T0191769

Internet of Things

Technology, Communications and Computing

The series Internet of Things - Technologies, Communications and Computing publishes new developments and advances in the various areas of the different facets of the Internet of Things. The intent is to cover technology (smart devices, wireless sensors, systems), communications (networks and protocols) and computing (theory, middleware and applications) of the Internet of Things, as embedded in the fields of engineering, computer science, life sciences, as well as the methodologies behind them. The series contains monographs, lecture notes and edited volumes in the Internet of Things research and development area, spanning the areas of wireless sensor networks, autonomic networking, network protocol, agent-based computing, artificial intelligence, self organizing systems, multi-sensor data fusion, smart objects, and hybrid intelligent systems.

Indexing: *Internet of Things* is covered by Scopus and Ei-Compendex **

More information about this series at https://link.springer.com/bookseries/11636

Sahil Garg • Gagangeet Singh Aujla
Kuljeet Kaur • Syed Hassan Ahmed Shah
Editors

Intelligent Cyber-Physical Systems for Autonomous Transportation

Editors
Sahil Garg
École de Technologie Supérieure
Montreal, QC, Canada

Kuljeet Kaur
École de Technologie Supérieure
Montreal, QC, Canada

Gagangeet Singh Aujla
Department of Computer Science
Durham University
Durham, UK

Syed Hassan Ahmed Shah
California State University
Fullerton, CA, USA

ISSN 2199-1073 ISSN 2199-1081 (electronic)
Internet of Things
ISBN 978-3-030-92056-2 ISBN 978-3-030-92054-8 (eBook)
https://doi.org/10.1007/978-3-030-92054-8

This Springer imprint is published by the registered company Springer Nature Switzerland AG
The registered company address is: Gewerbestrasse 11, 6330 Cham, Switzerland

Foreword

The advent of autonomous transportation systems (ATSs) has drawn greater attention to deal with the challenges faced by traditional transportation systems such as traffic congestion, energy utilisation, security, and parking. The concept of ATSs is not new in the market; it has undergone numerous laboratory and field testing to work successfully in real-life applications. However, the introduction of artificial intelligence (AI) in the ATSs is what attracts researchers the most. The advancement in technological research and development in the area of AI has proven to give promising solutions to the challenges faced by modern society, such as rise in population, climate variability and spike in energy needs. Therefore, AI, along with the use of data-collection and smart technologies, drives the research on ATSs to unrivalled heights.

Cyber physical systems (CPSs) have revolutionised further the autonomous transportation systems by achieving greater efficiency and reliability as compared to the traditional transportation systems. The CPS technology is based on integration of physical systems (sensors, actuators etc.) along with the data-driven cyber systems (software). It employs the use of IoT-based technology to design driverless vehicles that enables more safety and security and is a boon to the hustled human lives. The CPS for autonomous transportation is not dependent on human control and thereby delivers efficient results in areas where humans fail (such as construction sites, agriculture, and mining).

This book comprises a number of chapters that focus on the theoretical and practical concepts regarding the need for intelligent cyber-physical systems for autonomous transportation. Apart from covering the topics ranging from CPS to AI, it also sheds light on different use cases and security issues in intelligent transportation systems. It is a very good incubation for those who are novice to the area and at the same time helps to provide useful insights for researchers who wish to dig deep into the concepts of CPS utilising the intelligent algorithms. The "Intelligent Transportation" highlighted in this book employs the collaborative use of artificial intelligence technology and real-time data collection from the physical systems along with data analytics.

It is fascinating to mention here that the editors and authors of this book represent distinct regions in the world, thereby covering different aspects of ATS in those regions. Irrefutably, this geographic fusion of technical minds would provide a more comprehensive view in this emerging field.

Durham University, UK
July 2021

Anish Jindal

Preface

Cyber-physical system (CPS) is referred to as an engineered mechanism wherein the physical processes are controlled through computations and computer-based algorithms. CPS intertwines the software and physical elements (depicting different and discrete behavioral characteristics) working at spatial as well as temporal scales to interact with each other based on the context. Among others (smart grid, smart cities, health systems), one of the most challenging examples of CPS is the transportation system (including vehicles and drones). The orchestration of computing, things, and vehicles in a complex urban transportation ecosystem ends up in a complex environment. In such a scenario, the characterization of individual systems may provide a deep insight into the potential methods to interact with the surroundings. Due to these reasons, the future CPS will have limited dependence (or reliance) on human control and will focus on embedded intelligence in the form of artificial-intelligence (AI)-based processors and control. Such systems will operate vehicles and transportation key elements as a prevalent landscape of future transportation systems. Nowadays, the adoption of AI and machine learning (ML) to solve the problem of uncertainty along with the usage of probabilistic models and algorithms to handle predictive analytical problems is prevalent in a wide range of environments. On similar lines, the intelligent CPS also relies on AI and learning algorithms for approximation and employs statistical training and probabilistic algorithms running at the intelligence edge devices embedded with inference engines.

Although the automobile manufacturers are pushing hard towards the growth of this future intelligent CPS for the transportation sector, several challenges are new and increasingly complicated. The stringent need for coordination, amalgamation, and unification among the virtual (including the computational elements) and the physical worlds is essential for the success of intelligent CPS for smart transportation. The expanding smart cities require an instrumented transportation infrastructure and intelligent CPS to control and coordinate the same. The advent of connected and autonomous vehicles adds situational awareness in vehicles using the networked infrastructure of smart cities. For this reason, there are several latest technologies (edge intelligence, deep learning, generative adversary network, software-defined networks, intent-based networks, and context-aware computing) that are trying to act as enablers for future smart transportation. The main reason for the confluence of these smart technologies, adaptive systems, and runtime models is because of the increasingly instrumented world enabled through pervasive sensing and actuating capabilities, adaptive real-time and networked control, analytical and cognitive capabilities, intelligent edge-fog osmosis, and compute and storage clouds. The availability of cognitive intelligent assistants in the vehicles human-out-of-the-loop CPSs is proliferating in smart transportation. Hence, the enabling technologies and intelligent applications emerging from the integration of the CPS and embedded intelligence have a potential for innovation and incubation engine for a broad range of research, academia, and industrial ventures.

This book provides a comprehensive discussion on some key topics related to the usage or deployment of AI in urban transportation systems. It presents intelligent solutions to overcome the challenges of static approaches in the transportation sector to make them intelligent, adaptive, agile, and flexible. This book showcases different AI-deployment models, algorithms, and implementations related to the intelligent CPS along with the pros and cons of the same. Even more, this book provides deep insights into the CPS, specifically about the layered architecture and different planes, interfaces, and programmable network operations. The deployment models for AI-based CPS are also a part of this book with an aim towards the design of interoperable and intelligent CPS architectures. Uniquely, the readers will find practical implementations, deployment scenarios, and use cases related to vehicular, traffic management, underwater transportation, and many more.

The current and future application of AI lies in the CPS-based urban transportation use cases like traffic lights, emergency vehicles, traffic management, underwater transportation, signal processing, high-speed trains, subway track vibration, complex data analysis, vehicle safety, secure information transmission, and mobile crowdsensing. These applications cover almost every domain related to transportation where computational decision making and networking is partially or fully required, and so the future of this area is bright. The current level of research and literature in this area of focus is limited to specific and smaller segments; therefore, this book will make us understand the applicability of AI-based CPS in a wide range of transportation applications that drive life's routine. The current advances in smart cities require intelligent control and management of urban transportation infrastructure with the vision of autonomous cars to provide global visibility. As intelligence empowers the connected vehicles and transportation infrastructure and everything, AI-based CPS architecture will open various bottlenecks encountered by the connected transportation sector.

This book is divided into five parts. The first part provides an overview of transportation systems and future transportation systems. This part discusses the historic evolution of transportation systems alongside presenting the key challenges and applications. The second part focuses on AI and its need and application in transportation systems. This part also covers the application deployment of AI in transportation systems. The third part covers the historic evolution of CPS and its role in future transportation systems. The fourth part provides application use cases of autonomous transportation systems in traffic lighting, underwater transportation, advanced signal processing, high-speed train induced subway track vibrations, complex data analysis, and vehicle safety systems. The last part of the book covers the security perspective in intelligent transportation systems considering blockchain and privacy-preserved mobile crowdsensing.

Montreal, QC, Canada — Sahil Garg
Durham, UK — Gagangeet Singh Aujla
Montreal, QC, Canada — Kuljeet Kaur
Fullerton, CA, USA — Syed Hassan Ahmed Shah
July 2021

Contents

Contributors

Mansoor Ahmed Department of Computer Science, COMSATS University Islamabad, Islamabad, Pakistan

Muhammad Waseem Akhtar School of Electrical Engineering and Computer Science, National University of Sciences and Technology (NUST), Islamabad, Pakistan

Dongliang Chen College of Computer Science and Technology, Qingdao University, Qingdao, China

Yueyue Dai Research Center of 6G Mobile Communications and School of Cyber Science and Engineering, Huazhong University of Science and Technology, Wuhan, China

Amil Roohani Dar Department of Computer Science, COMSATS University Islamabad, Islamabad, Pakistan

Anju Devi Jaypee University of Information Technology, Waknaghat, India

Zhiwei Guo School of Artificial Intelligence, Chongqing Technology and Business University, Chongqing, China

Jinkang Guo College of Computer Science and Technology, Qingdao University, Qingdao, China

Syed Ali Hassan School of Electrical Engineering and Computer Science, National University of Sciences and Technology (NUST), Islamabad, Pakistan

Yi He Macquarie University, Sydney, NSW, Australia

Jia Hu University of Exeter, Exeter, UK

Sidra Iqbal School of Electrical Engineering and Computer Science, National University of Sciences and Technology (NUST), Islamabad, Pakistan

Alireza Jolfaei Macquarie University, Sydney, NSW, Australia

Gurpreet Kaur Chandigarh University, Mohali, India

Uswah Ahmad Khan School of Electrical Engineering and Computer Science, National University of Sciences and Technology (NUST), Islamabad, Pakistan

Muazzam A. Khan Khattak Quaid-e-Azam University, Islamabad, Pakistan

Hui Lin College of Computer and Cyber Security, Fujian Normal University, Fuzhou, China

Engineering Research Center of Cyber Security and Education Informatization, Fujian Province University, Fuzhou, Fujian, China

Haibin Lv North China Sea Offshore Engineering Survey Institute, Ministry of Natural Resources North Sea Bureau, Qingdao, China

Zhihan Lv College of Computer Science and Technology, Qingdao University, Qingdao, China

Huihui Ma University of Electronic Science and Technology of China, Chengdu, China

Qinyang Miao College of Computer and Cyber Security, Fujian Normal University, Fuzhou, China

Engineering Research Center of Cyber Security and Education Informatization, Fujian Province University, Fuzhou, Fujian, China

Fazeela Mughal School of Electrical Engineering and Computer Science, National University of Sciences and Technology (NUST), Islamabad, Pakistan

Liang Qiao College of Computer Science and Technology, Qingdao University, Qingdao, China

Geetanjali Rathee Department of Computer Science and Engineering, Netaji Subhas University of Technology, New Delhi, India

Bhawna Rudra National Institute of Technology Karanataka, Karnataka, Mangaluru, India

Hemraj Saini Jaypee University of Information Technology, Waknaghat, India

Munam Ali Shah Department of Computer Science, COMSATS University Islamabad, Islamabad, Pakistan

Sumit Sharma Chandigarh University, Mohali, India

Jakub Siłka Faculty of Applied Mathematics, Silesian University of Technology, Gliwice, Poland

S. Thanmayee National Institute of Technology Karanataka, Karnataka, Mangaluru, India

Abdul Wahid School of Electrical Engineering and Computer Science, National University of Sciences and Technology (NUST), Islamabad, Pakistan

Xiaoding Wang College of Computer and Cyber Security, Fujian Normal University, Fuzhou, China

Engineering Research Center of Cyber Security and Education Informatization, Fujian Province University, Fuzhou, Fujian, China

Muhammad Waqas School of Electrical Engineering and Computer Science, National University of Sciences and Technology (NUST), Islamabad, Pakistan

Michał Wieczorek Faculty of Applied Mathematics, Silesian University of Technology, Gliwice, Poland

Marcin Woźniak Faculty of Applied Mathematics, Silesian University of Technology, Gliwice, Poland

Jingyi Wu College of Computer Science and Technology, Qingdao University, Qingdao, China

Keping Yu Global Information and Telecommunication Institute, Waseda University, Shinjuku, Tokyo, Japan

Xi Zheng Macquarie University, Sydney, NSW, Australia

Part I
Overview of Transportation Systems

Chapter 1
Transportation Systems

Sidra Iqbal, Uswah Ahmad Khan, and Abdul Wahid

1.1 Background of Transportation Systems

Persistently occurring revolutionary changes in the transportation systems continues to enhance the consumers experience and awareness. These transportation systems are deeply embedded in presently commercial and private businesses all around the globe. It has led the transportation industry to a subsequent level in terms of speed, services, time, and quality. This section focuses on the vicissitudes in the transport systems over the years. The need to evolve the former systems arose from the rapid growth and demand of advanced technologies as depicted in Fig. 1.1. This was due to the fact that new transportation systems were adaptive in nature and were adjustable to the unceasing changes in social, economic, and environmental factors. The foremost changes, which encapsulate the gradual substitution of old systems by the innovative and improved systems, are discussed below:

1.1.1 Roads

In the year 1700s, the early modern highways were constructed by John Loudon McAdam (1756–1836), around the time of Industrial Revolution by using cheap soil and stone concrete paving material (macadam). The roads were embanked a few feet higher than the surrounding terrain to allow drainage of waterway off the surface. The demand for such roads increased in the forthcoming years to eliminate wash

S. Iqbal (✉) · U. A. Khan · A. Wahid
School of Electrical Engineering and Computer Science, National University of Sciences and Technology (NUST), Islamabad, Pakistan
e-mail: siqbal.mscs19seecs@seecs.edu.pk; ukhan.mscs19seecs@seecs.edu.pk; abdul.wahid@seecs.edu.pk

S. Garg et al. (eds.), *Intelligent Cyber-Physical Systems for Autonomous Transportation*, Internet of Things, https://doi.org/10.1007/978-3-030-92054-8_1

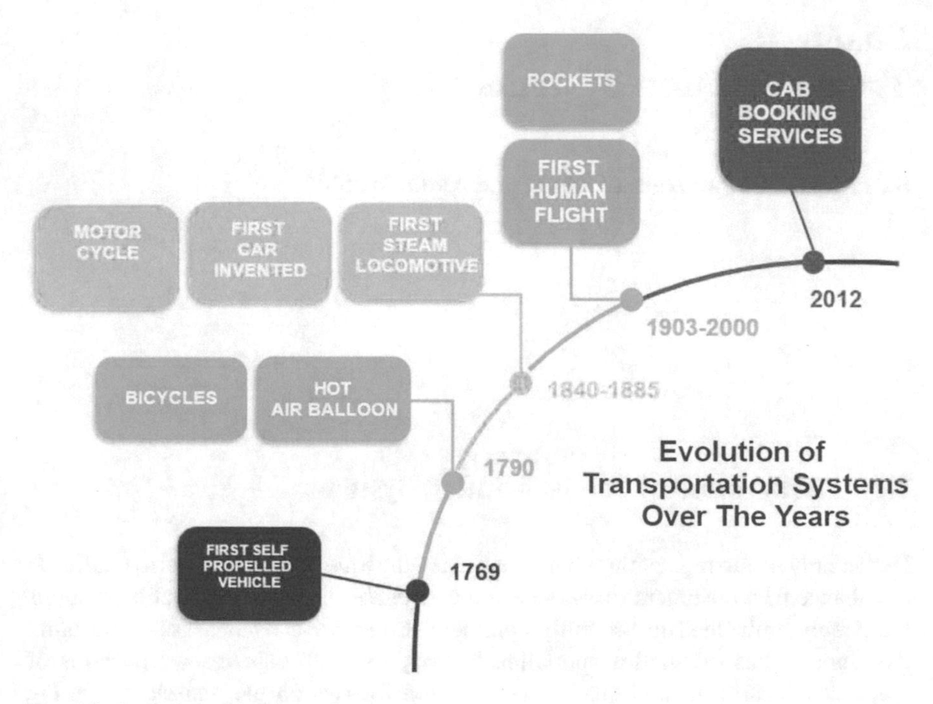

Fig. 1.1 Evolution of transportation systems

ways, bogging systems and dust on both municipal and agricultural roads. With the growth of motor transport in the early twentieth century, major western cities constructed roads initially by using cobblestones, wood paving, tar-bound macadam (tarmac), and concrete. In 1902, the world's first tarmac road was built then and named as Nottingham's Radcliffe Road [6].

1.1.2 Rails

In 1802, the first-ever ("Puffings Devils") steam-powered locomotive operated on smooth rails. It was designed and constructed by Richard Trevithick, a Cornish engineer, referred to as "father of railways" alongside, George Stephenson. Five years later, in 1807, Swansea and Mumbles Railway at Oyster mouth initiated the first passenger-carrying public railway. It was designed on existing tramlines by utilizing horse-drawn carriages. In England, in the 1820s, modern rail transport systems initially appeared [6].

1.1.3 Air

In 1783, the Montgolfier brothers invented the first air mode transport known as the hot air balloons. Jean-François Pilâtre de Rozier and François Laurent d'Arlandes were the chief explicitly confirmed individuals which took the human flight in Paris. They traveled 5 miles (approximately 8km) in the hot air balloon. Later, over a gap of 120 years, a ground-breaking change arose in the air transportation system. The Wright brothers made the first continuous, operated and powered heavier-than-air flight, and named it the Wright Flyer [6].

1.2 Growth of Transportation Industry

The transportation industry maintains a strong hold on the GDP growth of a country. It is one of the largest sectors that has been accumulating labor since its first-ever advancements in the sector. Transport industry pays a major surcharge to the governments which enhance the growth of countries globally. As millions of people, use multiple means of transport on a daily basis, the economic sector is forever to prosper. Different modes of transportation system include subways, metro buses, automobiles, air flights, and delivery of shipments through sea cargo. There is an evident relationship between quality and quantity of foundations of transportation, the stronger the relation, the better the economy [7].

Concentrated infrastructure of transportation and highly dense connected networks are deeply connected to the levels of development. An efficient transport system increases the provision of social and economic opportunities for the general public which highly impact the country nationally and internationally. Not only that, it also facilitates towards a better society by increasing the employment rate, accessibility in markets, and pave ways for investors to invest in the profitable industry of transportation. On the contrary, deficiency in production units of transportation system leads to inflation in a country and it lowers the living standards.

At an aggregate level, a productive transportation system decreases costs, while poor transportation increases such costs. In addition, transportation impacts are not always intended and can have unexpected or unintentional effects. For example, when supplying customers with free or low-cost transit networks, congestion is always an unexpected result. Nevertheless, congestion can also be taken as a good indicator of a rising economy where it is difficult for capacity and infrastructure to keep up with the increasing demands for mobility. Transport bears a major social and environmental load that cannot be overlooked.

1.2.1 Importance of Transportation Industry

Transportation industries play a vibrant role in our day-to-day lives. Without the convenience of transportation, the world would not be able to function effortlessly and smoothly. Over the years, the significance of transportation industry has expressively amplified. Transportation industry provides assistance in moving objects of daily usage from one place to another. Around the globe, millions of citizens are dependent on the availability of different modes of transportation systems in order to transit to offices, schools, and other places of interest. The importance of transportation system assists in several aspects depicted in Fig. 1.2. These different aspects are discussed below.

- **Quick Marketing:** Transportation industry provides considerable assistance in the growth of businesses that contains items which demand "quick marketing" (also known as fast marketing).Short-lived items like fish and green vegetables are swiftly delivered to different customers even in inaccessible market sectors through vehicles thus raising the demand of merchandise [9]. With the rapid advancement in transportation systems, there is a need to keep up with the latest, affordable, and reliable modes of transport to benefit the business. Availability of safe and fast transportation modes such as motorbikes and automobiles (nation-

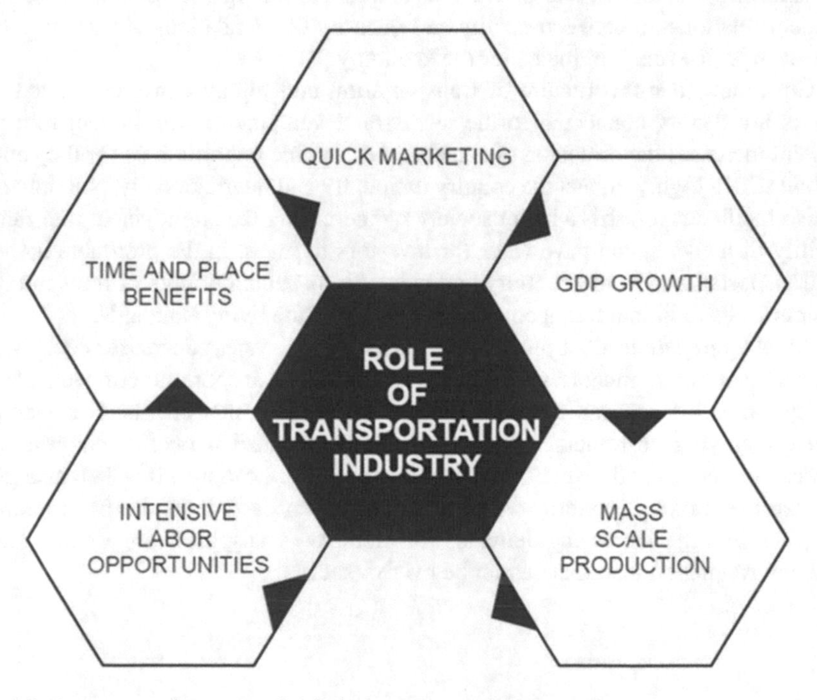

Fig. 1.2 Importance of transportation industry

ally), air cargo, and railway transport (internationally) has enabled effective collaboration between customers and the manufacturers.

- **Benefits of Time and Place:** A transportation system has eased up businesses by shortening the gap between time and place. Due to the climatic factors and the geographical conventions, industries are coerced to situate businesses far from the country local markets where they have little to no demands of the goods. Transport overcomes the barrier between the manufacturing and the utilization forces. Transportation industry has significantly increased the speed of transport by virtue of increased developments in the transportation modes. It enables the items to be disseminated in the minimum conceivable time.
- **Price Stability and Large-scale production:** One of the significant benefits imparted by transportation industry is their ability to dismantle any distinctions in item costs, i.e., products prices remain the same everywhere. When the production of goods is sufficiently enlarged, the transportation helps in delivering massive amounts of these products in the market from where the clients can purchase the products at lower costs. This results in mass production of items and goods. Transport places an evident impact upon the pricing of items by moving wares from surplus to shortage zones. This in turns levels the organic market factors and makes the cost of items stable. Transportation guarantees the progression of products under the control of the buyers throughout the period of utilization. Costs are likewise diminished due to the facility offered by transport for productions of goods at a larger scale.
- **Increases Employment:** Transport expands portability of work and capital. It causes individuals of one spot to relocate to different spots searching for occupations. Indeed, even capital, hardware, and other sort of gears are imported from far off nations through vehicles alone.
- **Economic Growth:** A country around the globe that lacks self-sufficiency relies on other nations to satisfy their necessities. This contributes to the economic growth by increasing employment and labor. Transportation has brought the nations together due to which it is considered to be the largest employment-intake sector. It also encourages the development and empowerment of business. The correlation between economic and transport industry is depicted in Fig. 1.3

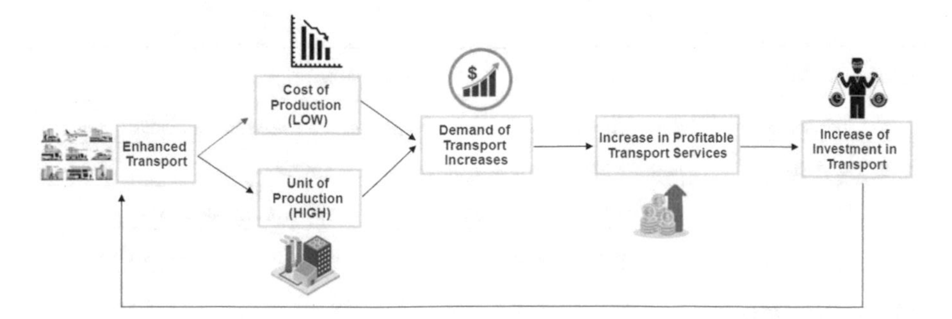

Fig. 1.3 Correlation between economic and transport industry

1.2.2 Growth Factors in Transportation Industry

Transportation systems are significantly concomitant with geography as it is concerned with measuring and assessing the relations between different areas and territories. For a nation's economic and social development, mobility is the elementary concern. Rapid advancements and unceasing emergence of technological aspects to transportation industry can swiftly ease the humans which in turns assist the financial sector, thus improving the country's civilization. The existing transportation industry cannot be pronounced by one factor, but there are several other factors contributing towards the growth of the transportation systems depicted in Fig. 1.4. Some of the factors are illustrated below:

- **Economic Factor:** Economic activities and the movement of all types of supplies from one destination to another are the two sides of a coin that collectively strengthen and reinforce a country's economy. Resourceful transport systems with advanced foundations can expand the business between the clients and the manufacturers. Thus, a proficient transport system can then benefit the overall economy. Reliable transport systems that offer low cost and less time benefits can provide services to markets on a larger scale. This will enable to impose "just in time" services to markets situated at distant places. As the delivery of goods and services has been improved due to the advancement in the transport systems, it enables the consumers to have access to varieties of goods from all across the world. This enables the clients to obtain quality products at lower costs. Efficient transportation with effective communication enables the production of goods and services in regions where it can attain the uppermost advantage. The production of goods in such region can improve the economic productivity as long as the proper transport is accessible for trade. So, for a country to grow economically, it is imperative that it should possess a virtuous transportation system.
- **Political Factor:** The growth of transportation systems is highly dependent on the political factors. Political facets include the safety and security conditions that are imposed on the transport system. It also defines the means in which different operations performed. The political aspect also outlines the trade agreements and contracts with other adjoining entities for the purpose of transit of goods and

Sr no.	Factors	Description
1.	**Economic**	Efficient transit of goods and services improves the economic productivity.
2.	**Political**	Implying safety conditions, trade agreements with other nations.
3.	**Social**	Providing healthcare facilities, social interaction through transport system.
4.	**Technological**	Using technology to make the transport system enhanced, inexpensive and cheaper.

Fig. 1.4 Growth factors

services. It defines the transnational connectivity and the interaction with other nations which can overall improve the country's economy.

- **Social Factor:** Transport system plays a very vibrant role in providing the necessities that improves and enhances the livelihoods. The need of providing healthcare facilities to people has made the use of different modes of transport systems unavoidable. The upsurge participation in cultural events and social interaction has also led to the growth of the transport system. Social activities have been stimulated to an immense extent due to the ease in mobility [10]. Now traveling from one place to another for any kind of societal collaboration is very stress-free. The more is the interaction, the better chances are there to improve and intensify the transit of goods and services among nations or also individuals. It thus humanizes the nation economic conditions.
- **Technological Factor:** Technological advancement has conveyed transportation services that are enhanced, inexpensive, and faster. People are now more likely to move efficiently from one endpoint to another. Similarly the transition of good and services is also enhanced and has become proficient over the period of time. Technology has led to the development of a friendly ecosystem. Use of GPS has headed people to get to their anticipated destination in a shorter period of time [8]. The hi-tech progression has also managed to lessen the fatal traffic accidents. Headlights in the dark assist the driver to drive safely, similarly the driver less automobiles have also led to the reduction of humanoid faults so that accidents can be reduced to an immense extent. Telecommunication has also endorsed the influence of "work from home" due to which the traffic mobbing on the roads can be reduced. Correspondingly, technology has directed the road administrators to highlight the highly congested areas so the traffic can be managed in an efficient way. This is owing to the technological innovations that the prevailing transport systems have made life easier and thought-provoking.

1.3 Challenges to Transportation Industry

With the significant evolution and improvements in the transport system, comes the challenges as well. As the transport system is becoming more saturated and complex, certain other sophisticated innovative and revolutionized mechanisms are required to overcome the challenges. These challenges can be categorized as the social, political and the technical challenges. Each of these is explained in detail in the following sections:

1.3.1 Social and Political Challenges to Transportation Industry

There are different social and political challenges in the transportation industry that needs to be addressed and resolves. Some of them are discussed below:

- **Social Challenges:** Some of the social concerns related to the transportation systems are described below:
  - **Limited Communal Interactions as a Social Challenge:** The transport system has eased the process of people to transit. It has also upgraded the transport of goods and services thus improving trade. But there is a certain level of inequity in the traveling of people belonging to different social groups. Most of people have ownership of personal vehicles and thus have the benefit of moving stress-freely and reaching the destination is shorter period of time and in a comfortable environment. As opposed to this, a person that cannot afford the ownership of vehicles therefore uses public transport which consumes time and is not comfortable. This results in limited social communication. The social interactions between people are also less due to the dichotomy in the modes of transportation available for people belonging to different demographics. Therefore, it is the responsibility of the transport sectors to offer a comfortable and cost-effective environment for people who are using public transport to transit. People living in the urban areas have a variety of transportation modes available for going to their offices, schools, shopping malls, parks, etc. In contrast people living in rural area are more subjected to limited public transport.
  - **Environmental Factor as a Social Challenge:** The social challenges of the transport system can also be considered from the environmental point of view. The transport system has affected the lifestyle of people. There are certain negative health impacts of transport on humans. Pollution is one such distress that has become an important social concern as it risks the human health. The emission of pollutants from the vehicles has polluted the air thus causing breathing issues and other various diseases. Similarly the noise pollution has caused annoyance and psychological instabilities. Similarly, the water pollution has contaminated the rivers and streams leading to different health concerns. This pollution has affected the human health and is therefore a social challenge in the transportation system that needs to be addressed and resolved.

    There should be supervisory agencies that should define standards in order to retain these social concerns. They should monitor the conditions and should impose regulation or penalties if any of the standards are violated. A good transport system is required, which will increase the amount of people using transport thus, improving the country economy.

- **Political Challenges:** There are different political concerns associated with the transport system. Some of them are discussed below:
  - **Inefficient collaboration:** The collaboration between different organizations that are accountable for the transportation management systems lacks efficiency. Due to the limited economic and financial development, initiating the expansion programs and enhancement campaigns for improving the transportation system is usually very less. The need of different transport modes is not properly gratified due to the deficiency in inter modal scheduling and planning. These consequences are due to the inefficient communication between the transport sector and the economic sector. Political pressure is one of the main factors responsible for the ineffectiveness in collaboration between different sectors. This is owing to the partial investment being made in projects that are concerned with the maintenance and preservation of the transport infrastructures. The Regimes are least considerate and sympathetic towards the improvement of transportation systems, which is one of the reasons why the transport systems are not competently enhancing and advancing in most of the countries.
  - **Less Prioritized:** Transport agencies ordinarily propose many stimulating and innovative projects schemes to the government corporate sectors. But still such projects are least prioritized. The initiation of such projects requires full fledge funding. The successful completeness of such projects can revolutionize and transfigure the entire transportation system, thus benefiting the nation from different aspects. One such aspect can be tourism; this can contribute to the economic development as well. As a matter of fact, it is observable that countries with dynamic and robust transport systems are economically strong. Hence, for a country to economically cultivate, vigorous transport systems are probably imperative.

1.3.2 Technical Challenges to Transportation Industry

Due to the extraordinary consumption of different modes of transport, the transportation system is thus becoming multifaceted and saturated. Therefore to efficiently carry out the transit of population, services, and goods, it is necessary that the mobility system should function applicably and appropriately. There are certain technical challenges that the transportation system is facing. Some of these issues are discussed below:

- **Traffic Congestion:** Due to the intensification in use of automobiles, traffic congestion has become the most predominant challenge. Increased quantity of vehicles on the roads has led to the problems of overcrowding and mobbing. Overcrowding leads to traffic jams and roadblocks, thus disturbing the normal day-to-day routine. This congestion also leads to fatal roadside accidents.

Therefore, in order to reduce the traffic congestion, it is highly endorsed to use public transports so that jamming on the roads can be reduced thus saving time.

- **Maintenance Expenditures:** Time is continually evolving and the movement of people as well as services and goods is rapidly mounting. This calls for a need to upgrade the transport system on continuous basis. Upgradation and maintenance of the infrastructures of the transport system requires high costs.
- **Hazardous Accidents:** As the transportation system is constantly developing, high congestion on the roads leads to fatal and hazardous accidents. This is a technical challenge that entails the road administration to persistently monitor the congestion so that the overcrowding should be decreased and people can safely transit. These mobbing also leads to delay and postponement in the transfer of goods from one place to another thus adversely affecting the business as well.
- **Safety:** The safety of the people is the main concern. The transport system requires alerting mechanisms that activates when certain speed limit is crossed. The existing transport system needs upgradation. Additionally, there is a need for latest and hi-tech sensors and cameras to constantly monitor the movement of vehicles so that any abrupt change in the traffic flow can be identified at hand and thus safety measures can be timely taken to avoid any unfortunate circumstances.
- **Parking:** The ever-increasing need for the transport has resulted in the increased number of personal vehicles. Comparing to the older system when there was only the public transport and very few people owned personal vehicles, in today's era almost every family owns a personal vehicle for traveling purposes. This increase in the number of vehicles has led to the problem of parking. People do not find parking spaces easily which causes them to search for vacant spots thus wasting time and fuel. An advance real-time parking is thus needed that will notify the drivers about the vacant parking slots so that the drivers can park their vehicles while saving their time and fuel.

1.4 Impact of Intelligent Transportation Systems

Previously, the transportation systems were more prone to subjective errors. Congestion, poor traffic monitoring, and insufficiency in transport management resulted in disastrous accidents and other mishaps. These challenges called for an upgradation in the existing transport systems and from here the need of incorporating technology to the transport system arose. This gave rise to the intelligent transportation system which overcame most of the challenges that were occurring in the conventional transportation.

The Intelligent transportation system (ITS) has gained prominence as it offers an effective real-time traffic managing and monitoring systems [12]. It has enhanced the quality of transit, provided safety measures to users, helped in reaching the destination in shorter time with less fuel consumption, decrease in traffic overcrowding and most importantly reducing the roadside calamities and disastrous accidents.

The impact of ITS on the lifestyle of people as well as its effect on the nation is huge. Advancement in ITS has led people to make constant use of different modes of transport thus improving their routines. Robust ITS promotes more trade and businesses all across the world contributing to more financial and economic development. A detailed explanation about the ITS is given in the following subsections.

1.4.1 Overview of ITS

Emerging countries are persistently improving their transportation systems. Due to the concentration in economic growth and population, transport systems are facing a number of challenges which includes overcrowding, hazardous accidents, and poor traffic management [3]. These situations call for immediate consumption of the ITS so that such extreme concerns could be fixed to the possible extent. To introduce ITS in a country, it is imperative that the country should know how to efficaciously incorporate technologies into the transportation systems.

ITS is concerned with refining the productivity and proficiency of the services provided by transport system. It involves the transmission of instantaneous information in order to advance the traffic management, reduce congestion, and lessen the road accidents. All these benefits of ITS can be yielded by applying the advance technologies of communication, sensing devices, computers into the transportation system. At the communal level, ITS applications are concerned with reducing the road congestion, minimizing the road accidents thus improving the traffic management. For those managing the road, ITS helps in determining the areas that remains highly congested so that the traffic can be managed in a competent way.

ITS assists in overcoming the socioeconomic issues as well as helps in the economic and financial development of the nations. Intelligent Transportation System supports the development of industrialized countries as well as the developing nations. Proper implementation of ITS can effectively reduce the fuel consumption enabling people to travel in a secure and inexpensive mode. ITS consists of various components. One such component is the Vehicular ad hoc network (VANET) which is considered to be the most significant [4]. VANET uses ITS techniques to impart reliable information regarding the vehicle's location, its speed, upcoming headings, and road conditions [13]. It is hence predictable that further advancement and progression in the Intelligent Transportation System will result in a more user convenient, secure, and efficient transport system.

1.4.2 Applications of ITS

The foremost intention of intelligent transportation systems is to enhance the existing transportation system and improve the quality of transport [2, 5]. Different

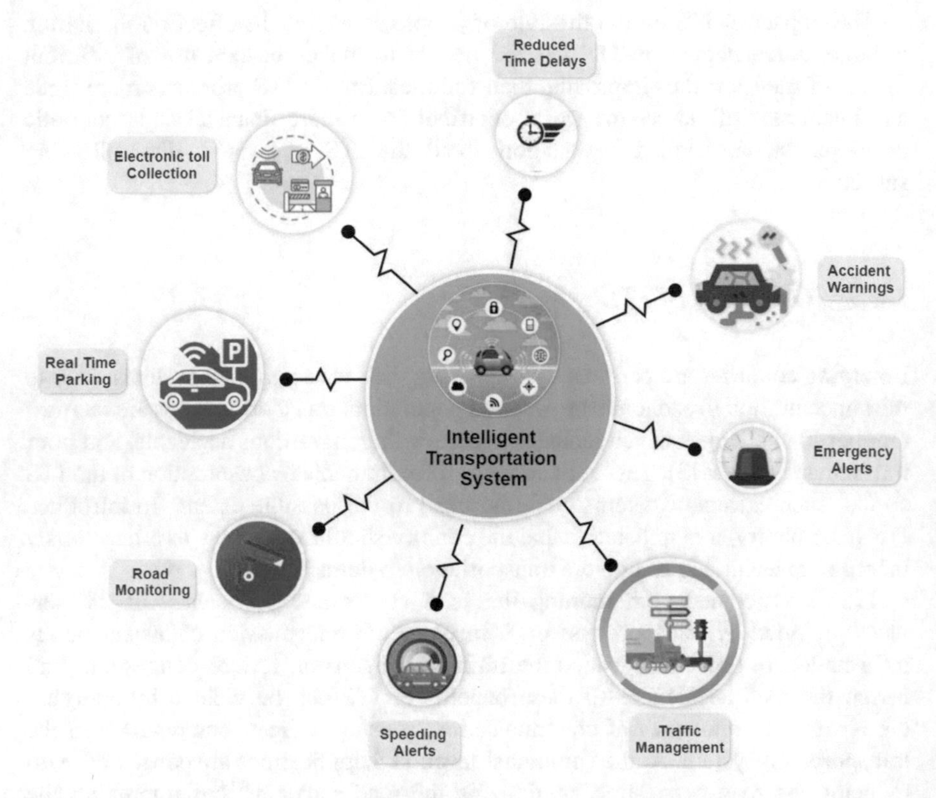

Fig. 1.5 Application of intelligent transport systems

monitoring and managing mechanism are used to enhance the transportation system. There is no doubt that, with the appearance of technology, the transport system has made rapid advancement and improvement. A colossal transformation can be observed that has changed the whole depiction of the transport. This is due to the constant emergence of technologies that has made the transport systems intelligent which are also shown in Fig. 1.5.

On a broader level, the functionalities offered by ITS can be categorized into two aspects, i.e., the Advance Traveler Information Systems (ATIS) and Advance Management System (ATMS). ATIS help people in making decisions regarding traveling whereas AMS focuses on the monitoring and controlling the transportation systems. Progression in ITS has led to the development in both ATIS and ATMS. There are voluminous applications of the intelligent transport systems in both of these categories.

- **Advance Traveler Information Systems:** This system collects, analyzes, and then delivers information to travelers so that they can easily transit from one location to another. It focuses on providing different mechanisms so that people can certainly make decisions about their traveling journeys. It imposes several models that help in deciding the routes, means of transports, lane restrictions, available parking places, expected time to reach the destination, recommended speed, etc. These applications are discussed below:

- **Emergency Notification System:** This is one of the most significant benefits offered by intelligent transportation system. It is a real-time alerting system that notifies the drivers that an emergency vehicle is passing. This directs the drivers to slow down and give ways so that the emergency vehicles can pass easily.
- **Reduced time and fuel consumption:** Mobility is an advance application of Intelligent Transportation Systems (ITS) that aims at providing the ease of transit. It enables people to move effortlessly from one destination to another. The travel time prediction models help in specifying how much time is required to reach the destination. Amalgamation of vehicles with mobiles, Global Positioning System (GPS), and Media Access Control (MAC) has intended to deliver services such as taking the shortest paths. This saves the time as well as reduces the fuel consumption because we can now reach our desired destination in shorter time by selecting the optimal path amongst different paths.
- **Route Guidance Systems**: Route Guidance Systems (RGS) are used to notify the drivers about the environment such as traffic congestion on the roads, weather conditions, any mishap or accidents on the roads ahead, etc. The route guidance system uses various systems to identify the conditions of the roads and then notifying the drivers. This helps the travelers to timely decide on whether to switch or not to another route. The guidance systems use different algorithms such as inter-vehicular communication, sensor networks, P2P, VANET, neural networks for operating, etc.
- **Real-time parking:** The intelligent transportation system has paved ways for the real-time parking management. It uses cameras, sensors, payment systems, and mobile apps to determine the vacant slots in the parking areas. It then provides information to the drivers about the unoccupied and available parking spaces so that the drivers can easily park their vehicles. This real-time parking has also reduced the traffic jamming and mobbing as shown in Fig. 1.6.
- **Obstacle Alerts:** This warning technology uses radars, sensors, and cameras to warn the driver about any hindrance that is sensed through these devices. This prevents the vehicle from collision. This application warns the drivers about the obstacles approaching either from the forward or backward direction as the in-vehicle sensors are capable of sensing in both the direction.

• **Advance Traffic Management System (ATMS):** This system makes use of real-time information collection, processing, analyzing, and then using it for the management of transport system. The data collected from enriched sources with quality information helps in managing and controlling the operations of the transport systems. It makes use of different mechanism such as emergency warnings, road conditions, transit services, and the transfer of goods and services

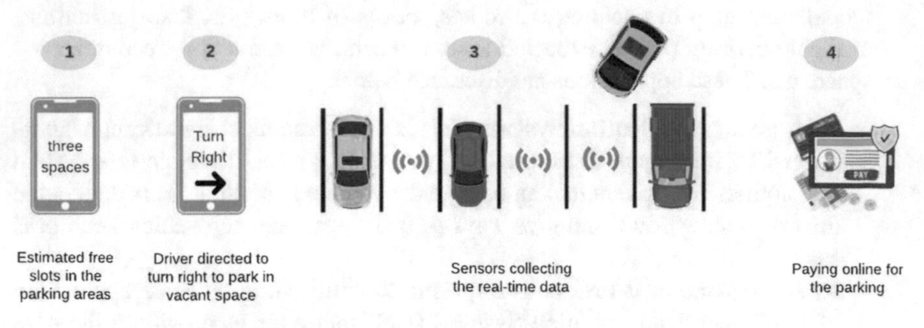

Fig. 1.6 Real-time parking system

for trade and business purposes, etc. It consists of some of the following applications:

- **Smart Traffic Management:** It involves tools, amenities, and services associated with different modes of transport and the traffic management system which enables people to have better understanding about the traffic system. The key objective is to proficiently control the traffic congestion, supervising the traffic lights, managing the flow of vehicles. The traffic system uses camera, radars, different sensing devices, and advanced networks to regulate the traffic congestion. Thus saving time, energy and reducing the road accidents.
- **Speed imposition:** Installation of speed monitoring cameras has upgraded the transportation system to an immense extent. It monitors the speeds of the vehicles where the over-speeding vehicles are caused penalties. The enforcement to follow a certain speeding limit has inestimably reduced the road accidents and due to which the traffic congestion has also been condensed. In addition to this, Adaptive Cruise Control (ACC) technology is largely used which concentrates on adjusting the speed between vehicles. Some examples of this technology include the drive-by-wire regulators in the car and the in-vehicle distance cautionary systems.
- **Electronic toll collection:** It is a very stimulating application of the intelligent transportation systems where the drivers are charged with a toll without stopping. Conventionally, the divers would entail to stand is long queues and wait for their turn to pass through the toll plaza thus consuming users time. The electronic toll collection system has supplementary ease the user's life. Now, the users have registered accounts for the billing procedures. The vehicles do not have to stop or stand in lines; rather there are sensors and cameras installed at the toll plaza that recognizes the license plates of the cars and the users can later pay the toll. This has competently reduced the time consumption, thus letting the users to reach their destination in shorter period of time.

- **Transit Management:** The transit organizations are responsible for monitoring the transit operations. This is done through the surveillance cameras, the Computer-Aided Dispatch (CAD) systems, and the automated vehicle location (AVL) systems [1]. CAD systems are used for dispatching the responder's personnel. It delivers services that are used by call-takers and 911operators for recording the calls in order to recognize the location of the responder and then dispatching its personnel. AVL is used by agencies to track the location of a vehicle by using the GPS over the internet. These surveillance's mechanism provides safety and security to public transport systems.

1.4.3 Challenges Overcome by Intelligent Transportation Systems

Transportation and traveling plays a very essential part in our day-to-day lives. We use different modes of transport to transit to offices, learning, shopping, and pleasure. Similarly, companies customize transport for their trade devotions. We have now become exceedingly reliant on transportation to accomplish our daily tasks. People are keen towards a transport system that is much safer, faster, reliable, and comfortable. There were a lot of challenges that people faced in the conventional transportation system. But after the immersion of technologies in the conventional transport, the transportation systems now are very intellectual. The intelligent transport system has eliminated most of the challenges previously faced and its prime emphasis is to provide people with resilient transit opportunities. Some of the challenges that have been overcome by these modern transport systems are:

- **Enhanced Mobility:** Previously the transport systems were not much reliable, due to the lack of smart solutions to congestion problems. People would get jammed in long traffic congestion. They would be late for offices; children would not reach the learning institution on time. This was adversely disturbing the overall day-to-day routine. It also resulted in the transit delays of delivering good and services thus affecting the trade and businesses. Nonetheless, with the emergence of advance technologies to this conventional transport system, the challenges have been overcome to a greater degree. The traffic congestion has now been excessively controlled. The drivers have in-vehicles and traffic management information that guides them about the congested network and notifying them if the road ahead is clear. Controlling and adjusting the traffic congestion has immensely reduced the travel time.
- **Finding appropriate routes:** Formerly people had no awareness about the conditions of the roads ahead on which they were traveling. Sometimes they may come across routes that are not apposite. For example, there is some ongoing road construction, road maintenance or road accidents ahead. People get stuck or they would require changing the route all cross and driving back from the place. This

result is delays and wastage of fuel. Route guidance systems have enabled the drivers to timely decide to take appropriate routes for reaching the destination. People use the smart navigation system that guide them about the circumstances on the road and thus saves travel time.

- **Safety:** The intelligent transportation system has condensed the roadside accidents and has provided a safer means of transport. Previously, traveling safely was a prevalent challenge which has been overcame by ITS. At the instant, the traffic flow is smoothly controlled and monitored. This fallout in reducing the chances of calamities and hazardous accidents. ITS has also reduced the risk associated with Vulnerable Road User (VRU) [11]. VRU is the term used for users that are at most risk than other on the road infrastructures which includes the cyclist, pedestrian, etc. They are broadly categorized as the non-motorized users. Intelligent transport system has enabled the automobile manufacturing companies to designed vehicles in such a manner to offer protection to the VRU. The vehicles have forward pre-installed cameras and other in-vehicle applications that constantly warn the driver about the collision. This assists the drivers to identify if there is a non-motorize users closer to his vehicle so that he can avoid collision.
- **Road Management:** Intelligent transportation system has provided massive benefits to the road management system. Previously it was a very cumbersome job for the management to monitor the roads. At present, ITS are constantly devising hi-tech mechanisms that monitors the roads and based on their usage identifying when most probably the roads needs maintenance. Incorporating advance technologies to the traffic management systems helps the administrators to easily manage the traffic signals and network.

1.5 Conclusion and Future Directions

Advancement in the transportation systems has eased the traveler's day-to-day routine. It has reduced the imprecision's and has made the transit effective and efficient. With constant incorporation of technologies, the transport system is persistently improving. It is further enhancing the consumers experience and awareness, thus increasing the use of different means of transport.

Currently, the transport systems are intelligent enough to perform tasks that were manually performed by humans. Amalgamation of hi-tech mechanisms, artificial intelligence and networks has completely revolutionized the world. ITS has immensely reduced the challenges which were previously faced by the conventional transport system. Smart traffic management, congestion control, emergency notification, reduced accidents, route guidance systems, real-time parking management and transit management are some of the applications of ITS. These improvements have contributed to the overall development of the nation, thus making countries economically strong and resilient.

Future of ITS lies in further advancing the information, communication, and technological mechanism to make transport system much more robust. It should provide consumers with more accurate and precise information. Data collection and analyzing mechanisms should be performed my models/algorithms that are less prone to errors and are more efficient. The traffic management systems should adopt more vigorous measures so that the safety and security of people should not be compromised at any cost.

Despite the various advantages provided by the existing transport system, there is always room for further improvements. The cellular devices and computers are extensively used for transferring and receiving data which is majorly used for the route guidance purposes. ITS in the future should use cars to transmit data instead of cellular phones. This will provide instantaneous benefits to the drivers and the travelers about the traffic conditions, route guidance, etc. This future aspect focuses on considering the cars as data points on the networks rather than the cellular phones or other devices.

References

1. (2012). Retrieved from 100PercentBronx: http://100percentbronx.blogspot.com/2012/05/liu-mismanagement-of-911-upgrade-picked.html?m=0.
2. Aujla, G. S., Jindal, A., & Kumar, N. (2018). EVaaS: Electric vehicle-as-a-service for energy trading in SDN-enabled smart transportation system. *Computer Networks, 143*, 247–262.
3. Chaudhary, R., Jindal, A., Aujla, G. S., Aggarwal, S., Kumar, N., & Choo, K.-K. R. (2019). BEST: Blockchain-based secure energy trading in SDN-enabled intelligent transportation system. *Computers & Security, 85*, 288–299.
4. Dua, A., Sharma, P., Ganju, S., Jindal, A., Aujla, G. S., Kumar, N., & Rodrigues, J. J. P. C. (2018). RoVAN: A rough set-based scheme for cluster head selection in vehicular ad-hoc networks. In *2018 IEEE Global Communications Conference (GLOBECOM)* (pp. 206–212). IEEE
5. Ercan, T. (2013). *Sustainability analysis of intelligent transportation systems*. University of Central Florida.
6. History of Transport (2021). Retrieved from https://en.wikipedia.org
7. Learning: Growth of the Transportation Industry and Importance of the Transportation Business (2018). Retrieved from desientrepreneurs: https://www.desientrepreneurs.com/
8. McTigue, K. (2019). News: Economic Benefits of the Global Positioning System. Retrieved from: https://www.nist.gov/news-events/news/2019/10/economic-benefits-global-positioning-system
9. Raut, S., Biswal, S., & Rout, S. (2015). Significance of Transportation and its Contribution to Economy. Retrieved from SlideShare Website: https://www.slideshare.net/sunil133/significance-of-transportation-55618349
10. Rodrigue, J. (2020). *The Geography of Transport Systems* (5th ed.). Routledge.
11. Safety of Vulnerable Road Users. (2020). Retrieved January 2021, from https://rno-its.piarc.org/en/network-operations-its-road-safety/vulnerable-road-users
12. What is Intelligent Transportation System(ITS): Applications and Examples (2019). Retrieved January 2021, from https://www.aindralabs.com
13. Yousefi, S., Mousavi, M. S., & Fathy, M. (2006). Vehicular ad hoc networks (VANETs): Challenges and perspectives. In *2006 6th International Conference on ITS Telecommunications*. Chengdu: IEEE.

Chapter 2
Future Autonomous Transportation: Challenges and Prospective Dimensions

Muhammad Waseem Akhtar and Syed Ali Hassan

2.1 Introduction

Autonomous vehicles (AVs) have tremendous potential to enable the time spent in a vehicle to be more efficient with low accidents, congestion costs, fuel consumption. Governments may also shift vehicle ownership models and land use patterns and may generate new industries and economic opportunities [23]. Lawmakers are faced with many political issues which would play a crucial role in shaping the adoption and impact of AVs. This covers everything from where and if this technology should be allowed on the roads under the required legislation. Technological developments establish a continuum between conventional, entirely human-driven cars, and AVs that are partially or completely driven themselves and may eventually not require a driver at all [11]. Technologies that allow a vehicle within this continuum are collision alert systems, lane warning systems, adaptive cruise control (ACC), technology for self-parking, and systems management.

AV technology deserves the urgent attention of lawmakers for many reasons. The efforts of Google, such as a fleet of vehicles that have collectively logged millions of autonomous miles, have gained extensive attention and prove that this technology has progressed significantly [28]. Each major commercial car manufacturer is engaged in research in this field and it is expected that full-scale commercial deployment of fully autonomous (including driverless) vehicles can be available within next decade [5]. Second, there are high stakes. In many countries

---

M. W. Akhtar · S. A. Hassan (✉)
School of Electrical Engineering and Computer Science, National University of Sciences and Technology (NUST), Islamabad, Pakistan
e-mail: engr.waseemakhtar@seecs.edu.pk; ali.hassan@seecs.edu.pk

S. Garg et al. (eds.), *Intelligent Cyber-Physical Systems for Autonomous Transportation*, Internet of Things, https://doi.org/10.1007/978-3-030-92054-8_2

around the globe, more than tens of thousands of people are killed in accidents every year, and millions are wounded, and the vast majority of these incidents are the result of human errors. Therefore, by implementing an autonomous transport system, the number of road accidents can be substantially reduced [5]. The use of an autonomous transport system can also prevent traffic congression, which can ultimately considerably reduce the costs associated with traffic congression, such as fuel and time. As per some estimates, with autonomous transportation, the road capacity (vehicles per lane per hour) can be enhanced three times compared to that of transpiration without any automation. Bagloee et al. [5]. It is also possible to save a lot of parking space with autonomous transport in suburban and urban areas. It could be possible to significantly lighten automobiles if collisions were increasingly rare incidents [16].

Human beings are uniquely conditioned, whether by smartphones, alcohol, rage, or simply looking at anything, to be distracted while driving [12]. Computers, on the other hand, pay attention all the time. They are not distracted and can therefore adapt to situations that might not be seen by humans. The widespread adoption of autonomous vehicles could also minimize deaths in motor vehicles by orders of magnitude, which would be "a big win for society." Benefits may also be evaluated in terms of efficacy. People spend vast quantities of time around the world waiting to change stoplights, inching through heavy traffic, or otherwise trapped behind their car's wheel. If cars could drive themselves, people could make more efficient use of that time. If vehicles can drive themselves, global economic productivity can increase quantifiably.

Total vehicle miles traveled (VMT) could be raised, which could lead to further congestion. If passengers travel ever further away from workplaces, they can raise sprawl. Similarly, to allow other activities, AVs can gradually adjust the preferences of users towards larger vehicles. This could also involve beds, sinks, baths, or offices, in principle. If AV software malfunctions, a single flaw can result in several accidents. It may be possible to hack malicious systems that are connected to the Internet. And the biggest threats are simply uncertain, perhaps.

Figure 2.1 depicts the level of automation from no automation (level 0) through fully automated vehicles (level 5) with a tentative timeline. It is anticipated that fully autonomous vehicles would be rolled out by the year 2030.

In the next section, we give an overview on the state-of-the-art in the field of autonomous transportation.

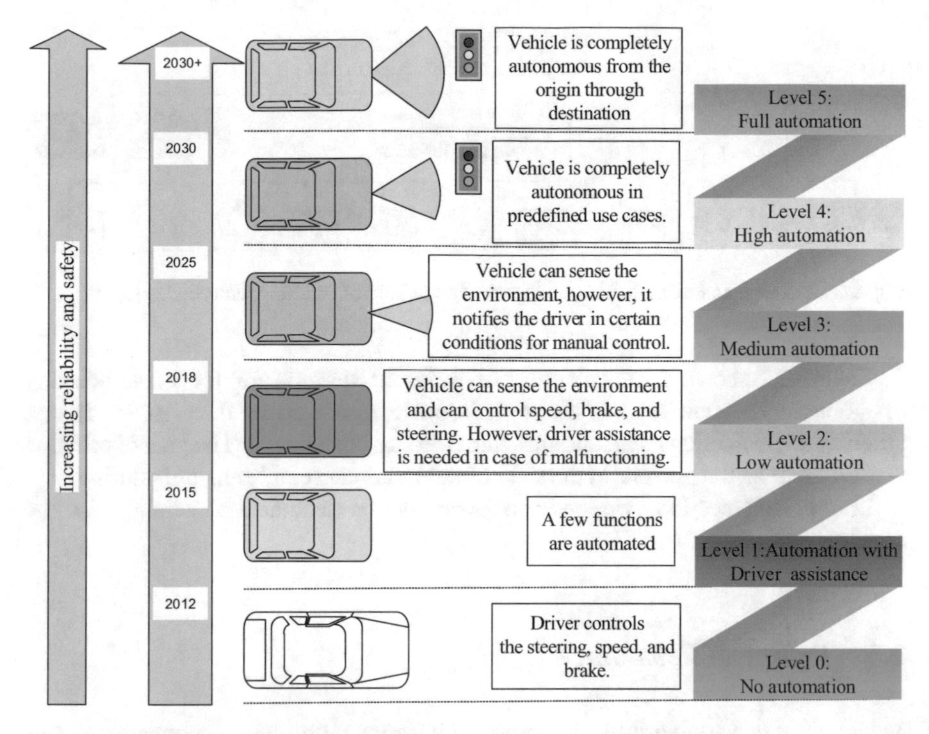

Fig. 2.1 A tentative timeline of the development of autonomous vehicles with respect to levels of automation [18]

2.2 State-of-the-Art System on Autonomous Vehicles

Land transport systems have evolved from basic electromechanical systems to complex, computer-controlled, networked electromechanical systems. Initially, cars were used for recreational purposes, but as quite practical means of transport, they are now an essential part of everyday life. As a result, over the years, road infrastructure has been developed and the development of large numbers of vehicles has contributed to a decrease in cost and availability for the general public [5]. These reforms, however, have resulted in numerous problems, such as road accidents leading to death or injury, pollution, traffic congestion, heavy cost competitiveness, depletion of fossil fuel reserves, etc.

In recent years, society has shifted the perception of transportation systems. Vehicles are seen as a means of comfort and social prestige that brought a lot of independence to business. However, today's transport networks are a source of growing concern as a viable mode of travel, due to the high rate of accidents, environmental constraints, high fuel prices, etc. [28]. Various propulsion systems are used for modern vehicles to allow them more computer-controllable, have numerous sensors for car navigation features, and nodes have become part of large communications networks.

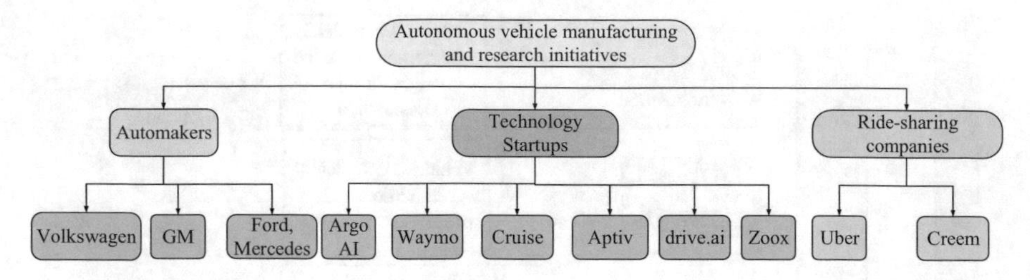

Fig. 2.2 Research and manufacturing initiatives in the field of autonomous vehicles [9, 19]

Figure 2.2 shows a tree diagram of different startups for the manufacturing of autonomous vehicles. Theses startups are categorized in three main classes, automakers, technology start startups, and ride sharing startups. The goal of all these startups is to be the pioneer in the field of autonomous vehicles manufacturing.

In the next section, we give an overview on technical feasibility for the development of ATS.

2.3 Technical Feasibility

Vehicles are developing into autonomous platforms with mobile connections. This is due to the political and financial can for a better life and to improvements in ICTs, which are gradually being introduced in new commercial vehicles. People are still worried about the security, safety, and comfort of modern travel [13]. Also, new vehicles can make it possible for people who are currently unable to drive, as well as promote the sustained mobility of the elderly population.

Progress is driven by initiatives by academia, the military, vehicle manufacturers, and by the production of sensors for autonomous transportation. Largely, the reasons why vehicles are designed with intelligent capabilities are defined by the motivators [16]. The introduction of wireless connections to vehicles could allow information to be shared and thereby increase drivers' peripheral vision as the anticipated environment expands. Sensors capable of detecting items in the road network are considered to dictate the speed of innovation. The advent of sensing devices, advanced digital maps, and wireless communications technologies, together with the availability of electric vehicles, could make it possible to deploy autonomous vehicles on public roads without changing the climate [22].

Over the past few years, the autonomous vehicle has gained a lot of interest and numerous manufacturers have built concept models. The industrial realization of autonomous vehicles, however, has often been an obstacle. The autonomous car is designed at a very simplistic level with a wide range of sensors and actuators that produce a lot of real-time data that must be interpreted and evaluated in order to make the appropriate decision. The size, speed, consistency, diversity, and real-time nature of data must also be incorporated in the development of the autonomous

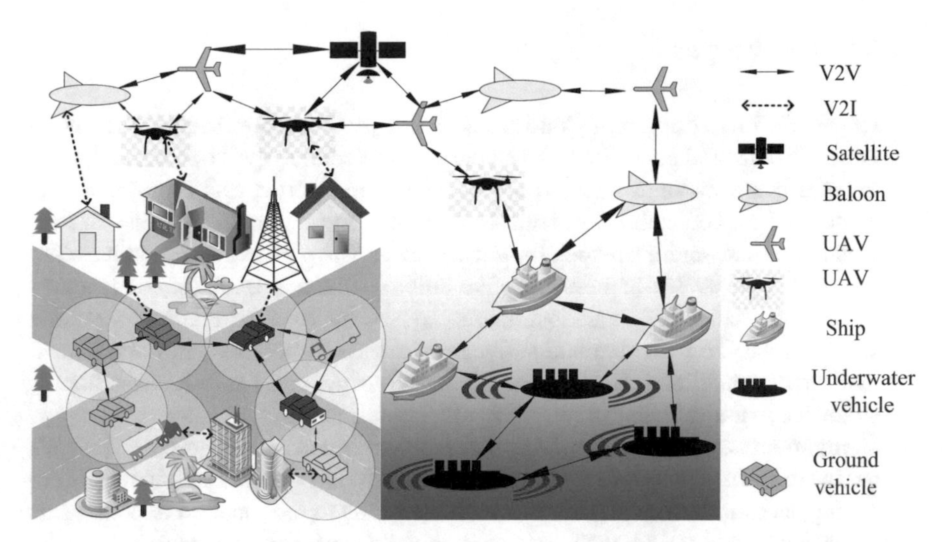

Fig. 2.3 A depiction of autonomous transportation in air, ground, and sea

vehicle. It should be remembered that the on-board sensor and actuator technologies (lidar, radar, global positioning system (GPS), video, ultrasound, etc.) are leveraged by various car companies for various forms of tailored requirements. However, the key prerequisite of being designed to function autonomously is at the heart of the autonomous car concept. In other words, the autonomous vehicle needs features that allow it to safely and accurately predict, evaluate, and drive-by any plan.

A depiction of autonomous air, land, and sea transport is shown in Fig. 2.3. It is shown that future network can be an integrated and connected network, in which vehicle could be equipped with sensors to sense the environment in their vicinity. An array of sensors including light (cameras), radio (radar), laser (lidar) and positioning systems are equipped for each vehicle (aerial, ground, and underwater). Autonomous cars, taxis, trucks, trains, and buses interact with each other on the ground via the vehicle-to-vehicle (V2V) interface, while they can interact via the vehicle-to-vehicle (V2I) interface with infrastructure such as buildings, roads, footpaths, trees, etc. Likewise, aircraft, UAVs, satellites, air taxis, flying cars, and balloons operate autonomously in the air and interact with each other via the V2V interface. Vehicles are equipped with sonar, lidar, and other sensors at sea and underwater and communicate with each other via the V2V interface.

2.4 Autonomous Transportation on Earth

In this section, we highlight practical challenges for autonomous transportation on earth.

2.4.1 Challenges

- **Uncertain Environmental Conditions:** Many problems are faced due to uncertain environmental conditions. A vehicle can immediately pull out, obstacles can fall over in a work zone, intense rain, and snow can obstruct visibility. An infinite amount of unpredictable incidents can happen on the road. By using reason, common sense, and even social norms, such as ways of interacting with other drivers, people deal with these and other unforeseen circumstances. To maneuver through unpredictable scenarios, autonomous vehicles must use hardware and software, but the hardware and software in a car can be tested in only a small number of conditions. It can be a while before we can make an assertion or guarantee that a vehicle can be safe in all circumstances.
- **Components malfunctioning:** Another problem is that an autonomous vehicle's hardware can malfunction, so redundancy needs to be built into the vehicle. From an engineering viewpoint, this is a solvable problem, but economic constraints are a consideration as well. At least initially, autonomous vehicles can be much more expensive than other vehicles, however, automation functions can ultimately be produced in massive quantities. However, until then, the advantages of autonomous vehicles may only be open to certain people [12].
- **Cyber Attacks:** Cybersecurity is another potential challenge in autonomous transportation. Hackers could intrude into the command and control system of self-driving vehicles, and it could be disastrous if multiple self-driving vehicles are taken over by criminals [2]. Therefore, cybersecurity for autonomous transportation should be made certain before the roll-out of self-driving vehicles [8].
- **Social and Psychological Challenges:** There are social and psychological problems further than technical barriers to autonomous vehicles. When you and your family are in the car, particularly driving at highway speeds, or traveling through an urban area with dense groups of pedestrians, cyclists, motorcycles, and so on, no one as a person would be willing to give up wheel control [1]. In such a situation, people, their friends, and the rest of the people around them are not going to be safe. Also, the media can be all about it when an event occurs, and no one knows what the social pushback can be. It is also probable that autonomous vehicles would create profound social dislocations.

  Autonomous vehicles can change the mindset of the people and the outlook of everything. At the individual level, the elderly and special persons can be given freedom back to their life. In the health sector, autonomous drones can transport medical goods to remote areas where other transportation systems are not available. An autonomous robot can be used for medical operations. Restaurants and hotel can change their dining in moving vehicles. Car accidents and airplane crashes can be reduced to a minimal level with the help of ATS. Companies are coming up with the idea of innovation, growth in technology in everyday life.

Connecting vehicles can have weather-related and safety issues too, e.g., when the weather is apparently not harsh, however, the road is slippery or icy. In such a scenario, the data collected from the multiple connected vehicles can be used to guide the upcoming traffic and warn the driver of a potential hazard. Real-time data from the connected vehicle can be transferred to the central traffic management system for early warning. Connected vehicle applications can help vehicles to avoid unnecessary stops by communicating with smart traffic signals. Autonomous vehicles (AV) can occupy less space and spare more space for our community. AVs can automatically repower and recharge. ATS can ensure seamless, secure, and safe mobility. Most long-haul trucking, for instance, is a very reasonable candidate for automation. From a technological point of view, technology is almost ready, while training and regulatory issues still need to be addressed. Similarly, an additional 70–75 cents of every dollar earned could be saved if companies like Uber were able to exclude drivers from vehicles [17].

- **Unemployment and re-employment:** At some point, moving to autonomous trucks, buses, taxis, and other vehicles can cause significant job losses, but it can probably take years to eliminate these jobs, providing time for preparation and adjustment. A substantial fraction of the world population worked as farmers, as did agriculture at the beginning of the nineteenth century, and a significant number worked in factories throughout the twentieth century [17]. The majority of those jobs are no longer open, yet the overall unemployment rate is less than 5%, which is a good sign.

A summary of impact of autonomous transportation on different aspects of humans life is given in Table 2.1. In the next section, we shed light on the challenges in autonomous transportation in air.

Table 2.1 Impact of autonomous transportation on different aspects of human's life

Impact	Description
Cost	• Cost of autonomous vehicle can be higher as compared to non-autonomous vehicle • Cost of trip may significantly be reduced due to no cost of driver • Cost incurred due to fines and damage to the vehicle parts, and waiting cost can significantly be reduced
Environmental impact	• There would be positive impact on the land saving, emissions, urbanization, food, and drinking water quality
Transportation	• The road quality, comfort/convenience during traveling can be improved • Journey time would be minimized with high degree of safety
Employment	• Drivers could be unemployed. However, at the same time the employment opportunities for auto-part manufacturing/repairing could be enhanced
Social impact	• People can build beds, rooms, and other luxurious items inside the vehicle. They may even start living inside the vehicles and leave their houses

2.5 Autonomous Transportation in Air

Air traffic control is about to become even more challenging when companies implement new technology that uses airspace. Unmanned aerial vehicles (UAVs), for example, could significantly add to the requirements of the air traffic control system, but in certain cases, they are very useful. For example, in a manner that is much better than sending first responders to the scene, they can collect information for rescue operations, disastrous reliefs. Giving more control and functionality to such systems would reduce risks to individuals [20].

Personal air transportation, or "flying cars," is another emerging technology that would allow individuals to travel in a single autonomous vehicle both on the air or on the ground. How should we understand the implementation of these cars, which are autonomous by nature, so the person in the cockpit is just a normal person [25]. As is the case with military UAV flights, the pilot in charge of a personal aircraft may be on the ground checking up on the control of the vehicle. A human controller could serve as a backup plan regardless of its role if the signals or on-board sensors of an aircraft were hacked [25].

The UAVs with fixed wings currently operating in the military and commercial sectors are costly, large, and have less autonomy. For the most part, small UAVs designed by universities in their research laboratories have a different architectural design than large UAVs, which need to be tackled.

2.5.1 Challenges

In this section, we discuss the key challenges in the realization of autonomous transportation in air.

- **Varying Channel Conditions:** For any application involving autonomous UAVs, channel state information (CSI) and link-budgeting is a basic requirement. Platform design, payload, and mobility limitations, however, impose significant limitations on sensing and available computational capacity. The vehicle's 3D stability, and agility require sensors that provide information on the 3D environment surrounding the vehicle with low latency [10]. However, varying channel condition may affect the information received through various sensors. The performance of V2V and V2I interface may be degraded due to hash environmental conditions. Therefore, to meet the challenges of varying channel conditions for autonomous vehicles, a robust mechanism should be designed.
- **High Mobility:** In comparison to the speed of a car traveling on a highway, another difficulty is related to the maximum speed a standard UAV can achieve. This can lead to problems for some applications, such as the flying police vehicle, since a UAV's maximum speed may be higher than a tracking vehicle's speed. By allowing the UAV to fly at a high altitude with certain speed limits can be beneficial for air traffic control and management [20, 24]. Similarly, high

mobility can cause the Doppler shift to increase, which can affect inter-UAV and UAV to ground communications.

- **Power Limitations:** Limited power availability is another issue. A standard UAV has the hover time of less than thirty minutes, which poses challenges in the flight operations with UAVs [25]. On the other hand, recent advances in battery technology, such as hydrogen fuel cells, and improved lithium-ion batteries, more energy-efficient UAV designs, and the use of alternative sources of energy, such as solar energy, to extend flight missions, would enable UAVs to fly for several hours in the future.

In the next section, we discuss the autonomous transportation in sea and related challenges.

2.6 Autonomous Transportation in Sea

Historically, people have used ships to explore the ocean. However, over two-thirds of the earth is occupied by the ocean, and ship-based exploration is costly and often risky. As a result, much of the ocean remains undiscovered, unpaved, and poorly defined in a scientific sense. The ocean is huge, deep, and dark, full of wonders. Cameras on underwater vessels, for instance, suggest that they are being surrounded by fish colonies, even in whales' waters in which such activity would appear to pose risks.

The ocean is a dynamic and shifting world below the sea, marked by tides, eddies, and large amounts of life, much of which is poorly understood. For example, until the 1970s, organisms living off the chemical and heat energy from hydro thermal vents were not discovered [30]. Today, robots are used to complement and supplement ships from above, on, and below the ocean floor. Instead of needing a ship to follow them, robots can run for weeks underwater, travel tens of kilometers, learn from unpredictable circumstances, and work from radio waves. We would also like to be able to anticipate the condition of the ocean, and using lots of robots is the only way we know how to do it.

Using robots, autonomous submarines, and ships in the ocean can eliminate the need to be in dangerous conditions for humans, as shown in Fig. 2.4. The ocean environment is similar to that of space in many respects. Externally, maintaining human protection at a water depth of thousands of meters or tens of thousands of pounds per square inch is very costly. For a group of individuals who do little other than operating the platform, one must construct a very big, complicated platform supported by a reasonably sophisticated surface vessel. Robots can lower costs, time, and risks by replacing individuals on oceanic platforms. Searches for ships or aircraft lost at sea, for example, can continue with robots seven to ten times faster than with ships.

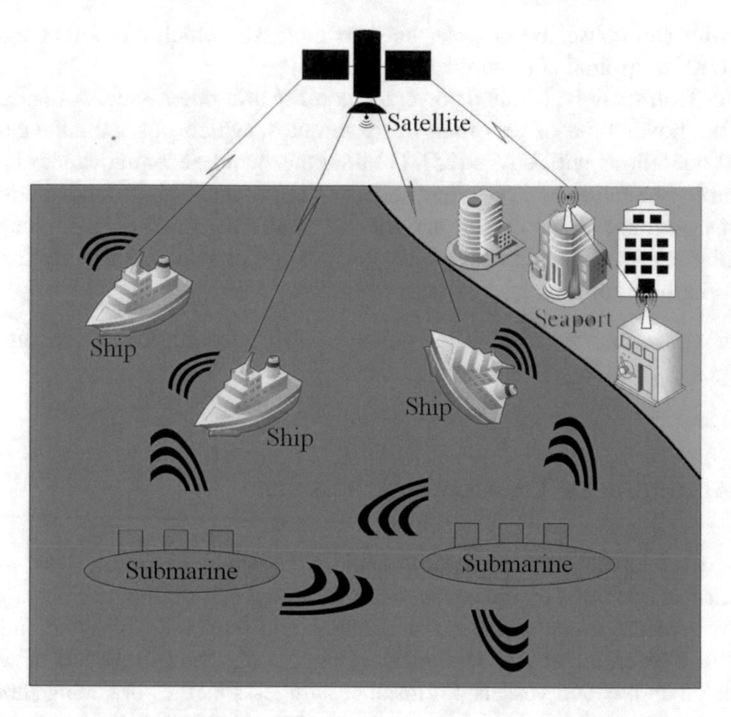

Fig. 2.4 Autonomous transportation in sea

2.6.1 Challenges

In this segment, we discuss the real-time issues of autonomous maritime transportation.

- **Different Radio Environment:** A fundamental challenge is faced by robots built to operate underwater is different channel conditions for radio waves [6]. It is a conductor since the ocean contains salt, ensuring that the radio waves used above ground to power vehicles cannot be used underwater. Consequently, untethered vehicles must be autonomous in the ocean climate that would be different from that of autonomous vehicles on the ground or in the air.
- Sensing Difficulties: The underwater atmosphere also presents major sensing problems. Cameras are not able to see very far, sonar systems are costly and consume fuel, and navigation is difficult. To communicate with a GPS, some machines have to surface; human guidance is unavailable in other situations, as when a robotic system searches under an ice cap that stretches to the sea coast [14].

2.7 Potential Technologies for Autonomous Transportation Systems

Imagine a future where vehicles talk to each other and the infrastructure to form an ATS. This ATS can become the reality with the aid of a safe and reliable wireless network that can make communication among the vehicles, infrastructure, and human beings. These connected vehicles can provide the drivers with 360-° awareness of all of its surrounding up to a distance of approximately 300 km. The personal information of the passenger can be kept private and hackers cannot be allowed to track or get sensitive information. Multiple warning applications can guide the driver to navigate the vehicle accordingly. In the this section, we discuss key enabling technologies for autonomous systems.

2.7.1 Blockchain

Blockchain, commonly regarded as one of the revolutionary technologies developed in recent years, is rapidly evolving and has the full potential to revolutionize applications with increasingly centralized ATS [2, 26, 31]. To build a stable, trustworthy, and decentralized autonomous intelligent transportation system (ITS) ecosystem, Blockchain can be used to create better use of the traditional ITS resources and infrastructure, especially efficient for crowdsourcing technology.

2.7.2 Artificial Intelligence and Machine Learning

In various applications, such as ATSs, automation with underlying artificial intelligence (AI) is evolving rapidly. Despite their use, however, there is still significant doubt about such applications in society [29]. The principle of trust provides a practical basis for explaining the connection between man and machines, making an analogy from social behavior. The implementation of intelligent automation includes autonomous driving. Automation is generally defined as a technique that effectively collects data, transforms information, and processes decision-making or control. Inherent artificial intelligence (AI) is the foundation of the decision-making method used in technology, hence the word 'intelligent automation.' Among many industries that are significantly affected by automation including underlying AI, the transportation sector is just one.

2.7.3 Edge/Fog Computing

The continuous growth of Autonomous Transport and the recent Internet of Vehicles (IoV) is aimed at addressing existing and new problems in the field of transportation systems [7]. Edge computing provides the Internet of Vehicles with natural support, enabling rapid response, sensing capabilities and reducing data transfer to centralized servers, both of which are facilitated by the availability of edge computing near mobile vehicles [3, 4, 15, 27]. These candidates for supporting IoV are mobile edge computing, fog computing, cloudlets, etc. Their frameworks and technologies have overlapping features, but also differences in approach, a complete convergence still has not been established between them [21].

2.8 Conclusion

The concept of a self-driving car seems like a distant dream; however, autonomous technology is advancing rapidly and on current car models several semi-autonomous features already are provided. This chapter discussed various levels of transportation automation with their immense benefits in human life, such as their ability to minimize accidents, ease traffic, reduce fuel consumption, minimize parking requirements, add independence to those unable to drive, and reduce waiting time. We also shed light on the state-of-the-art of autonomous vehicles. In addition, the issues of automated air, land, and sea transportation are addressed. Finally, as core innovations in automated transportation, we explore the promise of blockchain, artificial intelligence, edge computing, network virtualization.

References

1. Acheampong, R. A., Thomoupolos, N., Marten, K., Beyazıt, E., Cugurullo, F., & Dusparic, I. (2018). Literature review on the social challenges of autonomous transport. In *STSM Report for COST Action CA16222" Wider Impacts and Scenario Evaluation of Autonomous and Connected Transport (WISE-ACT)*.
2. Akhtar, M. W., Hassan, S. A., Ghaffar, R., et al. (2020). The shift to 6G communications: Vision and requirements. *Human-centric Computing and Information Sciences, 10*(53).
3. Aujla, G. S., Singh, A., Singh, M., Sharma, S., Kumar, N., & Raymond Choo, K.-K. (2020). BloCkEd: Blockchain-based secure data processing framework in edge envisioned V2X environment. *IEEE Transactions on Vehicular Technology, 69*(6), 5850–5863.
4. Aujla, G. S., Kumar, N., Garg, S., Kaur, K., et al. (2019). EDCSuS: Sustainable edge data centers as a service in SDN-enabled vehicular environment. *IEEE Transactions on Sustainable Computing*.
5. Bagloee, S. A., Tavana, M., Asadi, M., & Oliver, T. (2016). Autonomous vehicles: challenges, opportunities, and future implications for transportation policies. *Journal of Modern Transportation, 24*(4), 284–303.
6. Board, N. S., National Research Council, et al. (2005). *Autonomous Vehicles in Support of Naval Operations*. National Academies Press.

7. Borcoci, E., Vochin, M., & Obreja, S. (2018). Mobile edge computing versus fog computing in internet of vehicles. In *Proceedings of the 10th International Conference on Advances in Future Internet* (pp. 8–15).
8. Cheema, M. A., Shehzad, M. K., Qureshi, H. K., et al. (2020). A drone-aided Blockchain-based smart vehicular network. *IEEE Transactions on Intelligent Transportation Systems*.
9. Csiszár, C., & Földes, D. (2018). System model for autonomous road freight transportation. *Promet-Traffic&Transportation, 30*(1), 93–103.
10. Elliott, D., Keen, W., & Miao, L. (2019). Recent advances in connected and automated vehicles. *Journal of Traffic and Transportation Engineering (English Edition), 6*(2), 109–131.
11. Fagnant, D. J., & Kockelman, K. (2015). Preparing a nation for autonomous vehicles: opportunities, barriers and policy recommendations. *Transportation Research Part A: Policy and Practice, 77*, 167–181.
12. Finn, A., & Scheding, S. (2010). Developments and challenges for autonomous unmanned vehicles. *Intelligent Systems Reference Library, 3*, 128–154.
13. Földes, D., & Csiszár, C. (2016). Passenger handling functions in autonomous public transportation. In *3rd International Conference on Traffic and Transport Engineering (ICTTE)*.
14. Foresti, G. L. (2001). Visual inspection of sea bottom structures by an autonomous underwater vehicle. *IEEE Transactions on Systems, Man, and Cybernetics, Part B (Cybernetics), 31*(5), 691–705.
15. Giang, N. K., Leung, V. C. M., & Lea, R. (2016). On developing smart transportation applications in fog computing paradigm. In *Proceedings of the 6th ACM Symposium on Development and Analysis of Intelligent Vehicular Networks and Applications* (pp. 91–98).
16. Hancock, P. A., Nourbakhsh, I., & Stewart, J. (2019). On the future of transportation in an era of automated and autonomous vehicles. *Proceedings of the National Academy of Sciences, 116*(16), 7684–7691.
17. Jameel, F., Chang, Z., Huang, J., & Ristaniemi, T. (2019). Internet of autonomous vehicles: architecture, features, and socio-technological challenges. *IEEE Wireless Communications, 26*(4), 21–29 (2019).
18. Johnson, C. (2017). Readiness of the road network for connected and autonomous vehicles. *RAC Foundation: London*.
19. Kim, T. J. (2018). Automated autonomous vehicles: Prospects and impacts on society. *Journal of Transportation Technologies, 8*(03), 137.
20. Kumar, V., & Michael, N. (2012). Opportunities and challenges with autonomous micro aerial vehicles. *The International Journal of Robotics Research, 31*(11), 1279–1291.
21. Liu, S., Liu, L., Tang, J., Yu, B., Wang, Y., & Shi, W. (2019). Edge computing for autonomous driving: Opportunities and challenges. *Proceedings of the IEEE, 107*(8), 1697–1716.
22. Lu, M., Wevers, K., & Van Der Heijden, R. (2005). Technical feasibility of advanced driver assistance systems (ADAS) for road traffic safety. *Transportation Planning and Technology, 28*(3), 167–187.
23. Menouar, H., Guvenc, I., Akkaya, K.,Uluagac, A. S., Kadri, A., & Tuncer, A. (2017). UAV-enabled intelligent transportation systems for the smart city: Applications and challenges. *IEEE Communications Magazine, 55*(3), 22–28.
24. Muntaha, S. T., Hassan, S. A., Jung, H., et al. (2020). Energy efficiency and hover time optimization in UAV-Based HetNets. *IEEE Transactions on Intelligent Transportation Systems*.
25. Pines, D. J., & Bohorquez, F. (2006). Challenges facing future micro-air-vehicle development. *Journal of Aircraft, 43*(2), 290–305.
26. Saranti, P. G., Chondrogianni, D., & Karatzas, S. (2018). Autonomous vehicles and blockchain technology are shaping the future of transportation. In *The 4th Conference on Sustainable Urban Mobility* (pp. 797–803). Springer.
27. Singh, A., Aujla, G. S., & Bali, R. S. (2020). Intent-based network for data dissemination in software-defined vehicular edge computing. *IEEE Transactions on Intelligent Transportation Systems*.
28. Smith, B. W. (2012). Managing autonomous transportation demand. *Santa Clara Law Review, 52*, 1401.

29. Varlamov, O. O., Chuvikov, D. A., Aladin, D. V., Adamova, L. E., & Osipov, V. G. (2019). Logical artificial intelligence Mivar technologies for autonomous road vehicles. In *IOP Conference Series: Materials Science and Engineering* (vol. 534, pp. 012015). IOP Publishing.
30. Yoerger, D. R., Jakuba, M., Bradley, A. M., & Bingham, B. (2007). Techniques for deep sea near bottom survey using an autonomous underwater vehicle. *The International Journal of Robotics Research, 26*(1), 41–54.
31. Yuan, Y., & Wang, F.-Y. (2016). Towards blockchain-based intelligent transportation systems. In *2016 IEEE 19th International Conference on Intelligent Transportation Systems (ITSC)* (pp. 2663–2668). IEEE.

Part II
Artificial Intelligence

Chapter 3
Artificial Intelligence

Zhiwei Guo and Keping Yu

3.1 AI Conception

3.1.1 Cognitive AI

The concept of "artificial intelligence (AI)" dates back to the mid-twentieth century in the United States, proposed by some computer scientists, information scientists, neurophysiologists, and so on. Nowadays, AI has a history of more than 60 years, and involved in many fields. Generally speaking, artificial intelligence is based on exploring the activities of human intelligence, using intelligence technology to create artificial systems, in order to endow human intelligence with computer systems to replace traditional labor. The tasks can only be completed by human intelligence in the traditional mode; it can be handled efficiently with computer hardware and software in the era of artificial intelligence.

AI is a discipline that learns, understands, and simulates human intelligence, intelligent behavior and its laws. It seeks to understand the nature of intelligence and produce a new kind of intelligent machine that can respond in a similar way to human intelligence. In the past thirty years, it has achieved rapid development and has been widely used and has achieved fruitful results in many disciplines such as machine translation, intelligent control, expert systems, robotics, natural language processing, and image recognition.

Z. Guo
School of Artificial Intelligence, Chongqing Technology and Business University, Chongqing, China
e-mail: zwguo@ctbu.edu.cn

K. Yu (✉)
Global Information and Telecommunication Institute, Waseda University, Shinjuku, Tokyo, Japan
e-mail: keping.yu@aoni.waseda.jp

S. Garg et al. (eds.), *Intelligent Cyber-Physical Systems for Autonomous Transportation*, Internet of Things, https://doi.org/10.1007/978-3-030-92054-8_3

People had different understandings of artificial intelligence in the early days. Some people think that AI is synonymous with any form of intelligence realized through non-biological systems, and it does not matter whether the realization of intelligence is the same as that of AI. Others believe that AI systems must be able to imitate human intelligence. With the development and application of AI technology, the definition of AI is more inclined to the first statement.

AI can be divided into "strong AI" and "weak AI." Those who hold a strong AI view believes that, it is possible to create an intelligent machine that can actually reason and solve problems, and that is sentient and self-aware. Strong AI is often divided into two categories: one is human-like AI, in which machines thinking and reason just like humans; the other is non-human-like AI, in which machines generate completely different perceptions and consciousness from humans and use completely different reasoning methods from humans. And the Weak AI is the opposite of strong AI, they do not believe that they can create an intelligent machines, which can realize thinking like humans. They think the machines created may be seems to intelligent, but they are not intelligent like we thought, not to mention being self-aware.

Professor Nelson defined AI as "The discipline of knowledge, the science of how knowledge is represented and how knowledge is acquired and used." And another Professor Winston of Massachusetts Institute of Technology thinks: "Artificial intelligence is to study how to make the computer to do the intelligent work that only human can do in the past." The definition of AI on Wikipedia is: Artificial intelligence refers to the intelligence displayed by machines made by humans.

3.1.2 Machine Learning AI

Artificial intelligence is the intelligence represented by systems made by humans. Its branch domain mainly focuses on solving specific practical problems, one of important thing is how to complete specific problems by using different tools. The core of AI including reasoning, knowledge, planning, learning, communication, perception, and the ability is to operate objects.

Machine learning (ML) is a means to achieve AI, and deep learning is a technology that implements ML, the relationship between the three is shown in Fig. 3.1. ML is to enable machines to automatically learn the relationships in the data, then make predictions and decisions about future events based on the learning. Deep learning is a technology that uses a series of "deep-level" neural network models to solve more complex problems [14, 16].

ML is a multidisciplinary subject that studies algorithms from data. It studies human behaviors and how can computers simulate to be like that. According to the existing data or past experience, select algorithms, build models, predict new data, and reorganize the existing knowledge structure to continuously improve its performance. ML usually refers to a kind of problem and the method to solve such problem, how to find rules from observed data (samples), and use the learned rules

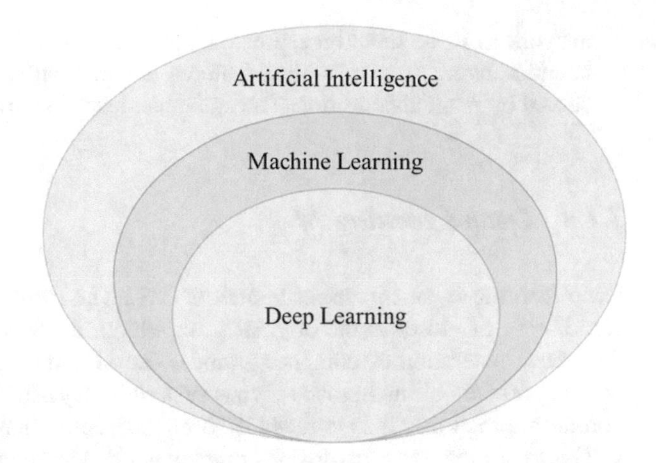

Fig. 3.1 The relationship between AI, ML, and deep learning

(models) to predict unknown or unobservable data. ML includes the following four categories:

- **Supervised Learning (SL):** SL refers to the process of using labeled samples to optimize the parameters of the algorithm and improve its performance. Use the samples (x, y) with a known characteristic as the training set, build a mathematical model, solve $f : x \rightarrow y$, and predict the unknown samples. The dataset used by SL contains not only features but also labels. Based on these labels, design a learning strategy (loss function) to optimize the model.
- **Unsupervised Leaning (UL):** The training dataset has only features x and no labels y. The purpose is to try to extract the structure and laws implicit in the data. Common algorithms include K-means, dimensionality reduction, and text processing (feature extraction). UL is generally used as the preliminary data processing of SL, and its function is to extract the necessary label information from the original dataset.
- **Semi-Supervised Learning (SSL):** SLL is a combination of SL and UL, using a small number of labeled samples and a large number of samples without label to training and testing. SSL includes three basic assumptions:

  1. Smoothness Assumption: If two samples are joined by an edge in a dense data region, then there is a high probability that they will have the same class label.
  2. Cluster Assumption: If two samples are in the same cluster, then there is a high probability that they have the same class label.
  3. Manifold Assumption: The high-dimensional data is embedded in the low-dimensional manifold. If two examples are located in a small local neighborhood in the low-dimensional manifold, then they have similar class labels.

- **Reinforcement Learning (RL):** RL is performed by the agent in a trial and error manner. The agent continuously interacts with the environment to obtain rewards or punishments. The goal is to maximize the reward value that the agent finally obtains. Deep reinforcement learning is essentially a kind of method using neural network as value function estimator. The main advantage is that it can use deep neural network to extract state features automatically, avoiding the inaccuracy caused by manual definition. The agent can learn in a more primitive state.

3.1.3 Deep Learning AI

Deep learning is an important branch of ML. The introduction of deep learning could bring us closer to the original goal, which means artificial intelligence[15]. As a new generation of computing mode, deep learning attempts to simulate the working process of the human nervous system by layering and combining multiple nonlinear functions. It is a method based on representation learning of data in machine learning. By constructing complex nonlinear structures formed by multiple intermediate hidden layers, it can better deal with the complex real problems. Its technological breakthrough has set off a new wave of development in artificial intelligence.

Deep learning is to learn the internal rules and representation levels of sample data. The ultimate goal is for machines to be as analytical and learning as humans, able to recognize data such as words, images, and sounds. Deep learning is a complex machine learning algorithm that has far outperformed previous technologies in speech and image recognition.

With the rise of deep learning in recent years, the prediction accuracy and classification accuracy achieved by the latest deep learning algorithm have far exceeded that of the traditional machine learning algorithm. Deep learning does not need feature selection by manual, it can extract the features from the data automatically and ignore the information unimportant. Compared with the general method of supervised learning in traditional machine learning, feature engineering is less, and a lot of working time is saved.

The Artificial Neural Network (ANN) algorithm in deep learning is different from the traditional computing model. Essentially it is a multi-level artificial neural network algorithm, imitating the neural network of the human brain. The operating mechanism of the human brain is simulated from the most basic unit. It can spontaneously sum up rules from a large amount of input data, and then draw inferences about it and apply it to other scenarios. Therefore, it does not need to manually extract the characteristics of the problem to be solved or summarize the rules for programming.

Deep learning and ML are inclusive relationships, and deep learning is a subcategory of ML. For beginners, deep learning can be understood as a "multi-layer neural network." Strictly speaking, deep learning is a learning mode, which refers to the use of a "deep" model for learning, and it is not a model in itself. The essence of deep learning is the learning of complex nonlinear models. From the perspective of the development history of machine learning, the rise of deep learning represents the natural evolution of machine learning technology. In 1957, Rosenblatt proposed the Perceptron model, which is a linear model and can be regarded as a two-layer neural network. In 1986, Back Propagation (BP) algorithm was proposed, which is used in a three-layer neural network, and it can represents a simple nonlinear model. In 1995, Support Vector Machine (SVM) was invented. The Radial Basis Function (RBF) kernel SVM is equivalent to a three-layer neural network actually, and it can also be thought of a simple nonlinear model. Since 2006, deep learning actually uses more than three layers of neural networks, also known as deep neural networks, which are complex nonlinear models.

There are several variants of deep neural networks, such as Convolutional Neural Network (CNN) and Recurrent Neural Network (RNN). The deep neural network is actually a complex nonlinear model with a complex structure and a large number of parameters. It has a very strong representation ability and is especially suitable for complex pattern recognition problems. But deep learning is not a panacea. First, deep learning is not suitable for all problems. If the problem is simple, such as linear problems and simple nonlinear problems, deep learning is at best as accurate as support vector machines. If learning falls into a local optimum, it may not be as excellent as other methods. In addition, if the amount of training data is not large enough, the deep neural network cannot be fully learned, and the effect will not be very outstanding.

3.2 Need and Evolution of AI

3.2.1 Origin and Development About AI

The timeline of AI from origin to the full development is shown in Fig. 3.2.

- **Birth about AI: 1943–1956**: In the 1940s and 1950s, a group of scientists from different fields (mathematics, psychology, engineering, economics, and politics) began to explore the possibility of making artificial brains. In 1943, psychologist McCulloch and mathematical logician Pitts published mathematical models of neural networks in the "Journal of Mathematical Biophysics." In 1945, John von Neumann proposed the concept of stored procedures. American mathematician

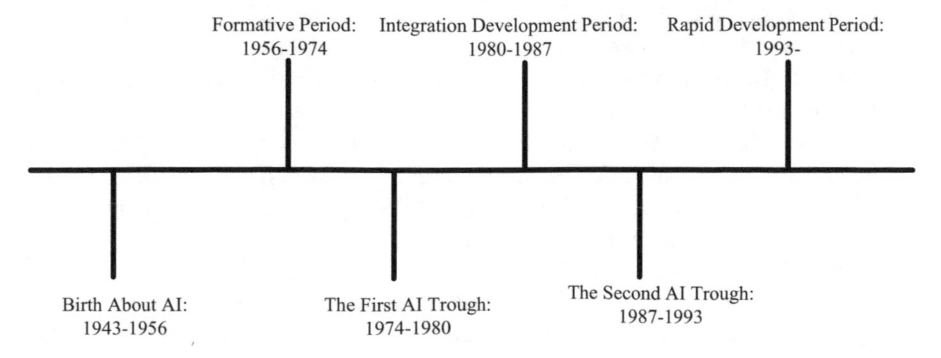

Fig. 3.2 The timeline for the development of AI

and pioneer of electronic digital computers, Mokley and others developed the world's first general-purpose electronic computer in 1946. In 1948, Shannon published "The Mathematical Theory of Communication," marking the birth of information theory. In the same year, Wiener founded Cybernetics, which is the subject of studying and simulating automatic control of biological and artificial systems. In 1950, Turing published a famous paper "Computer Machines and Intelligence," which clearly put forward the viewpoint of "machines can think." At this point, the embryonic form of artificial intelligence has initially taken shape, and the conditions for the birth of artificial intelligence are basically available.

- **Formative Period: 1956–1974** The years following the Dartmouth Conference in 1956 were the era of great discoveries. For many people, the programs developed at this stage are magical: computers can solve algebra problem, prove geometric theorems, learn and use English. At the time, most people could hardly believe that machines could be so "smart." In 1956, Samuel successfully developed a chess program with self-learning, self-organization, and self-adaptation capabilities. In 1957, Rosenblatt developed the perceptron, a system that uses artificial neurons for recognition. From 1958 to 1960, McCarthy established an action planning consulting system and AI language. In 1965, Robinson proposed the resolution principle, which brought a major breakthrough in the research of machine proof. In 1968, a research team led by Feigenbaum at Stanford University proposed the chemical expert system DENDRAL. In 1969, the first International Joint Conference on Artificial Intelligence (IJCAI) was held, marking that AI has entered the international academic arena as an independent discipline.
- **The First AI Trough: 1974–1980** After the rapid development of AI in the formation period, it soon encountered a lot of troubles. In terms of the game, the chess program designed by Samuel lost four out of five games when playing against the world champion. In the proof of theorem, Robinson proposed the resolution principle and it was found that the proof ability was limited. In terms of problem solving, since most of the problems studied in the past are problems with good structures, but in the real world, they are mostly with poor structures. If we still use the original method to deal with it, there will be a combinatorial explosion problem. In the 1970s, AI began to meet skepticism, and with it there were financial difficulties. The researchers failed to judge the research subjects correctly. The previous optimism raised expectations for AI too high, and funding for AI began to be reduced or withdrawn when promises were not met. At the same time, Marvin Minsky launched a fierce criticism of the perceptron, leading to the disappearance of connectionism (neural networks) for ten years. At the time, even the best AI could only solve the simplest part of the problem. AI research ran into insurmountable obstacles.
- **Integration development period: 1980–1987** At the beginning of 1980, research on ANN began to recover. A class of AI programs called "expert systems" began to be adopted by companies all over the world, and "knowledge processing" has become the focus of mainstream AI research. In 1982, the

breakthrough of Hopfield network proposed by Hopfield once again aroused people's enthusiasm for neural network research. In 1984, Hinton developed Boltzmann, using the "simulated annealing" method to enable the system to jump out of the local minimum state and tend to the global minimum state. In 1986, Rumhardt and others proposed a new generation of multi-layer perceptron Back Propagation (BP) neural networks. It breaks through the limitations of simple perceptron and improves the recognition ability of multi-layer perceptron. In 1987, the first International Neural Network Society (INNS) set off the second climax of ANN research.

- **The Second AI Trough: 1987–1993** In the late 1980s, the pursuit and neglect of AI of commercial organization was in line with the classic pattern of economic bubbles, also the bursting was observed by government agencies and investors of AI. The term "AI Winter" was coined by researchers who experienced financial difficulties in 1974. They note the craze for expert systems and predict a turn to disappointment in the near future. Unfortunately, they stated the truth. Despite all the criticisms, this field is still advancing.
- **Rapid Development Period: 1993–** Since the beginning of the twenty-first century, with AI as the core, the integrated development model of natural intelligence, AI, integrated intelligence and collaborative intelligence has attracted great attention. During this period, the main characteristics of intelligent science and technology research mainly included the following four aspects: (1) From a single research on artificial intelligence to a research on collaborative intelligence integrating natural intelligence, artificial intelligence, and integrated intelligence; (2) From the independent research of artificial intelligence disciplines to the intersection of brain sciences, cognitive sciences and other disciplines; (3) From separate studies of multiple different schools to comprehensive studies of multiple schools; (4) From the study of individual and centralized intelligence to the study of group and distributed intelligence. The Internet provides an important research, popularization and application platform for the development of AI.

3.2.2 Common Approaches and Technologies

3.2.2.1 Natural Language Processing (NLP)

NLP is a technology for studying computer processing of human language, and it is an intersecting research field combining computer science and linguistics. Generally speaking, the tasks faced by natural language processing mainly include three aspects: language perception, language understanding, and language generation. Language perception is equivalent to the computer's "listening" and "reading" abilities, and is the main information input part in human-computer interaction; Language understanding is the main task and core challenge of natural language processing research, which is equivalent to human language "thinking" and "understanding" ability; Language generation is the study of how to let the computer express the extracted information in a smooth and fluent language. This expression can be text or speech.

Natural language processing tasks cover a relatively large range and are widely used at the same time. Here, according to the complexity of the objects involved in the language understanding process from low to high, the main content of natural language processing research is introduced in the following aspects.

- **Voice and Text Recognition:** Text is the main object of natural language processing. Therefore, in order to facilitate processing, it is usually necessary to use speech recognition to convert speech or pictures into text, and then do further analysis and understanding. This process is usually processing at the character and word level.
- **Text Analysis:** A sentence usually includes words of different parts of speech, phrases, and clauses of different functions and attributes. Text parsing includes lexical and syntactic parsing. Its main task is to correctly segment a given text, word division and part-of-speech tagging, fixed phrase recognition, and sentence relationship judgments to facilitate subsequent analysis and understanding. This process is to process the text at the level of words and sentences.
- **Text Analysis and Mining:** This process is to classify and cluster text, topic recognition, sentiment analysis, abstract compression, and visual representation from the full text content level. Text analysis and mining are often no longer limited to a single word or sentence, but a deeper processing on the overall content of the text.
- **Machine Translation:** Similar to human translation, machine translation converts text or speech from one language to another. At the same time, try to accurately and completely convey the information contained in the source language. This process involves the accurate extraction of the information contained in the source language, the appropriate way of conversion between languages, and the correct and smooth generation of the target language.
- **Question and Answer/Dialogue System:** The ultimate goal of natural language processing is to design a system that can complete a natural communication or dialogue similar to that between people, so as to achieve efficient and smooth interaction between humans and machines. This process is a comprehensive language processing process, including the entire process of language perception, language understanding, and language generation. At the same time, it is required to be able to correctly understand the meaning of the language according to the dialogue situation.

3.2.2.2 Artificial Neural Networks (ANN)

ANN abstracts biological neuron networks from the perspective of information processing, establishes neuron models, and composes artificial neural networks with different functions according to different neuron connection methods and learning strategies.

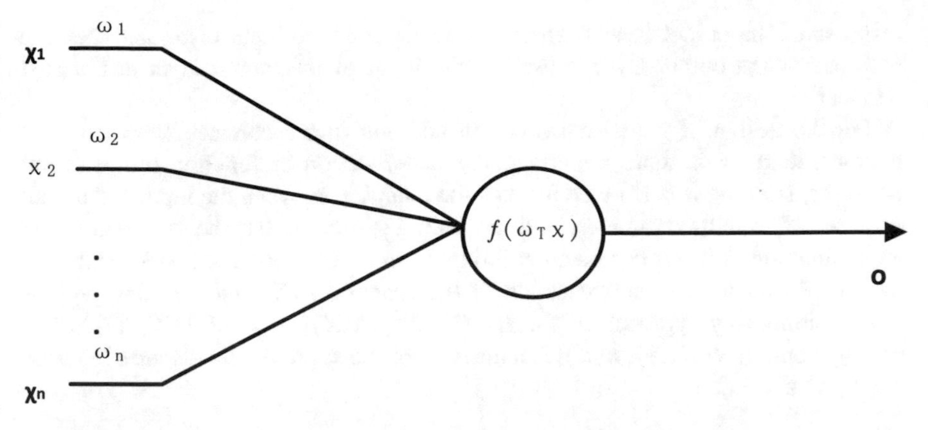

Fig. 3.3 General artificial neural network model

The main functions of artificial neural networks include pattern classification, function fitting, associative memory, optimization problem solving, pattern matching, and pattern clustering. Among them, classification is the most important function of neural networks. For the recognition of different patterns (such as faces, fingerprints, etc.), after feature extraction, classification is required to realize pattern recognition. ANN has been widely studied and applied due to its self-learning and self-adaptation, fault tolerance and parallelism. Classical ANN have been successfully applied in many fields such as pattern recognition (such as character recognition), intelligent robot planning, and intelligent control.

Figure 3.3 is a general artificial neural network model. Among them, $x = [x_1, x_2, ...x_n]$ is the input vector, $w = [w_1, w_2, ..., w_n]$ is the weight vector, o is the output of the neuron, and f is the response function.

$$o = f\left(\sum_{i=1}^{n} w_i^T x_i\right)$$

The response function usually selects the threshold function and the continuous sigmoid function, as shown below.

$$f(\omega) \triangleq \begin{cases} 1, \omega > 0 \\ 0, \omega < 0 \end{cases}$$

$$f(\omega) \triangleq \frac{1}{1 + \exp(-\omega)}$$

Neural network is divided into forward network and feedback network according to the connection mode of neurons. The forward network obtains the response of the output layer from the input of the input layer, and there is no feedback between the output layer and the input layer node. The feedback network obtains the initial response of the output layer from the initial input of the input layer, and then the response of the output layer is used as the input of the output layer at the next moment.

For the design of neural networks, in addition to the connection structure of neurons, learning is more important. For classification or function fitting neural networks, learning is performed through the mapping between the input and output of a set of training examples, and the learning process can be regarded as an approximation process. For function fitting, suppose the neural network $H(W, X)$, then $H(W, X)$ as the approximation of the function $h(X)$, the learning process is to continuously approach $h(X)$. $d[H(W, X^*), h(X)] <= d[H(W, X), h(X)]$, among them, $d[H(W, X), h(X)]$ is a measure of the approximate distance between $H(W, X)$ and $h(X)$.

3.2.2.3 Computer Vision

Computer vision also known as machine vision, is a discipline that "teaches" computers how to "see" the world, further, machine vision of identifying, tracking, and measuring targets in the place of the human eye [9]. And further doing image processing by computer into images more suitable for human eye observation or instrument detection. As a scientific discipline, computer vision research related theories and techniques, trying to establish artificial intelligence systems that are capable to obtain information from images or multi-dimensional data. Learning and computing to enable machines better understand the picture environment and build truly intelligence visual systems. There are a lot of picture and video content in the current environment, which urgently need scholars to understand and find patterns to reveal details that we have not noticed before. In recent years, the relatively basic and popular directions in computer vision mainly include: object recognition and detection, motion and tracking, visual question and answer. In addition, there are semantic segmentation, 3D reconstruction, action recognition, and so on.

- **Object Recognition and Detection:** Object recognition and detection, that is, given an input picture, the algorithm can find out the common objects in the picture automatically, and the category and location of its output. Of course, it also derived such as Face Detection, Vehicle Detection, and other subclassification detection algorithms. Object detection has always been a very basic and important research direction in computer vision. Most of the new algorithms or deep learning network structures are first applied in object detection, such as VGG-net, GoogLeNet, Resnet, etc.
- **Motion and Tracking:** For now, the evaluation criteria for tracking in academia is that the position and scale of the tracked object are given in the first frame of a given video. In the subsequent videos, the tracking algorithm needs to find the position of the tracked object from the video and adapt to various illumination transformation, motion blur and apparent changes. Tracking is also one of the basic problems in the field of computer vision. In recent years, it has been developed in a very sufficient way. The method has been transferred from the

past non-deep algorithm to deep learning algorithm, and the accuracy is getting higher and higher.

- **Visual Question and Answer:** Visual question and answer (VQA), is a very popular field in recent years. The purpose of the research is to ask questions based on an input image, and the answer is answered by an algorithm automatically. In addition to question and answer, there is an algorithm called Caption Generation, where the computer generates a paragraph of text describing an image without asking questions automatically. Algorithms that span two data forms, such as text and image, are sometimes referred to as multi-modal, or cross-modal problems.

3.2.2.4 Expert System

Expert system is defined as an interactive and reliable computer-based decision system that uses factual and heuristic methods to solve complex decision problems. An expert system is a computer program that uses the facts that exist in a particular problem domain. To draw similar conclusions from these facts in a manner similar to that of human experts who have the same facts. Such programs need to access all the facts in the domain and draw conclusions from the same facts. Such expert systems are sometimes referred to as rule-based or knowledge-based systems.

3.2.3 Successive Cross-Field Solutions

The successive cross-field solutions are shown in Fig. 3.4 and discussed below.

Smart Manufacturing

Smart Cities

Smart Healthcare

Smart Education

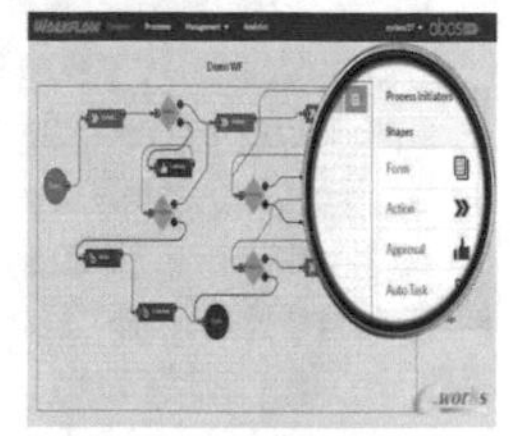

Smart Workflow

Fig. 3.4 The successive cross-field solutions

3.2.3.1 Smart Manufacturing

Smart manufacturing is a man-machine integrated intelligent system composed of intelligent machines and human experts. It can carry out intelligent activities in the manufacturing process, such as analysis, reasoning, judgment, conception, and decision-making. To enlarge, extend and partially replace the mental work of human experts in the manufacturing process through the cooperation of human and intelligent machines. It extends the concept of manufacturing automation to flexibility, intelligence, and high integration. The integrated application of AI in the manufacturing industry is the focus of promoting the development of the real economy and the key to the digital, networked, and intelligent transformation of the manufacturing industry. Common smart manufacturing includes smart factories, smart equipment and smart robots.

Smart manufacturing is to realize the intelligence and innovation of the whole manufacturing value chain, and it is a further improvement of the deep integration of informatization and industrialization. Smart manufacturing combines information technology, advanced manufacturing technology, automation technology and artificial intelligence technology. Smart manufacturing system has the ability of data acquisition, data processing and data analysis. It can accurately execute instructions and realize closed-loop feedback. The trend of smart manufacturing is to truly realize "Intelligent." It includes the smart factories, smart equipment, smart robots, and so on. Finally, intelligent decision-making is realized.

- **Smart Factories:** Smart production, smart factory, and smart manufacturing are three different stages of enterprise intelligence, which are different but closely connected to each other. Smart factories use various modern technologies to realize factory office, management and production automation, achieve standardization and strengthen corporate management, reduce work errors, plug various loopholes, improve work efficiency, ensure safe production, provide decision-making reference, strengthen external contacts, and broaden the purpose of the international market.
- **Smart Equipment:** Smart manufacturing equipment is defined as: manufacturing equipment with the functions of perception, analysis, reasoning, decision-making, and control. It is the integration and deep fusion of advanced manufacturing technology, information technology and intelligent technology. Its main technical capabilities include: the ability to perceive, process, and analyze the equipment operating status and environment in real-time; the ability to independently plan, control, and make decisions based on changes in equipment operating status; the ability to self-diagnose and self-repair faults; the ability to actively analyze and maintain the degradation of its own performance; the ability to participate in network integration and network collaboration.
- **Smart Robots:** The fundamental purpose of smart robots is to simulate human thinking with computers. Smart robots simulate human intelligent activities by using mechanical, electronic, optical, or biological devices to create a device or machine. During its inception, many researchers are committed to studying and

summarizing the patterns and laws of human thinking, and through the simulation of intelligent activities by smart robots, they can understand human thinking and motivation at a deeper level.

3.2.3.2 Smart City

With the help of emerging technologies like cloud computing, the Internet of Things (IoT), big data. Smart city provides cities with more convenient and efficient decision-making tools and more humanized service models through the collection, sorting, analysis and storage of massive data, as well as data interconnection, exchange, and sharing, so as to realize more efficient and green production processes [23]. Smart City include smart transportation, smart car, and smart city management.

- **Smart Transportation:** Intelligent transportation effectively integrates advanced information technology, data communication transmission technology, electronic sensing technology, control technology and computer technology in the entire ground transportation management system, so as to achieve real-time, accurate, and efficient dynamic control [12]. Intelligent transportation can helps to reduce traffic load and environmental pollution, can effectively ensure traffic safety and improve transportation efficiency [4]. With the help of intelligent transportation, vehicles, roads, and management personnel can achieve dynamic balance, traffic flow is automatically adjusted to the best state, and management personnel have a clear grasp of the road and the whereabouts of vehicles.
- **Smart Car:** The emergence of automobiles has not only changed the concept of time and space and transportation methods of human beings but also promoted the transformation of people's lifestyles and the progress of social science and technology [22]. However, with the increase in traffic volume and the acceleration of vehicle speeds, more and more traffic accidents caused by automobiles. Smart car based on computer technology, microelectronics technology and intelligent automation technology can respond in a short time and can effectively reduce the occurrence of traffic accidents [7].
- **Smart City Management:** The application of information technology in smart city construction can help government departments better understand the situation, analyze and predict, and make accurate decisions, so as to realize urban management [5]. The whole process of food safety supervision from field to table, real-time monitoring of ecological environment indicators such as air and water, automatic identification of traffic violations, and upper and lower connectivity of networked management are all aspects of smart city management [10]. In order to provide a stronger city innovation development, sustainable and harmonious development ability. Based on the comprehensive interconnection of things and things, people, human mutual perception. Smart cities obtain more scientific and efficient analysis and prediction, intelligent control, safe and

stable information resources integration, and more harmonious multi-level cross-department cooperation ability.

3.2.3.3 Smart Care

Under the trend of interdisciplinary and cross-field development, there are many integrations between artificial intelligence technology and the medical field, and it has made great progress in many fields such as medical imaging, clinical decision support. With the continuous advancement of intelligent technology, AI has become more and more widely used in the medical industry, and has gradually become an important promoter of promoting the progress of the medical industry and improving the quality of medical treatment, which can effectively improve the accuracy and efficiency of medical diagnosis. There are three most critical technologies in the development of smart care: computer computing power, computer computing methods, and large database establishment.

3.2.3.4 Smart Education

At the current stage, artificial intelligence is often used to replace teachers to complete tedious and repetitive tasks. For example, to help teachers correct test papers, to help analyze and count the data of each learning link of the students, and to recommend the next work plan, and to recommend personalized exercises and review plans for each student's learning situation and characteristics. There are five main applications of artificial intelligence and related technologies in the field of education: adaptive learning systems, virtual tutors, educational robots, technological education based on programming and robots, and scene-based education based on virtual reality/augmented reality.

3.2.3.5 Smart Workflow

Smart workflow is a software tool for process management, which integrates work performed by people and machines. Allows users to start and track the state of end-to-end process in real-time, so as to facilitate the management of switching between different groups, including between robots and human users, also provide bottleneck phase statistics. With the continuous progress of society and technology, all fields have begun to develop rapidly towards automation and intelligence. The research of workflow related technology is also paid more and more attention, and is widely used in different fields such as manufacturing, software development, banking finance, biomedicine, and so on. The workflow can not only automate related activities and tasks, reduce the potential errors in the process of human-computer interaction. And precise each processing step, maximize the generation efficiency, apply the workflow to dynamic, variable, and flexible application scenarios.

3.3 AI for Transportation Systems

3.3.1 Motivations

Recent years, the application of artificial intelligence in a number of industries has sparked a wide discussion around the world, and people have higher expectations for the application of AI in production and life. In terms of technology, artificial intelligence is extremely active in the past few years, largely because of the development of deep learning. Under the promotion of deep learning, the intelligent level of the machine has been significantly improved. The intelligent machines applying deep learning algorithms can accomplish something that humans could not do before [20].

Artificial intelligence has brought a subversive changes to human economy and society. Machines are no longer just cold tools [21]. Enabled by artificial intelligence, it can interact with people, systems and the environment, and have a profound impact on various industries. Improving the efficiency and quality of work through intelligent machines is the key to research by relevant practitioners. Under the impact of disruptive technology, all industries need to make corresponding adjustments, including the transportation industry. The construction of intelligent transportation is an important foundation for building modern cities, and artificial intelligence technology will provide strong support for this.

Traffic optimization has gradually become the focus of urban construction, and the intelligence transportation has huge development potential. With the rapid construction of urbanization, the scale of automobile is expanding rapidly, many cities are faced with serious traffic congestion, frequent traffic accidents and serious environment pollution problems, which increase the burden of urban construction and development. In this case, more and more regions began to build intelligent transportation system (ITS). With the help of neural network and deep learning technology, artificial intelligence has improved its own learning and understanding ability. With the rapid development of this field in recent years, artificial intelligence will gradually infiltrate into daily life, improve work efficiency and have an impact on the way of thinking of people. The application of artificial intelligence can accelerate the development of social economy, promote the transformation and upgrading the whole, and play an increasingly important role in the field of road traffic management.

ITS uses advanced communication technology, control technology, sensor technology, computer technology and system integration technology to present the interaction between people, vehicles, and roads in a new way, so as to realize real-time, accuracy, efficiency, safety and energy saving goals. The emergence of ITS has effectively alleviated traffic problems and is the development direction of the future transportation system. ITS can effectively use the existing infrastructure, reduce the traffic's pollution to the load and the urban environment, and ensure the safety of drivers and passengers, thereby improving traffic efficiency.

The socialization of automobile development, the sustainability of the human environment, and the intelligence of information technology are the background and motivation for the development of ITS. With the rapid development of economy and the rapid increase of the number of cars, traffic is also facing greater challenges. Therefore, ITS is valued by various countries. For example, the smart road system designed by Japan, the green smart transportation proposed by Europe, and the smart driving strategy proposed by the United States are all effective practices for the development of smart transportation.

The ITS that people mainly use is a particularly advanced comprehensive traffic monitoring system. In such a system, the vehicles on the highway are all driving on the intelligently planned road, and the highway can reach a particularly good state according to the regulation of the traffic situation. With the help of the system, it can help traffic management staff to grasp the latest road condition information.

3.3.2 Application Status

The application of AI for transportation systems is shown in Fig. 3.5 and discussed below.

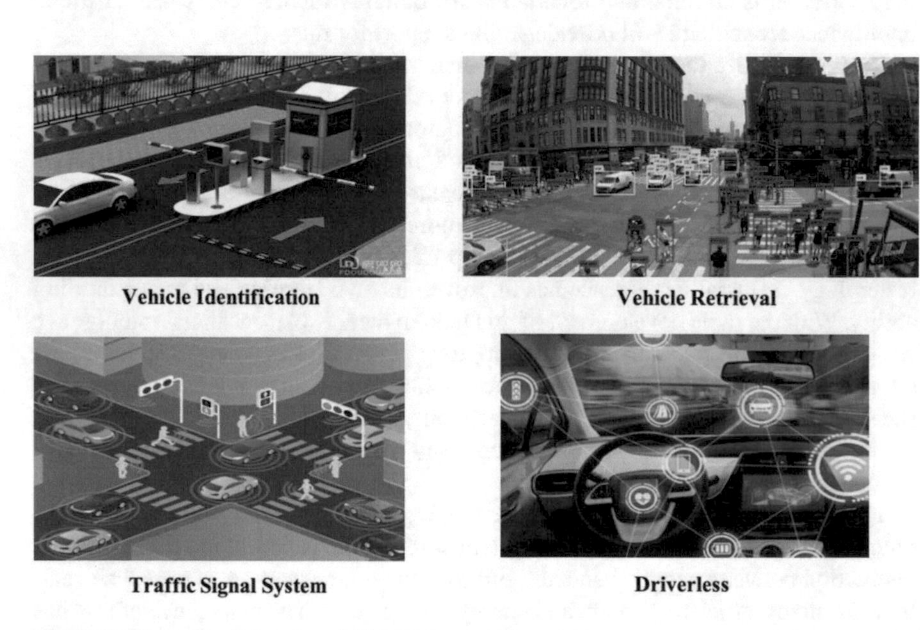

Fig. 3.5 The application status of AI for transportation systems

3.3.2.1 Vehicle Identification

At present, license plate recognition is the most ideal application of artificial intelligence in the field of intelligent transportation. It is understood that the license plate recognition rate can reach 99% under the video condition of the standard checkpoint and some preset conditions. Deep learning is used for image detection and recognition, it does not need to set artificial features. As long as sufficient images are prepared for training, better results can be achieved by continuous iteration. Judging from the current situation, as long as new data are constantly added and sufficient time and resources are available, the network level of deep learning will continue to increase and the recognition rate will continue to improve. Compared with traditional image processing and machine learning methods, the application effect of this method is much better.

- **Vehicle Color Identification:** In the past, different lighting conditions, camera hardware errors and other factors have led to vehicle color changes [11]. Nowadays, with the assistance of artificial intelligence technology, the problem of error recognition caused by the change of image color has been effectively solved. The recognition rate of the color of vehicles at checkpoints has reached 85%, and the recognition rate of the main color of electric police vehicles has exceeded 80%.
- **Vehicle Manufacturer Identification:** In the past, the traditional Histogram of Gradient (HOG), Local Binary Patterns (LBP), Scale Invariant Feature Transform (SIFT), and so on were generally used for vehicle manufacturer mark recognition. Using machine learning technology based on support vector machine to develop a multi-level classifier for recognition, the error rate is relatively high. Now, with the introduction of big data and deep learning, the recognition rate of the vehicle manufacturer mark has increased from 89 to 93% or more.

3.3.2.2 Vehicle Retrieval

In the aspect of vehicle retrieval, the vehicle images may be overexposed or underexposed and the size of the vehicle may change under different circumstances[18]. In this case, if continue to use the traditional method to extract the vehicle features, there will be errors, resulting in a bad impact on the retrieval rate of the vehicle. After the introduction of deep learning, the system can obtain relatively stable vehicle characteristics, more accurately search for similar targets, with the search rate of some devices exceeding 95% [2]. In face recognition, the face will undergo some changes due to the influence of light, expression, posture, and other factors. At present, many applications require that the scene and posture of face recognition be fixed. After the introduction of deep learning algorithm, the face recognition rate of fixed scene can be increased to 99%, and the requirements for light, posture, and other conditions are also relaxed.

3.3.2.3 Traffic Signal System

Traditional traffic light conversion uses default time, which is updated every few years. But with the continuous development of traffic mode, the application time of traditional system is getting shorter [3]. The intelligent traffic signal system with artificial intelligence uses radar sensors and cameras to monitor traffic flow, and then uses artificial intelligence algorithm to determine the conversion time. By integrating artificial intelligence and traffic control theory, the traffic flow in urban road network is optimized reasonably.

3.3.2.4 Driverless

To achieve driverless, object recognition and detection is very important, which can effectively identify vehicles, pedestrians, obstacles, roads, traffic signs, signal lights and other objects in front of them [8]. It can effectively improve the travel experience of people, reshape the transportation system and make people truly enter the era of intelligent transportation.

3.3.3 Challenges

With the rapid development of the modern economy, the rapid increase in the number of vehicles, the continuous improvement of people's requirements for travel, as well as more data types and sources and new application fields, all have put forward unprecedented challenges and requirements on ITS. We summarize the challenges faced by ITS into the following six aspects:

- **Large Amount of Data and Instability:** Although ITS has brought us great convenience, due to the huge amount of data to be processed in the past, some data processing instability will also occur [1, 13]. That is to say, if the data is too large, some unavoidable problems and difficulties will appear when the system is processing. It is precisely because of the instability of these data that it brings great challenges to ITS.

  Due to the large amount of data that needs to be integrated and processed at the same time, some small errors will cause great inconvenience to people's travel. For example, the latency of data. When a certain road section is congested due to a car accident or some other accidents, if the large amount of data not processed or analyzed in time, or the analysis is wrong, the information will be delayed. And many people may receive the wrong information: the road is smooth, causing more vehicles to enter and unnecessary congestion. Or the analysis of the route is wrong, the people who do not know the road drive into the wrong section of the road and go to the wrong place. There was also a problem of misreporting traffic information due to data processing errors. The information was reported as a green light at an intersection that was originally a red light, and there were also erroneous reports about the speed limit on certain road sections.

- **Information Security:** After entering the era of big data, intelligent transportation has indeed brought great convenience to our lives. However, it is precisely because of the popularization and application of the Internet that the security of our personal information has become a worrying issue.

  When using the information system, you need to fill in personal information to log in. Anything related to personal information is related to the issue of information security, and the same for ITS [19]. However, this will inevitably lead to user information security issues. Only on the premise of ensuring the security of user information, can people be assured of using ITS, and people can further recognize the intelligent transportation system. Now that personal information is connected to the Internet, there have been many instances of information theft. Various information leakage problems occur on the Internet, causing many people to question the security of the Internet. ITS is also facing such pressure.
- **Efficient Information Technology Platform:** Intelligent transportation systems should develop available information technology platforms. Improving the comprehensive utilization of traffic data resources, and quickly and accurately find the required resources with the technical support of big data, reduce memory consumption, and increase the speed of use are the key to current research [17]. In the era of big data, the requirements for intelligent transportation are accurate data resources, real-time calculations and feedback. However, how to quickly identify and calculate in the massive information to achieve the purpose of correct use, and how to achieve the reasonable sharing and use of data, and to allocate permissions according to the different needs of different people, is the problem that ITS faces when it wants to develop better.
- **Re-employment of Industry Personnel:** The emergence of any new technology will not only bring positive factors such as efficiency improvement, increase experience but also bring some negative factors. With the intelligence of the transportation system, the employees of the industry will be greatly streamlined. How to ensure and rearrange their employment will be a new problem.
- **Cost:** The up-front development and manufacturing of related equipment and systems itself requires a large amount of cost. How to control and coordinate the cost is an important problem in the integration of artificial intelligence and transportation industry.
- **Public Acceptance:** The combination of artificial intelligence and transportation industry in some fields is acceptable to the ordinary people. But some areas are not necessarily recognized by everyone, such as driverless vehicles [6]. Take the United States, for example, the United States attaches great importance to the construction of transportation channels. The government invests about 88 billion dollars every year to improve infrastructure, with the purpose of increasing the vitality and competitiveness of the economy. At present, automatic driving technology has attracted wide attention in the United States. The emergence of this technology has undoubtedly expanded the freedom of travel for vulnerable

groups. However, there are also some problems. According to the survey, 70% of American people are afraid of self-driving cars.

The same is true for other vehicles combined with artificial intelligence. So, on the one hand, we should improve relevant laws and regulations. And on the other hand, strengthen the popularization of science and positive publicity of unmanned means of transportation. So as to eliminate the fears and concerns of people, enhance confidence in unmanned vehicles, and promote the long-term development of artificial intelligence and transportation.

References

1. Aujla, G. S., Singh, A., Singh, M., Sharma, S., Kumar, N., & Raymond Choo, K.-K. (2020). BloCkEd: Blockchain-based secure data processing framework in edge envisioned V2X environment. *IEEE Transactions on Vehicular Technology, 69*(6), 5850–5863.
2. Brambilla, M., Nicoli, M., Soatti, G., & Deflorio, F. (2020). Augmenting vehicle localization by cooperative sensing of the driving environment: Insight on data association in urban traffic scenarios. *IEEE Transactions on Intelligent Transportation Systems, 21*(4), 1646–1663.
3. Chavhan, S., Gupta, D., Chandana, B. N., Khanna, A., & Rodrigues, J. J. P. C. (2020). Iot-based context-aware intelligent public transport system in a metropolitan area. *IEEE Internet of Things Journal, 7*(7), 6023–6034.
4. Chen, C., Liu, B., Wan, S., Qiao, P., & Pei, Q. (2021). An edge traffic flow detection scheme based on deep learning in an intelligent transportation system. *IEEE Transactions on Intelligent Transportation Systems, 22*(3), 1840–1852.
5. Chen, L.-W., & Ho, Y.-F. (2019). Centimeter-grade metropolitan positioning for lane-level intelligent transportation systems based on the internet of vehicles. *IEEE Transactions on Industrial Informatics, 15*(3), 1474–1485.
6. Chowdhury, A., Karmakar, G., Kamruzzaman, J., & Islam, S. (2021). Trustworthiness of self-driving vehicles for intelligent transportation systems in industry applications. *IEEE Transactions on Industrial Informatics, 17*(2), 961–970.
7. Dairi, A., Harrou, F., Sun, Y., & Senouci, M. (2018). Obstacle detection for intelligent transportation systems using deep stacked autoencoder and *k* -nearest neighbor scheme. *IEEE Sensors Journal, 18*(12), 5122–5132.
8. Han, B., Wang, Y., Yang, Z., & Gao, X. (2020). Small-scale pedestrian detection based on deep neural network. *IEEE Transactions on Intelligent Transportation Systems, 21*(7), 3046–3055.
9. Hussain, T., Muhammad, K., Ser, J. D., Baik, S. W., & de Albuquerque, V. H. C. (2020). Intelligent embedded vision for summarization of multiview videos in IIoT. *IEEE Transactions on Industrial Informatics, 16*(4), 2592–2602.
10. Lin, C., Han, G., Du, J., Xu, T., Shu, L., & Lv, Z. (2020). Spatiotemporal congestion-aware path planning toward intelligent transportation systems in software-defined smart city IoT. *IEEE Internet of Things Journal, 7*(9), 8012–8024.
11. Lin, C.-T., Huang, S.-W., Wu, Y.-Y., & Lai, S.-H. (2021). Gan-based day-to-night image style transfer for nighttime vehicle detection. *IEEE Transactions on Intelligent Transportation Systems, 22*(2), 951–963.
12. Naufal, J. K., Camargo, J. B., Vismari, L. F., de Almeida, J. R., Molina, C., González, R. I. R., Inam, R., & Fersman, E. (2018). A2CPS: A vehicle-centric safety conceptual framework for autonomous transport systems. *IEEE Transactions on Intelligent Transportation Systems, 19*(6), 1925–1939.

13. Raja, G., Anbalagan, S., Vijayaraghavan, G., Dhanasekaran, P., Al-Otaibi, Y. D., & Bashir, A. K. (2021). Energy-efficient end-to-end security for software-defined vehicular networks. *IEEE Transactions on Industrial Informatics, 17*(8), 5730–5737.
14. Sharma, S., Dudeja, R. K., Aujla, G. S., Bali, R. S., & Kumar, N. (2020). DeTrAs: deep learning-based healthcare framework for IoT-based assistance of Alzheimer patients. *Neural Computing and Applications*, 1–13. https://doi.org/10.1007/s00521-020-05327-2
15. Shukla, A., Bhattacharya, P., Tanwar, S., Kumar, N., & Guizani, M. (2020). Dwara: A deep learning-based dynamic toll pricing scheme for intelligent transportation systems. *IEEE Transactions on Vehicular Technology, 69*(11), 12510–12520.
16. Singh, A., Aujla, G. S., Garg, S., Kaddoum, G., & Singh, G. (2019). Deep-learning-based SDN model for internet of things: An incremental tensor train approach. *IEEE Internet of Things Journal, 7*(7), 6302–6311.
17. Srinivas, J., Das, A. K., Wazid, M., & Vasilakos, A. V. (2021). Designing secure user authentication protocol for big data collection in IoT-based intelligent transportation system. *IEEE Internet of Things Journal, 8*(9), 7727–7744.
18. Thomas, S. S., Gupta, S., & Subramanian, V. K. (2018). Event detection on roads using perceptual video summarization. *IEEE Transactions on Intelligent Transportation Systems, 19*(9), 2944–2954.
19. Wang, J., Jiang, C., Han, Z., Ren, Y., & Hanzo, L. (2018). Internet of vehicles: Sensing-aided transportation information collection and diffusion. *IEEE Transactions on Vehicular Technology, 67*(5), 3813–3825.
20. Xie, J., Zheng, Y., Du, R., Xiong, W., Cao, Y., Ma, Z., Cao, D., & Guo, J. (2021). Deep learning-based computer vision for surveillance in its: Evaluation of state-of-the-art methods. *IEEE Transactions on Vehicular Technology, 70*(4), 3027–3042.
21. Xu, X., Liu, Y., Wang, W., Zhao, X., Sheng, Q. Z., Wang, Z., & Shi, B. (2019). Its-frame: A framework for multi-aspect analysis in the field of intelligent transportation systems. *IEEE Transactions on Intelligent Transportation Systems, 20*(8), 2893–2902.
22. Zhou, Z., Gao, C., Xu, C., Zhang, Y., Mumtaz, S., & Rodriguez, J. (2018). Social big-data-based content dissemination in internet of vehicles. *IEEE Transactions on Industrial Informatics, 14*(2), 768–777.
23. Zhu, F., Lv, Y., Chen, Y., Wang, X., Xiong, G., & Wang, F.-Y. (2020). Parallel transportation systems: Toward iot-enabled smart urban traffic control and management. *IEEE Transactions on Intelligent Transportation Systems, 21*(10), 4063–4071.

Chapter 4
Artificial Intelligence: Evolution, Benefits, and Challenges

Fazeela Mughal, Abdul Wahid, and Muazzam A. Khan Khattak

4.1 Introduction to Artificial Intelligence

Artificial intelligence (AI) is delineated as, "The science and engineering of making intelligent machines, especially intelligent computer programs" by John McCarthy (the father of Artificial Intelligence). AI consists of several aspects such as learning, logical reasoning, problem solving, perception, and linguistic intelligence. Learning in AI can be carried in a variety of ways, amongst which the simplest one is "Learning by Trial & Error," whereas logical reasoning is grounded on deductive and inductive reasoning. In order to solve wide range of problems, general purpose methods are ordinarily used while solving a particular type of problem, special purpose methods are preferred. Different sensory organs are typically used for scanning in perception. Hence it can be concluded that, the decision-making power of machines working on Human Behavior through perception, learning, and reasoning can be collectively referred to as Artificial Intelligence. AI effortlessly attempts to eliminate the barrier between the human and robots by acting as an expert machine. Artificial Intelligence is therefore valuable and beneficial in time saving and increasing the productivity of work. AI goes on automating the routine tasks that were formerly executed by humans, thus the cost of hiring human resource has been effectively reduced. Furthermore, the goal of future AI is to build machines that would perform human tasks smartly and in much faster rate.

F. Mughal (✉) · A. Wahid
School of Electrical Engineering and Computer Science, National University of Sciences and Technology (NUST), Islamabad, Pakistan
e-mail: fmughal.mscs19seecs@seecs.edu.pk; abdul.wahid@seecs.edu.pk

M. A. K. Khattak
Quaid-e-Azam University, Islamabad, Pakistan
e-mail: muazzam.khattak@qau.edu.pk

S. Garg et al. (eds.), *Intelligent Cyber-Physical Systems for Autonomous Transportation*, Internet of Things, https://doi.org/10.1007/978-3-030-92054-8_4

4.1.1 Types of Artificial Intelligence

There are mainly two ways in which artificial intelligence is classified, one is on the basis of classification of technology and the other one is AI-enabled machines which are based on their likeness to human mind and their ability to think, feel, and react according to given situation. According to this type of classification following are the four types of AI-based systems as depicted in Fig. 4.1:

- **Reactive Machines:** Reactive machines works on the basis of current scenario through perception. They do not use past experience and also are not able to predict future. Example of reactive machine is “Deep Blue.”
- **Theory of Mind:** By social intelligence, machine would be able to understand the emotions, predict behavior and infer intensions. Example of theory of mind is “Autonomous Cars.”
- **Limited Memory:** In this type of AI systems have memory and it uses past experience to predict future. Example of limited memory is “Self-driving Car.”
- **Self-Aware:** Machines have artificial consciousness and are so akin to human brain, e.g., Robot arm made by group of people from Colombia University. The alternative system of classification is generally based on the classification of technology into Artificial Narrow Intelligence (ANI), Artificial General Intelligence (AGI), and Artificial Super Intelligence (ASI).
- **Artificial Narrow Intelligence:** ANI has very limited competencies because it performs a singles task autonomously using human-like capabilities. It falls under weak AI category because it does not replicate human intelligence but only simulates human behavior based on limited range of parameters. It corresponds to all types of reactive machines as well as limited memory systems. Also those

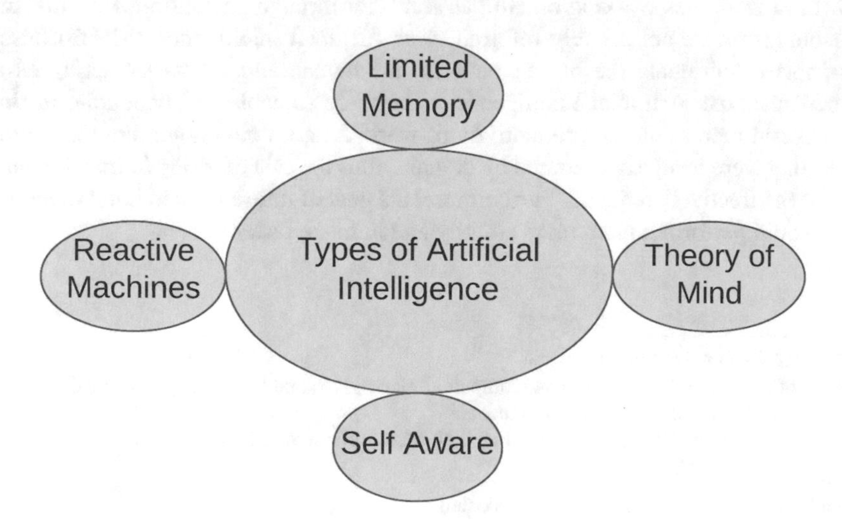

Fig. 4.1 Types of artificial intelligence

machines which teach itself by using machine learning and deep learning falls under ANI. Examples of ANI are smooth voice of Siri and Cortana, Google assistant, Apple's face recognition technology, Google photos application, and DeepMind's board game.

- **Artificial General Intelligence:** AGI replicates the human behavior by learning, perceiving, and understanding. It will be able to build multiple competencies and form connections across multiple domains and cut down the time needed for training. AGI falls under the category of weak AI but it is close to strong AI because it replicates human behavior and has the ability to think. Examples of AGI are self-driving cars, IBM's Watsons supercomputer and expert systems.
- **Artificial Super Intelligence:** ASI is hypothetical artificial intelligence because it is imaginary AI which is not only able to interpret or understands human behavior but in actual if it is possible to achieve ASI then machines based on this technology will surpass the capacity of human intelligence and the ability of reacting on certain things. Furthermore, it has the ability to replicate multi-faceted intelligence of human beings. ASI based machines will not only focus on understanding human emotions and experiences but it must also has the capability of understanding the emotions, desires, and beliefs.

Currently, AI is used in almost all applications that perform tasks faster and efficiently such as chatbots, self-driving cars, Google maps, image recognition, speech recognition, ridesharing apps (Uber, Careem, and Lyft), autopilots, Turnitin, Amazon, etc. Advanced robotics is based on strong AI mechanisms whereas Siri and Alexa are based on weak AI.

4.2 Needs and Evolution

There were some needs in different fields to perform tasks smartly, efficiently and without any error and for that John McCarthy proposed the idea of AI. The main objectives of artificial intelligence was to explore the ways in which machine could reason like human (the power of decision making, learning, perception, problem solving, and understanding).

4.2.1 Needs of Artificial Intelligence

Need of Artificial intelligence arose due to the fact that most of work was performed by humans. Therefore, it was indeed needed to automate the routine tasks. Humans normally consume much amount of time in decision making and solving complex problems. But with the help of AI, these tasks are now performed by machines within seconds and milliseconds without any error. AI is needed in almost every field. Some of them are discussed below:

- **Healthcare:** Before AI in healthcare, manually maintaining patient's health data and keeping track of the availability of machinery, beds, rooms, doctors, and nurses on the spot was cumbersome and time consuming. Early diagnoses of serious diseases like heart attack, cancer was not probable which led to increase in death ratios. But now artificial intelligence has led to more accurate and timely diagnosis. CT scans, CT scanogram, ECG, MRI, etc. are some of the applications of AI that are used for timely diagnosis of some of the serious diseases [8]. As it can be seen from Fig. 4.2 that patient's medical data, patient's health data and other related data is given to AI application as an input after processing it will display disease type and treatment plan.
- **Education:** People would travel miles away to seek learning and education. It was time consuming and expensive. But with AI it is now possible for everyone

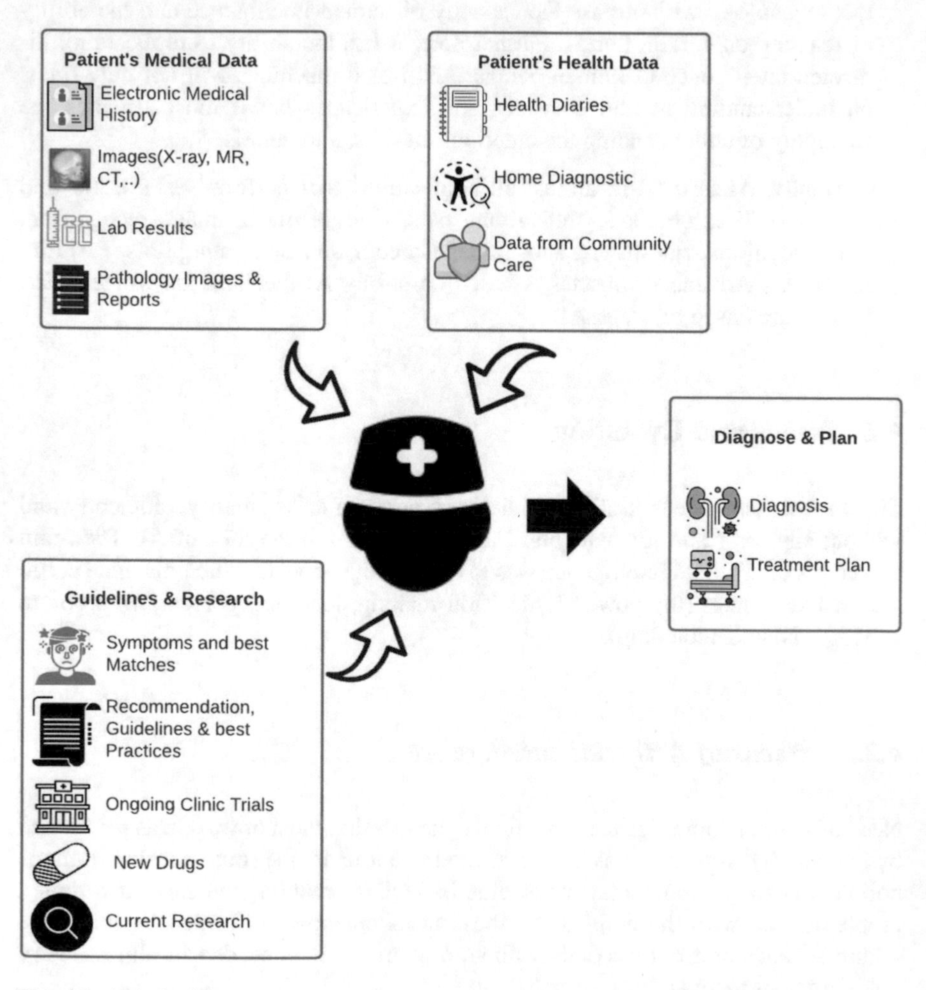

Fig. 4.2 AI in healthcare

to attend classes online via applications installed on cell phones or on laptops to get knowledge in lesser time, faster and on average expenses. Also paperless examinations (online exams) and autonomous grading is possible by AI.

- **Banking Sector:** As it was a complex task for the banks to maintain transactions record, provide faster and safe end-to-end transactions. But now with AI it is possible to maintain transaction record, provide faster and safe end-to-end transactions. Better understanding of customer and their behavior is developed by AI which identifies fraud and detect anti-money laundering pattern.
- **Business Sector:** In business it was a complex task to detect fraud, customer satisfactions, data analysis, more productivity and personalization but AI took it as a challenge in business sector and solved these problems. AI reduces the fraudulent transactions and does speedy transactions. It also provides real-time responses to their customers via telephone or chatbots. Data analysis is done by automated machine learning (AutoML) and H_2O.ai (an open source platform). H_2O.ai has the ability to analyze the data in accurate, transparent, and trustworthy manner. AI reduces the expenses and increases the productivity by predicting and preventing IT issues. AI has also solved the problem of personalization by personalized advertisement, recommendations on the basis of search and online shopping.
- **Manufacturing and Production:** Manufacturing and production industry faced a lot of issues such as; it took a lot of time to produce, poor quality of products, waste of materials and maintenance. But with artificial intelligence these issues were resolved. AI manages quality, shorten the design time, reduce materials waste, perform predictive maintenance and increase the concept of reusability of products.
- **Security and Surveillance:** Fraud detection, security of humans and valuable things was an issue. AI-based technologies like biometric and face recognition systems makes it possible to detect fraudulent activities, security of valuable things and make the places safe to live. Security cameras, recorders, and other surveillance equipment are used to keep track of humans.
- **Sports:** In previous sports system it was impossible to watch live matches. Decision-making process and training of players was a complex task. But now with the combination of different AI sensors, coaches improve the performance of players. Tactics of winners and losers is recorded while they are playing and analyzed by coaches. Live streaming is also possible. By the automation of media it is possible to pick out the highlights of a match.
- **Military:** Before applying AI in military, the wars were fought by soldiers were they would travel by foot towards attacking places but now with advancement in technology and by using AI it is possible for a military to attack other countries by drones.
- **Agriculture:** It was complicated to detect diseases in plants, the pest types and plants nutrition. With the help of AI and computer vision techniques, images of the farms are captured and analyzed which helps in identifying the diseases in plants and crops

Years	Evolution
1956	Idea of AI was presented in Dartmouth Conference
1956-1974	Considered as an imperative topic.
1974-1980	Winter (downfall) of AI.
1980-1987	Expert Systems
1997	Deep Blue
2004	DARPA Grand Challenge

Fig. 4.3 Evolution of AI

- **Transportation System:** Conventionally, transport system had a lot of issues like traffic management, increase in accident rates, scheduling and routing concerns, etc. But all of those issues are now resolved by ITS [3]. These aspects are further discussed in more detail in Sect. 4.3.2.

Apart from these fields there are a lot of other areas in which artificial intelligence was needed to solve problems faster and efficiently. Needs of AI let too many evolutions in the intelligent systems.

4.2.2 Evolution of Artificial Intelligence

The notion of modern AI was first presented by John McCarthy in Dartmouth conference held in 1956. The field of AI began to congregate its attention from 1956 to 1974 when it was considered to be an imperative topic for research. However, at the end of 1974 the first winter of AI started, as it was deliberated that AI would not be able to make valuable progress. Afterwards, another rebirth of AI occurred from 1980 to 1987 with the awareness about the Expert Systems. Based on AI skills, a supercomputer titled as Deep blue, was manufactured in the year 1997 which beat the world chess champion Gary Kasparov in a chess match. After that DARPA Grand Challenge in March 2004 occurred; which was about a self- autonomous vehicle drive through uncharted terrain. Hence, transportation system began to use Artificial Intelligence mechanisms. The brief summary of the evolution of AI is shown in Fig. 4.3.

4.3 AI for Transportation System

4.3.1 Introduction

Use of artificial intelligence has made all modes of transportation systems safer, smarter, and more comfortable. AI can be applied in vehicles such as self-driving cars, infrastructures for controlling traffic signals and for providing better transport services. Incorporating AI to transport systems helps in controlling traffic congestion, minimizing the road accidents and predicting the future moves (e.g., RATP dev AI engine works on it). In addition to this, AI can be used for scheduling, routing, and also for applying higher and lower traffic patterns. In Intelligent Transportation Systems (ITS), specifically VANET (Vehicular Ad hoc Networks) that consist of two modes of communications. These are V2V (vehicle to vehicle) and V2I (vehicle to infrastructure). Safe, reliable, fast, and efficient traffic management communication is done by V2V whereas in V2I communication is between vehicle and infrastructure via wireless exchange. In V2I, road conditions, accidents, parking availability, traffic congestion are captured by sensors and provide it to travelers. AI is also used in Swiss Federal Railways and Tesla self-driving cars. According to UITP's survey it is stated that 62% of public transport players are involved in Artificial Intelligence projects and 86% are engaged in partnership to develop and adopt AI.

There are six levels of vehicles' automation. From level zero to level two humans monitor the driving environment whereas from level three to level five it is automated system which monitors the driving environment. The detail of each level is mentioned in Fig. 4.4. Following are some major means of transportation system in which artificial intelligence is applied:

Level	Functionality
Level 0	**No Automation:** In this level control is manual. Driver performs all the functions/tasks such as: steering, acceleration, speed etc.
Level 1	**Function Specific Automation:** Function specific automation is the lowest type of automation in which one feature of automation is used to assist driver e.g steering or accelerating (cruise control).
Level 2	**Partial Driving Automation:** In this level advance driving assistance is given to the vehicle. Vehicles have the control of both steering and accelerating/decelerating e.g. Tesla Autopilots.
Level 3	**Conditional Driving Automation:** It is more like level 2 but the additional feature here is the capability of environmental detection.
Level 4	**Limited Self-Driving Automation:** Safety-critical functions can be performed by drivers.
Level 5	**Full Self-Driving Automation:** Vehicles perform all driving functions. There is no need of driver to sit in and take control for safety measures.

Fig. 4.4 Levels of vehicle automation

- **Artificial Intelligence in Road Transport:** AI is applied on road transport system in the form of street traffic signals, automated sign detection boards and self-driving cars (e.g., Tesla self-driving car). On a large scale, many research groups, manufacturers, and technology firms are studying AI technologies to develop automated vehicles for personal as well as commercial use [1]. These automated vehicles have sensors such as GPS, which receives real-time geographical data to navigate, radars, and cameras to detect speed and obstacles in the vicinity. With the combination of sensors, these vehicles have some pre-installed software, actuators (devices which transforms input signal into motion) and control units. Some of these technologies are used to take over certain driving functions such as parking and others are to replace the human driver.

  In European Single market vehicles that have adopted AI technologies to take over certain driving function are available whereas fully automated vehicles are under consideration. Truck platooning is also possible by Artificial intelligence as it has been seen by the Scania project (Autonomous truck platoon in Singapore). As it can be seen from Fig. 4.5 that there is a leading truck driver driving the truck and the other trucks are following the first one autonomously. The other trucks have drivers inside them which are ideal and will function only in case of an incident.

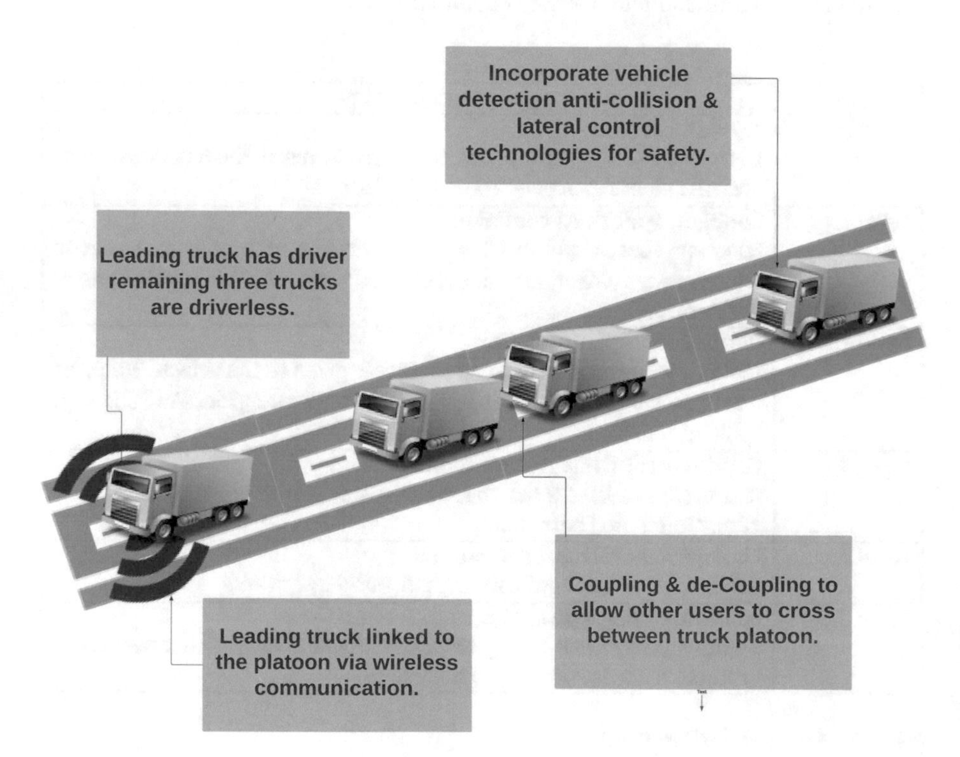

Fig. 4.5 Autonomous truck platooning

- **Artificial Intelligence in Aviation:** Artificial intelligence is widely used in aviation to modernize business and for safety measures. With advancement in machine vision, many applications are used to maximize system security and operating efficiency in airspace. As we know the global aviation industry is tremendously increasing, the growth aviation industry needs to update the current strategy of functioning and needs to adapt better opportunities to control cost factor and to optimize the resources. With the usage of artificial intelligence in aviation, it is easy to identify passenger by facial recognition technologies. Without any human intervention, the process of baggage screening is carried out effectively by AI technologies. In addition to this, responding to customers' queries (customer assistance) and data confidentiality is also management by AI.
- **Artificial Intelligence in Railway Transport:** In railways, AI can improve the manufacturing, operational, and maintenance processes. Automation of Train Operations (ATO) has given the responsibility of a driver to train control system for the managing operations timely and efficiently. Fully automated trains are not deployed yet but as SNFC (Société nationale des chemins de fer) planned to deploy semi-autonomous trains till 2023 and fully autonomous train till 2025. Sensors placed on critical trains provide the power data to recommended actions for maintenance. By doing so, the advantages are now casting and forecasting infrastructures, reduction of maintenance cost, faster repair and better customer satisfaction. AI used in railway transport can also improve train scheduling, avoiding accidents, managing speed of the train and predicting delays. There has been increase in the speed of train that is more than 300km/h by fast and convenient intelligent transportation systems.
- **Artificial Intelligence in Water Transport:** AI is also applied on maritime shipping and inland navigation. Electronic navigation charts, radars, and other sensors provide the data about other maritime ships and their performances. Automatic Identification System (AIS) conveys data such as ship's identification number, speed, position, and destination. AI detects the anomalies in marine operations to improve the safety and to reduce the chances of accidents by managing them.

Intelligent Transportation System (ITS) is an advanced application which provides innovative services to all means of transportations systems. ITS improves the safety factor and the efficiency of transport system. To reduce journey times and to increase the capacity of busy roads, ITS technologies are used. Various mechanisms are applied in ITS to perform management tasks such as car navigation, automatic number plate detection, traffic signal control systems, etc.

4.3.2 Benefits of AI in Transportation System

There are enormous benefits of artificial intelligence the transportation system. Following are some of them:

- **Make Transportation Safer:** By applying artificial intelligence in transportation system, the number of accidents can be reduced. As mostly accidents are happened due to drunk, distracted, or tired drivers. Since automated vehicles are not involved in these human factors and have the ability to reduce the chances of accidents.
- **Reduce Transportation Related Pollution:** By optimization of traffic flow and reduction of fuel will ultimately reduce the emission. As transportation is the important source of global warming emission, optimists' beliefs that automated vehicles reduce the pollution because they are based on electric power.
- **Save Time:** Autonomous vehicles save driver's time. According to current situation faster decisions are taken on the basis of prior knowledge and this can happen by limited memory type of AI.
- **Traffic Management:** AI can predict and analyze traffic conditions and also directs about alternative paths.
- **Reduce Congestion:** AI can control congestion by optimizing the traffic flow, traffic light management and by using CCTV to monitor the condition of roads. It also reduces congestion by scheduling public and private transportation services in improved manner.
- **Improve Access to Transportation:** The access of transportation can be improved by providing speedy shared autonomous vehicles or low capacity autonomous shuttles to fixed routes for travelling. It is also beneficial for people who are unable to drive or mobility impaired.
- **Real-Time Parking:** Cameras, sensors, and various applications are used to find the vacant place to park the vehicle.
- **Facility of Ridesharing:** It is beneficial because it saves time and cost. It provides alternative, secure (tracking feature) and better transportation facility which is available all the time.

4.3.3 Challenges and Future Research Directions

There are some challenges in artificial intelligence for transportation system. These challenges are related to:

- **Job Issues:** Researchers predict that autonomous vehicles will lead to unemployment of drivers because fully autonomous vehicles will completely replace the drivers' task. According to the research carried out by Stanford University in 2014, showed that there could be 70–90% automated vehicles on the roads which will lead to unemployment of drivers [2].
- **System Reliability and Cybersecurity:** AI-based automated vehicles need to access a lot of sensitive or protected data. If this data is accessed by third parties then vehicles and their users are endangered [4]. In 2016, researchers of Chinese company demonstrated that the malicious hackers can exploit autonomous vehicles for serious attacks [2].

- **Heavy Weather:** Another challenge for fully automated vehicles will be to ensure good interaction with other road users, recognizing obstacles properly under all weather conditions and dynamic environment [7].
- **Ethical and Legal Situations:** One important challenge that needs to be addressed is the liability of accidents. As the vehicles are semi-automated or fully automated then what are the rules and liabilities to be implied [6].
- **High Cost Automation Vehicles:** If the system fails it could be fatal to both passengers and other road users. Because all parts of autonomous vehicles need to meet high manufacturing, repair, installation, testing, and maintenance standards, this will be expensive for common people [5].

Future of Intelligent Transportation Systems may fall in multiple layers of social, cyber, and physical environment. This may include analysis of information by cyber sources and the models that flow in connected environment. One more future research direction can be the issue of cyber-attacks that is cyber security in AI in fully automated vehicles. Another important future research direction can be the energy consumption and emissions on the basis of the short-term and long-term impacts of automated vehicles.

References

1. Aujla, G. S. S., Kumar, N., Garg, S., Kaur, K., & Ranjan, R. (2019). EDCSuS: sustainable edge data centers as a service in SDN-enabled vehicular environment. *IEEE Transactions on Sustainable Computing*. https://doi.org/10.1109/TSUSC.2019.2907110
2. Bayyou, D. (2019). Artificially intelligent self-driving vehicle technologies, benefits and challenges. *International Journal of Emerging Technology in Computer Science and Electronics, 26*(3), 5–13. https://www.wavestone.com/app/uploads/2017/10/2017-driverlesscar.pdf
3. Gulati, A., Aujla, G. S., Chaudhary, R., Kumar, N., Obaidat, M., & Benslimane, A. (2021). Dilse: Lattice-based secure and dependable data dissemination scheme for social internet of vehicles. *IEEE Transactions on Dependable and Secure Computing, 18*(6), 2520–2534, https://doi.org/10.1109/TDSC.2019.2953841
4. How Automakers Can Survive the Self-Driving Era. AT Kearney https://www.atkearney.com/documents/10192/8591837/How+Automake
5. Litman, T. (2019). Autonomous Vehicle Implementation Predictions Implications for Transport Planning, Victoria Transport Institution, 18 March 2019. http://www.vtpi.org/avip.pdf
6. Maximizing the Benefits of Self-Driving Vehicles, Union of concerned scientists Maximizing-Benefits-Self-Driving-Vehicles.pdf (ucsusa.org).
7. Reina, G., Johnson, D., & Underwood, J. (2015). Radar sensing for intelligent vehicles in urban environments. *Sensors, 15*(6), 14661–14678.
8. Sharma, S., Dudeja, R. K., Aujla, G. S., Bali, R. S., & Kumar, N. (2020). DeTrAs: Deep learning-based healthcare framework for IoT-based assistance of Alzheimer patients. *Neural Computing and Applications*, 1–13. https://link.springer.com/content/pdf/10.1007/s00521-020-05327-2.pdf

Chapter 5
Artificial Intelligence: Need, Evolution, and Applications for Transportation Systems

Yueyue Dai and Huihui Ma

5.1 Overview of Artificial Intelligence

Nowadays, we have heard a plenty of AI stories. AI has permeated every aspect of our day-to-day life. When you ask “Siri” to call your friend Mike, you actually use Natural Language Processing (NLP), a branch of AI. When you do shopping online, the platform will recommend you several items that you might like by AI technique. When you enter some buildings by face identification, there is also AI. AI is everywhere even you have not noticed it. AI has been described as the fourth industrial revolution.

But what exactly is AI? Literally, it can be divided into two terms: artificial and intelligence. Artificial refers to anything that is made by human, not natural. Intelligence means the ability to understand and learn well and to form judgments and opinions based on reason. Combining the two terms, AI is a broad area of computer science that makes machine solve problems seem like they have human intelligence, in contrast to the natural intelligence displayed by humans. The official idea and definition of AI was first coined by John McCartey in 1956 at the Dartmouth conference. “Every aspect of learning or any other feature of intelligence can in principle be so precisely described that a machine can simulate it. An attempt will be made to find out how to make machines use language, form abstraction and concepts solve kinds of problems, now, reversed for humans and improve themselves.” After six decades of development, in 2019, the High-Level Expert Group on Artificial Intelligence produced a new AI definition.

Y. Dai (✉)
Research Center of 6G Mobile Communications and School of Cyber Science and Engineering, Huazhong University of Science and Technology, Wuhan, China
e-mail: yueyuedai@ieee.org

H. Ma
University of Electronic Science and Technology of China, Chengdu, China

S. Garg et al. (eds.), *Intelligent Cyber-Physical Systems for Autonomous Transportation*, Internet of Things, https://doi.org/10.1007/978-3-030-92054-8_5

"Artificial intelligence (AI) systems are software (and possibly also hardware) systems designed by humans that, given a complex goal, act in the physical or digital dimension by perceiving their environment through data acquisition, interpreting the collected structured or unstructured data, reasoning on the knowledge, or processing the information, derived from this data and deciding the best action(s) to take to achieve the given goal. AI systems can either use symbolic rules or learn a numeric model, and they can also adapt their behavior by analyzing how the environment is affected by their previous actions. As a scientific discipline, AI includes several approaches and techniques, such as machine learning (of which deep learning and reinforcement learning are specific examples), machine reasoning (which includes planning, scheduling, knowledge representation, and reasoning, search, and optimization), and robotics (which includes control, perception, sensors, and actuators, as well as the integration of all other techniques into cyber-physical systems)."

5.1.1 Evolution of Artificial Intelligence

In terms of time dimension, AI has gone through about three steps of development: the initial phase (1950s–1970s), the industrialization phase (1970s–1990s), and the explosion phase (1990s–2020s). Several important AI breakthroughs are shown in Fig. 5.1.

The first "AI period" began with the Dartmouth conference in 1956, where AI got its name and mission. From 1957 to 1974, AI flourished. People were excited for the computers were solving problems like human. But because of the limitation of computing and related techniques, the first AI winter started in the 1970s.

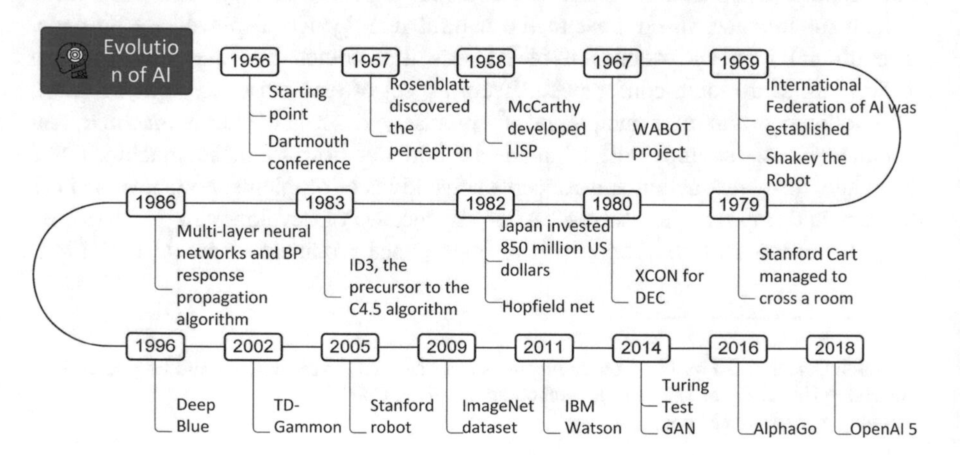

Fig. 5.1 Evolution of AI

In the 1980s, AI was reignited by two sources: an expansion of the algorithmic toolkit, and a boost of funds. The AI paradigm shifted to symbolic AI and the so-called expert systems or knowledge-based systems. As the expert system was increasingly being used for more and more fields, the limitation and errors came to the surface. However, the PC performance of Apple and IBM continued to increase and surpass the expensive specialized Lisp machines by 1987. From 1987 to 1993, the term "expert system" dropped from the IT lexicon, and it was referred to as the second AI winter.

The third major development in AI was from 1993 to the present. IBM's Deep Blue victory and the sensational event of Google's AlphaGo defeating world champion LiSedol let the world see some of the iconic results of AI. AI had addressed complex problems and provided solutions that were found to be useful in different application domains including data mining, industrial robotics, logistics, business intelligence, banking software, medical diagnosis, recommendation systems, and search engines. The AI evolution is also accelerated by the openness of code, frameworks, datasets, scientific publications, and overall knowledge sharing.

In terms of development level dimension, AI is divided broadly into three sections: artificial narrow intelligence (ANI), artificial general intelligence (AGI), and artificial super intelligence (ASI).

ANI, often referred to as "Weak" AI, is the type of AI that mostly exists today. ANI is limited in scope with intelligence restricted to only one functional area, e.g., those exploited by personal assistants Siri, language translations, recommendation systems, image recognition systems, face identification, etc. ANI can perform routine, repetitive, and mundane tasks that humans would prefer to avoid. While ANI is superior in some specialized domains, it is incapable of generalization, i.e., to re-use learned knowledge across domains, e.g., the ANI capable of image recognition cannot transfer its knowledge in the domain of speech recognition. The generalization problem is still an open question.

AGI, often referred to as "Strong" AI or "human-level" AI, is at an advanced level: it covers more than one field like power of reasoning, problem-solving, and abstract thinking. AGI is still entirely theoretical, with no practical examples in use today. AI will take a couple of decades to reach the AGI stage. However, the transition from ANI to AGI has begun.

ASI is the final stage of the intelligence explosion, in which AI surpasses human intelligence across all aspects, such as creativity, general wisdom, and problem-solving. At this moment, ASI belongs to science fiction.

5.1.2 Machine Learning

One of the common definitions of ML is the ability of systems to "make intelligent decisions without being explicitly programmed." Despite the fact that it is used interchangeably with AI by some people, ML is actually a subset of the field of AI. ML algorithms build a model based on sample data, known as "training data." In essence, ML approaches are data-driven. We are just truly in "The Information

Age." At the same time, computational capabilities continue to climb. They have opened the floodgates for the rise of ML.

According to the amount and type of supervision they get during the training, ML techniques are usually categorized into three, namely, supervised learning, unsupervised learning, semi-supervised learning, and RL.

Supervised Learning Uses labelled dataset $\mathbf{T} = \{(x_1, y_1), (x_2, y_2), ..., (x_n, y_n)\}$ to train algorithms to classify data or predict outcomes. Cases such as the digit recognition example, in which the aim is to assign each input vector to one of a finite number of discrete categories, are called classification problems. Cases such as predicting the house price or the stock market index, in which the desired output variable consists of one or more continuous variables, are regression problems. Common regression and classification techniques are linear and logistic regression, naive bayes, KNN algorithm, decision tree, random forest, support vector machine (SVM), and neural network. Supervised learning models can be used to build a number of business applications, including: image and object recognition, predictive analytics, customer sentiment analysis, spam detection, etc. Supervised learning techniques are more powerful than unsupervised techniques because the availability of labelled training data provides clear criteria for model optimization, but the computation time is vast and the pre-processing of data is no less than a big challenge.

Unsupervised Learning Uses unlabelled data $\mathbf{X}$ for training to learn the unknown hidden patterns or the structures. Unsupervised learning problems can be further grouped into clustering and association problems. A clustering problem is where you want to discover the inherent groupings in the data. An association rule learning problem is where you want to discover rules that describe large portions of your data. Typical unsupervised learning algorithms include: K-means clustering, KNN clustering, hierarchical clustering, self-organizing map, and principal component analysis (PCA). Some of the most common real-world applications of unsupervised learning are: news sections, computer vision, medical imaging, anomaly detection, customer personas, recommendation engines, etc. Unsupervised learning can handle large volumes of data in real time, but it might be less accurate than supervised learning for the lack of labelled data.

Semi-supervised Learning Is a happy medium, when the training data contains both labelled and unlabelled data. Typically, the amount of labelled data is small, and the amount of unlabelled data is large. Practical applications of semi-supervised learning are: speech analysis, Internet content classification, protein sequence classification, etc. The main advantage of this type of learning is that it reduces the errors of both supervised and unsupervised learnings. It consumes less computational power and is less time-consuming. But when the problem is complicated and your labelled data are not representative of the entire distribution, semi-supervised learning will not help.

Reinforcement Learning Is not given any labelled input/output pairs, in contrast to supervised learning, but must instead discover them by a process of trial and error.

RL is applicable to a wide range of complex problems that cannot be tackled with other machine learning algorithms. The real-world applications of RL are: robotics, autonomous driving, industry automation, delivery management, power systems, traffic control, etc. RL is closer to AGI, as it possesses the ability to learn policy autonomously. It has the potential to be a groundbreaking technology and the next step in AI development. Thus, we will introduce RL in detail in the next section.

5.1.3 Reinforcement Learning

The formal framework for RL borrows from the problem of optimal control of MDP, a probabilistic model of a sequential decision problem, represented by $< \mathcal{S}, \mathcal{A}, P, R, \gamma >$, where their definitions are given as follows:

State Set $\mathcal{S}$ At time step t, the agent observes state $s_t \in \mathcal{S}$ of the environment. The state is a unique characterization of all that is important in a state of the problem that is modeled. The state space can be discrete and continuous.

Action Set $\mathcal{A}$ At time step t, the agent takes the action $a_t \in \mathcal{A}$ to interact with the environment.

Transition Function P $P(s_{t+1}|s_t, a_t)$ records the probability of transitioning from state s_t to s_{t+1} after taking action a_t.

Reward Function R At time step t, the agent obtains a reward r_t according to $R(s_t, a_t)$. Reward is a signal of how well the learning algorithm is performing at achieving the global objective. It can be positive or negative to give direction in which way the system should be controlled.

Discount Factor γ Controls the importance of immediate rewards versus future rewards. The goal of the agent is to maximize some notion of cumulative reward over a trajectory τ, which is $R(\tau) = \sum_{t=0}^{\infty} \gamma^t r_t$.

Modeling a control task as an MDP is a key concept in RL. Given an MDP, a policy π is taken by the agent to determine the next action a_t based on the current state s_t. Solving a given MDP means computing an optimal policy π^*. To explain what is the optimality, the value functions are defined.

Value Function Usually means a state–value function, which maps a state to the expected reward, given the current state s, and follows the policy π. $V^{\pi}(s) = \mathbb{E}[R(\tau)|s_0 = s]$. The goal of MDP is maximizing the value function for all states $s \in \mathcal{S}$. An optimal policy, denoted π^*, is that $V^{\pi^*(s)} = \max_{\pi} V^{\pi}(s)$ for all states and all policies. Intuitively, $\pi^*(s) = \arg\max_{a} V^{\pi}(s)$. It greedily selects the best action using the value function.

Q-Value Function Also called action–value function, which maps a state–action pair to the expected reward of taking action a in the state s following the policy

π. $Q^{\pi}(s, a) = \mathbb{E}[R(\tau)|s_0 = s, a_0 = a]$. Its relationship with V is: $V^{\pi}(s) = \mathbb{E}[Q^{\pi}(s, a)]$.

The algorithms for MDP can be divided into model-based and model-free algorithms. Model-based algorithms exist under the general name of dynamic programming (DP). Model-free algorithms, under the general name of RL, do not rely on the availability of a perfect model. We discussed model-free algorithms in the next section, which refers to RL algorithms. They are further classified as on-policy or off-policy. An off-policy algorithm is an algorithm that, during training, uses a behavior policy that is different with the optimal policy it tries to estimate. An on-policy algorithm is an algorithm that, during training, chooses actions using a policy that is derived from the current estimate of the optimal policy, while the updates are also based on the current estimate of the optimal policy. Another way to classify the RL algorithms is by considering what component is optimized by the algorithm: policy-based, value-based, and actor-critic. Value-based RL aims to learn the value/action–value function in order to generate the optimal policy. Policy-based RL aims to learn the policy directly using a parameterized function. Actor-critic combines policy and value. The policy is called the actor and the value function is called the critic. The critic (typically a state-value function) evaluates, the actions executed by the actor. After action selection, the critic evaluates the action using the advantage function: $A^{\pi}(s, a) = Q^{\pi}(s, a) - V^{\pi}(s)$. Table 5.1 enumerates the most influential RL algorithms.

5.1.4 Other State-of-the-Art AI Algorithms

While the capabilities of AI improve rapidly, the algorithms behind AI models will also evolve. The advancements in the algorithm designs will enable AI to work more

Table 5.1 RL algorithms

Algorithm	Description	Policy	Action space	State space	Operater
Q-learning	State–action–reward–state	Off-policy	Discrete	Discrete	Q-value
SARSA	State–action–reward–state–action	On-policy	Discrete	Discrete	Q-value
DQN	Deep Q-network	Off-policy	Discrete	Continuous	Q-value
DDPG	Deep deterministic policy gradient	Off-policy	Continuous	Continuous	Q-value
A3C	Asynchronous advantage actor-critic	On-policy	Continuous	Continuous	Advantage
TRPO	Trust region policy optimization	On-policy	Continuous	Continuous	Advantage
PPO	Proximal policy optimization	On-policy	Continuous	Continuous	Advantage

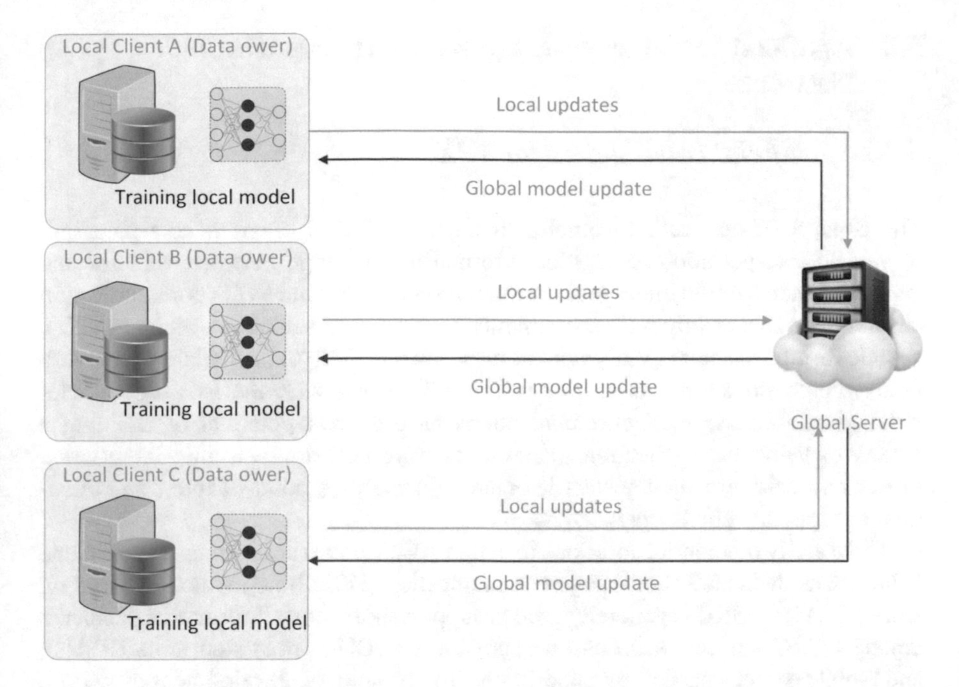

Fig. 5.2 Federated learning

efficiently and be available to more people with less amount of technical knowledge. Below the other state-of-the-art AI algorithms are introduced.

Federated Learning Brings ML models to the data source, rather than bringing the data to the model. It helps to solve privacy and cost considerations by using a centralized server to coordinate a federation of participating devices. Federated learning can be broken down into four steps as shown in Fig. 5.2.

Transfer Learning Is to use knowledge learned from tasks for which a lot of labelled data is available in other tasks where only little labelled data is available. Creating labelled data is expensive, so optimally leveraging existing datasets is key, which is one reason to use transfer learning. Another reason to use this technique is that some AI models are not easy to train and can take weeks to work properly. Using a trained model but not creating a new one for the new task is time-saving.

Meta-Learning Also known as "learning to learn," intends to design models that can learn new skills or adapt to new environments rapidly with a few training examples. Although transfer learning and meta-learning both contain the concept of pre-training, they are different. Meta-learning is a way of performing hyperparameter optimization and thus fine-tuning, but not in the sense of transfer learning, which can be roughly thought of as retraining a pre-trained model but on a different task with a different dataset.

5.2 Artificial Intelligence-Empowered Transportation Network

5.2.1 Artificial Intelligence for V2X

The concept of connecting vehicles to anything (V2X) opens a new paradigm in vehicle transportation to provide information exchange between vehicles and various elements of the intelligent transportation infrastructure with communication capabilities. The mainly V2X communications include vehicle-to-vehicle (V2V), vehicle-to-infrastructure (V2I), vehicle-to-pedestrian (V2P), and vehicle-to-network (V2N) communications, as shown in Fig. 5.3. Along with the existing vehicle-sensing capabilities, V2X communications have a great potential of enabling a variety of vehicular applications, such as forward collision warning, intersection movement assist, emergency electric brake light warning, point-of-interest notification, and remote vehicle diagnostics.

There are two potential solutions to support V2X communications. The first one is known as dedicated short-range communication (DSRC) [1], which is based on IEEE 802.11p. DSRC is generally used to support short-range information exchange among DSRC devices, such as on-board units (OBUs), road-side units (RSUs), and mobile devices carried by pedestrians. To promote the development of DSRC, 75 MHz of the spectrum has been allocated in 5.9 GHz frequency band by the U.S. Federal Communications Commission (FCC). The second one is known as cellular-based V2X communication (C-V2X) [2]. C-V2X is expected to operate in the ITS band and possibly co-exist with DSRC. According to 3GPP Release 14 [3], the physical layer of C-V2X has two new communication modes (Mode 3 and Mode 4). In Mode 3, vehicles communicate directly between them, but the cellular network selects and manages the radio resources for each direct V2V communication. In Mode 4, vehicles autonomously select and manage their radio resources for their direct V2V communications without any cellular infrastructure support. To this aim, Mode 4 defines a distributed resource allocation scheme, namely sensing-based

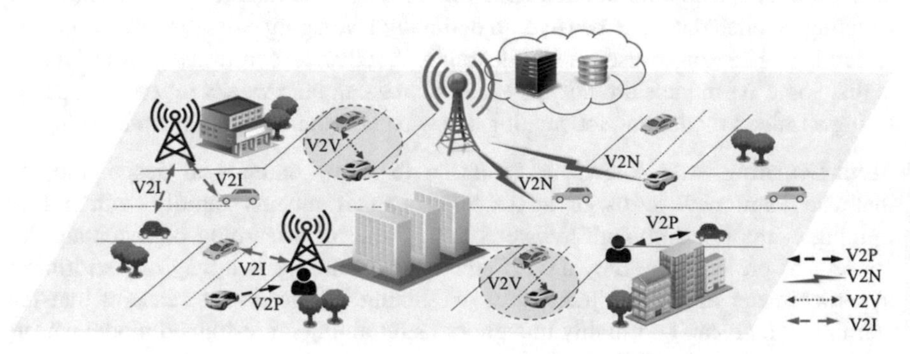

Fig. 5.3 General V2X communication

semi-persistent scheduling scheme, in that vehicles can autonomously select their radio resources and make distributed congestion control.

To promote safety and efficiency of transportation systems, V2X provides heterogeneous communications among vehicles, pedestrians, and infrastructures. The heterogeneous communication resources, such as channel, spectrum, transmission power, bandwidth, and time slots, are unevenly distributed in networks. As vehicular applications have diverse and strict requirements, the joint optimization of multi-resource allocation is a challenging task. In [4], the authors proposed a spectrum sharing and power allocation scheme to maximize the overall link throughput with the consideration of slow and large-scale fading information of wireless channels. In the proposed scheme, the authors considered that V2I connectivity is enabled by macrocellular link with large capacity and V2V connectivity is supported by localized D2D link with high reliability. The authors in [5] transformed the reliability and latency requirements of V2V links into optimization problem to maximize the overall data rate. A SOLEN algorithm was proposed to solve the formulated problem. In the first step of the SOLEN algorithm, the resource block allocation problem is transformed into the maximum weight matching problem, and the Hungarian algorithm is used to solve the transformed problem. In the second step, the power control allocation problem is solved with convex optimization method, i.e., dual decomposition method. However, these works did not consider the unique characteristic of vehicular network, that is, the high mobility of vehicles. Since vehicles are dynamically moving in vehicular networks, the wireless communication links are easily and frequently disconnected, which will deteriorate communication quality. Furthermore, the environment is spatial–temporal changing in a vehicular network, and the resource allocation scheme should consider the dynamic topology and QoS requirements from vehicular applications.

Deep reinforcement learning is a promising AI technology, which has the ability to interact with wireless environment to facilitate resource management and orchestration. To solve the mentioned challenges, some AI-based dynamic resource allocation algorithms have been proposed. In [6], the authors utilized DRL to jointly optimize D2D relay selection and a power-level allocation problem. In the proposed DRL-based method, there are two DRL agents: a transmission agent and a relay selection agent. According to the locations, velocities, shapes, antenna heights, of vehicles, the transmission agent is responsible for assigning transmission power levels to transmitters with the objective of transmission rate maximization. The relay selection agent selects relays for packets with the objective of transmission delay minimization. Simulation results show the transmission delay performance of proposed method is better than a link-quality-prediction-based method and close to a link-quality-known method. The authors in [7] investigated the joint optimization problem to make transmission mode selection and resource allocation for cellular V2X communications for maximizing the total capacity of vehicle-to-infrastructure users. A two-timescale federated DRL algorithm was designed. In the large timescale, the BS constructs undirected graphs based on the large-scale channel gains and makes resource allocation. In the small timescale, each V2V pair updates their current clustering.

5.2.2 *Artificial Intelligence for Vehicular Edge Computing*

To reduce traffic accident rate and improve traffic efficiency, vehicles are equipped with a plenty of intelligent transportation applications, which have strict requirements on resources, e.g., CPU, GPU, and memory. Although current powerful vehicles have a certain amount of local resources, it is still insufficient to run some machine learning-based applications with a low latency. Vehicular edge computing (VEC) is a new paradigm that aims to move computing and caching resources in close proximity of vehicular users, by combining mobile edge computing and vehicular networks. In a VEC, smart vehicles have the following features: (1) sensing: vehicles can sense surrounding traffic with on-board devices, such as cameras, radars, and GPS; (2) communication: vehicles can exchange traffic information with each other via V2X communication; (3) computing and caching: vehicles can execute parts of the computation-intensive tasks and store its own sensed information locally [8]. RSUs located along a road act as edge servers. Different from the existing RSUs, these RSUs are deployed with computing and storage resources and AI functions. As shown in Fig. 5.4, with VEC, vehicles can enjoy:

- Low-latency computing services: since computation resources and AI functions are widely deployed on RSUs, vehicles with limited resources can offload their computation-intensive and latency-sensitive tasks to nearby edge servers via V2R communications. Due to the short distance and powerful data processing abilities, RSUs can provide low-latency computing services, which can reduce task response time and alleviate the heavy burden on backhaul link [9]. As the state-of-the-art smart vehicles are equipped with certain computing resources, the vehicle with sufficient computing resources can also provide computing services to the resource limited vehicles via V2V communications.

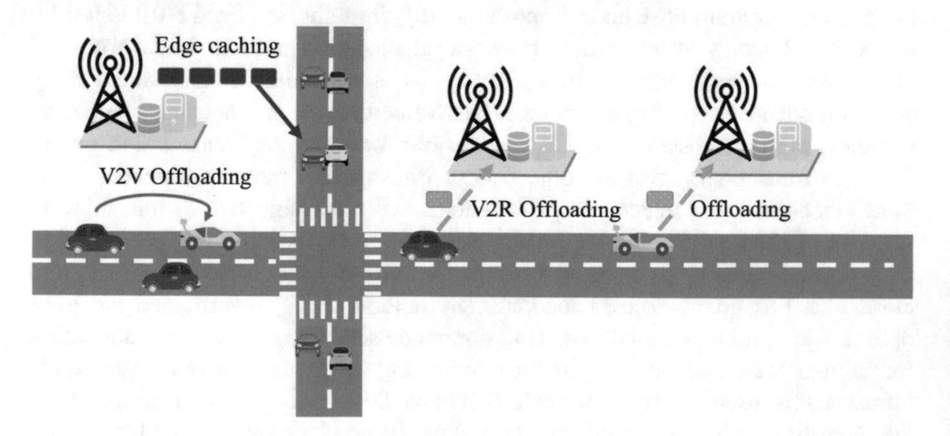

Fig. 5.4 The illustration of vehicular edge computing

- Ubiquitous edge caching services: since edge servers can cache some popular contents (e.g., videos, news, files), vehicles can directly obtain the interested content from caching nodes without accessing the core network, to reduce remote content delivery time and improve efficiency of network bandwidth usage.

There have been widely studied researches on how to investigate the performance improvement of edge computing and caching in vehicle networks. In [9], the authors integrated a joint multi-user multi-server computation offloading and load balancing problem in vehicular networks to maximize system utility. First, the authors analyzed the IEEE 802.11p protocol and modeled it into the task execution time. Then, a joint load balancing and offloading problem was formulated as a mixed-integer non-linear programming problem. To solve this problem, the authors decoupled the original problem into two sub-problems, developed a low-complexity algorithm to jointly make VEC server selection, and optimized offloading ratio and computation resource. In [10], the authors studied how to utilize edge caching to enable content caching and forwarding at the edge of vehicular networks. The authors first proposed a cooperative edge caching architecture to enhance network edge caching capability. Then, considering the mobility feature of vehicles, a mobility-aware hierarchical caching scheme was designed in which vehicles are taken as collaborative catching agents for sharing content cache tasks with edge servers.

The above methods to address the edge computing and caching problems are based on convex optimization and game theory. Nevertheless, these methods may suffer from the following issues: (1) The vehicular network requires immediate decisions to deal with the real-time situation; however, the above methods are often with high computational complexities. (2) Some key factors, such as wireless channel condition and content popularity, are assumed to be known. But in reality, they are time-varying and difficult to find the proper distributions to model them. (3) The VEC network is constructed with various edge servers, which is embedded with different resource capacities. For various kinds of QoS requirements in the heterogeneous vehicular network, jointly allocating suitable resources for each vehicle is challenging. Since machine learning can discover the in-depth feature of the network and can tackle complex optimization problems, recently it has been widely utilized in VEC computation offloading and caching. For example, in [11], the authors proposed an AI-empowered vehicular network architecture with cognitive and intelligent abilities for smart vehicular edge computing and caching. A joint vehicular edge computing and caching problem was formulated as an optimization problem to maximize computing and caching utility. Considering mobility parameters (i.e., location, speed, and direction), a mobility-awareness deep reinforcement learning-based resource allocation method was designed to make computing and caching resource allocation. In [12], the authors considered a VEC network with dynamic topologies, unstable connections, and unpredictable movements. In such a scenario, vehicles can offload computation tasks to neighboring VEC clusters with the objective of both minimizing energy consumption of systems and satisfying task latency constraints. To increase searching efficiency for large-scale networks, the authors proposed an imitation learning-enabled online task scheduling algorithm where an expert can obtain the optimal scheduling policy by solving the formulated optimization problem with a few samples offline. In [13], the authors considered a multi-access VEC network and proposed a DRL-based edge caching and content delivery approach to minimize the total content delivery latency of vehicles. The authors first presented a multi-access edge caching framework in which vehicles can obtain the requested contents via V2V, V2R, and V2I links. Then, a vehicular edge caching and content delivery problem was formulated, and a deep deterministic policy gradient-based algorithm was designed to solve the formulated problem.

5.2.3 Artificial Intelligence for Unmanned Aerial Vehicle

With the attractive characteristics of agility, versatility, and low cost, unmanned aerial vehicles (UAVs) can be utilized in diverse applications and purposes, such as environmental and natural disaster monitoring, border surveillance, emergency assistance, search and rescue missions, delivery of goods, and construction. Due to the easy-to-deploy and large coverage capabilities, UAVs also play an important role in future communication systems. That is, UAVs can be deployed as aerial communication relays or flying base stations to provide cost-effective wireless connectivity for devices without infrastructure coverage. According to [14], UAV communications have the following attributes:

- Line-of-Sight (LoS) Links: Compared with vehicles and infrastructures, UAVs without human pilots flying in the sky have a higher probability to connect ground nodes via LoS links, which facilitates highly reliable transmissions over long distances. Based on the real-time traffic information collected by embedded sensors, UAVs can adjust their hovering positions to maintain the quality of links, which can enhance the reliability of ground networks.
- Dynamic Deployment Ability: Different from traditional infrastructures fixed on dedicated locations, UAVs can be dynamically deployed and allocated to different users or controllers based on their real-time requirements. In addition, UAVs can be easily and cost-effectively re-deployed with the spatial and temporal changes of demands.
- UAV-based Swarm Networks: A swarm of UAVs can construct a scalable multi-UAV networks and offer ubiquitous connectivity to ground users. Due to its high flexibility and rapid provision features, the multi-UAV network can form a multi-hop network to recover and establish communication for devices without infrastructure coverage, which is quite useful for the communication unavailable scenarios.

There are a plenty of works investigating UAV communications to improve network connectivity and enhance information dissemination and data collection. In [15], to maintain the high quality of wireless connectivity, the authors formulated a UAV trajectory design problem to minimize UAV mission completion time. To solve the formulated problem, the authors proposed an effective approach to check the connectivity quality based on a graph verification method. By examining the GBS-UAV association sequence during the UAV mission, the optimal UAV trajectory can be obtained. The authors in [16] proposed how to improve the energy efficiency in UAV-aided information dissemination and data collection. First, the authors formulated the energy efficiency maximization problem in a UAV-aided network, where cluster heads have different SNR levels and different geographical locations. Based on the potential game, a game-theoretic game data collection method was proposed to optimize the allocation of time slots with the fairness constraint. Moreover, the authors also analyzed the price of anarchy of the proposed game. Extensive simulations exhibit the effectiveness of the proposed scheme under varying environments.

Recently, as AI can bring intelligence to machines and make them able to perform tasks better, the combination of AI and UAV can enhance the overall system level of intelligence, benefit the design of UAVs trajectory and deployment, and improve the quality of UAV-based communication. More specifically, the authors in [17] presented an idea of proactive deployment of cache-enabled UAVs for optimizing the quality-of-experience (QoE) of wireless devices in a cloud radio access network (CRAN). To optimize QoE, a novel machine learning-based framework, i.e., conceptor-based echo state networks (ESNs) was proposed, which can effectively predict each user's content request distribution and its mobility pattern. Based on the predictions of the users' content request distribution and their mobility patterns, the optimal user-UAV association, optimal locations of the UAVs as well as the content to cache at UAVs can be obtained. In [18], the authors considered a cellular-UAV network, where UAVs can execute sensing tasks through cooperative sensing and transmission to minimize the age of information. However, the cooperative sensing and transmission is tightly coupled with the UAVs' trajectories, which makes the trajectory design challenging. Leveraging DRL, the authors proposed a compound-action actor-critic algorithm to learn the optimal policy of the UAV trajectory design.

5.3 AI-Based Applications in Transportation System

5.3.1 Autonomous Driving

In 2019, traffic accidents resulted in 33,244 fatalities in the United States [19] and 22,800 fatalities in the European Union [20]. With the steady increase in the number of vehicles on the road, traffic congestion and road safety are becoming critical issues. Autonomous vehicles have gained significant interest as solutions to these challenges since 90% of all car accidents are estimated to be caused by human errors, while only 2% are caused by vehicle failures [21]. Early autonomous vehicle systems relied heavily on accurate sensor data with multi-sensor settings and expensive sensors, such as LIDAR, to enable accurate environmental perception. The control of these autonomous vehicles is handled by a rule-based controller, where the parameters are set by the developer and manually adjusted after simulation and field testing. However, the disadvantage of these methods is that the parameter time is manually adjusted, and the rule-based controller is difficult to extend to new scenarios. Moreover, the highly non-linear nature of driving makes it difficult to use linear vehicle model in control systems [22].

In recent years, since machine learning has obtained great achievements in fields such as image classification and speech recognition, it leads to the increasing use of machine learning in autonomous driving applications, including planning and decision-making, perception, as well as mapping and localization. For example, in [23], the authors proposed a Convolutional Long Short-Term Memory Recurrent Neural Network (C-LSTM), that allows end-to-end learning both visual and temporal dependencies of driving. In addition, the authors formulated the steering angle regression problem as a deep classification problem by imposing a spatial relationship between the output layer neurons. By introducing correlation between the class neurons, the proposed method can improve steering root mean square error by 35% over recent methods. In [24], the authors presented a deep convolutional neural net-based end-to-end controller for steering autonomous vehicles, in which deep neural net is served as a controller and trained with human guidance. Specifically, in the training phase, screen captures and human drivers steering angles were generated by the car simulator CARSIM, and CAFFE is used to build the neural net structure and train the net. In [25], the authors proposed an object-centric perception approach and provided a comparative evaluation on autonomous driving. Compared to the existing end-to-end holistic models, the proposed approach can provide object-level attention.

5.3.2 Traffic Prediction and Control

With the explosion of the number of vehicles, there are lots of traffic jams, collisions, and air pollution. Traffic prediction and control are of great important for transportation system to alleviate congestion and guarantee road safety. Traffic prediction can forecast the real-time traffic information according to the historical data. And then traffic control uses the output of the traffic prediction to make optimal control strategy to ease congestion. But traffic prediction is very challenging as it is affected by many complex factors, such as inter-region traffic, events, and weather. Conventional statistical methods, such as Kalman filtering (KF) and Auto-Regressive Integrated Moving Average (ARIMA), cannot completely capture traffic flow patterns; hence, they cannot provide high accuracy for prediction of traffic with diverse characteristics. For the case of autonomous driving, high accuracy is especially crucial. How to improve the prediction accuracy has drawn increasing attention in AI research field. The emergence of Internet of Things (IoT), vehicular network, sensor networks, and social media has surpassed traditional means of collecting data, by creating voluminous and continuous streams of real-time data. This fits with the data-hungry nature of AI. The authors in [26] proposed a deep-learning-based traffic flow prediction method. A stacked autoencoder (SAE) model was used to learn generic traffic flow features, and it was trained in a layer-wise greedy fashion. It was the first time that the SAE approach was used to represent traffic flow features for prediction. The spatial and temporal correlations were inherently considered in the modeling. The authors in [27] presented a deep architecture for traffic flow prediction, which had been implemented as a stack of restricted Boltzmann machines (RBMs) at the bottom with a regression layer at the top. The stack architecture at the bottom was a DBN, and it was effective for unsupervised learning. The authors in [28] proposed an expansive smart traffic management platform (STMP) based on the unsupervised online incremental ML, deep learning (DL), and DRL. The STMP integrated the heterogeneous big data streams, such as the IoT, smart sensors, and social media, traffic flow forecasting, and optimized traffic control decisions.

Real-time traffic forecasting is the useful input for intelligent traffic control to optimize traffic network performance. Traffic control includes traffic light signals control, dynamic speed limit, route guidance, ramp metering, etc. The typical approach that conventional transportation methods take is to cast traffic control as an optimization problem under certain assumptions about the traffic model. The key issue here is that the simplified models often deviate from the real world. Since model-based methods may not accurately describe the complex nature of traffic dynamics in all situations, model-free data-driven methods are more suitable for traffic control. Especially, DRL is a generic and flexible way to develop intelligent and adaptive traffic control methods. The authors in [29] proposed a DRL algorithm that combines several tricks to master an appropriate control strategy within an acceptable time for Urban Traffic Control (UTC). The simulation experiments had shown that DRL performed better than the traditional UTC approaches and could handle more complex environments while using fewer computing resources. The authors in [30] successfully trained a set of two autonomous vehicles to lead a fleet of vehicles onto a roundabout and then transferred this policy from simulation to a scaled city without fine-tuning. The authors in [31] classified the known approaches based on the RL techniques and provided a review of existing models with analysis on their advantages and disadvantages. It explored future directions in the area of RL-based traffic signal control methods. The existing RL methods for games usually require a massive number of update iterations and trial and errors for RL models to yield impressive results in simulated environments. How to learn efficiently is a critical question for the application of RL in traffic control. While RL methods learn from trial and error, the learning cost of RL could be critical or even fatal in the real world as the malfunction of traffic signals might lead to accidents. Discrepancies between simulation and reality confine the application of learned policies in the real world. Although AI-based traffic control systems still have some limitations to what they can achieve intelligently, the future still holds a lot of promise for these systems.

5.3.3 UAV Path Planning

Due to the large coverage and ease of installation, UAV has been used in cargo transfer, object detection, vision-assisted navigation, and disaster rescue. To realize these applications, the efficient and safe path planning schemes are required to guide UAVs traveling from an initial point to the goal point without obstacles and collisions. Generally, the path planning problem is often treated as a global optimization problem with various constraints from certain mission, environment (such as flight length and threat constraints), and UAV physical constraints (such as flight altitude and turning angle constraints). Currently, the path planning methods can be roughly divided into two types: graph-theory-based and non-graph-theory-based methods. The graph-theory-based approaches aim to find a reasonable path in a certain graph modeling the environment. Non-graph-theory-based approaches include rapidly exploring random tree, vector field histogram, genetic algorithm, which plan a path based on random sampling, potential filed theories, or bionics theories. The Voronoi diagram search method [32] is a typical graph-based method. In this method, the battle area is partitioned into a number of convex polygons, and in every polygon, only one threat is contained. The optimal path is decided using the Eppstein's k-best algorithm in the presence of simple disjoint polygonal obstacles. In [33], the genetic algorithm with the particle swarm optimization algorithm was proposed to design the UAV path in a complex 3D environment. The optimal path is produced by considering line segments, circular arcs, and vertical helices.

Although the above work has made progress in UAV path planning, dynamic environments still pose a great challenging as it may result in unexpected flying obstacles. To deal with this, recently, model-free RL methods have become popular in the field of path planning. Based on Q-learning, the authors in [34] proposed an Adaptive and Random Exploration approach (ARE) to design a UAV path planning and obstacle avoidance method. The basic idea of ARE is to let the UAV explore the environment itself and make actions according to the current evaluation. The proposed approach can balance the adaptive mechanism of self-learning and convergent random search so that the subjective UAV can possess both the ability of finding general directions and escaping the dilemma due to learning errors. In [35], the authors proposed a Deep Reinforcement Learning approach for UAV path planning based on the global situation information. First, a fast situation assessment model was proposed to translate the global environmental states into sequential situation maps to represent dynamic enemy threats. Second, a deep reinforcement learning UAV path planning was designed, which utilizes the dueling double deep Q-networks to predict Q-values and ϵ-greedy strategy to select path policy. The authors in [36] optimized the UAV trajectory in the UAV-mounted MEC network, where the UAV was deployed as a mobile edge server to dynamically serve the mobile TUs. First, considering the movement model of users, the UAV trajectory optimization problem was formulated as an MDP form to maximize the long-term system reward. Then, using double deep Q-network, a QoS-based ϵ-greedy selection policy algorithm was proposed. Simulation results demonstrated that the proposed algorithm can achieve almost 99% guarantee rate in QoS of each user.

References

1. *IEEE Draft Guide for Wireless Access in Vehicular Environments (WAVE)-Architecture*, IEEE Standard P1609.0/D9 (2017) (pp. 1–104).
2. Papathanassiou, A., & Khoryaev, A. (2017). Cellular V2X as the essential enabler of superior global connected transportation services. *IEEE 5G Tech Focus, 1*(2), 1–2.
3. Evolved Universal Terrestrial Radio Access (EUTRA); Physical Layer; Measurements (Release 14), document TS 36.201, 36.300, 36.211, 36.212, 36.213, 36.214, 3GPP (2017).
4. Liang, L., Li, G. Y., & Xu, W. (2017). Resource allocation for D2D-enabled vehicular communications. *IEEE Transactions on Communications, 65*(7), 3186–3197.
5. Sun, W., Ström, E. G., Brännström, F., Sou, K. C., & Sui, Y. (2016). Radio resource management for D2D-based V2V communication. *IEEE Transactions on Vehicular Technology, 65*(8), 6636–6650 (2016)
6. Zhang, H., Chong, S., Zhang, X., & Lin, N. (2020). A deep reinforcement learning based D2D relay selection and power level allocation in mmWave vehicular networks. *IEEE Wireless Communications Letters, 9*(3), 416–419.
7. Zhang, X., Peng, M., Yan, S., & Sun, Y. (2020). Deep-reinforcement-learning-based mode selection and resource allocation for cellular V2X communications. *IEEE Internet of Things Journal, 7*(7), 6380–6391. https://doi.org/10.1109/JIOT.2019.2962715
8. Liu, L., Chen, C., Pei, Q., Maharjan, S., & Zhang, Y. (2020). Vehicular edge computing and networking: A survey. *Mobile Networks and Applications*. https://doi.org/10.1007/s11036-020-01624-1
9. Dai, Y., Xu, D., Maharjan, S., & Zhang, Y. (2019). Joint load balancing and offloading in vehicular edge computing and networks. *IEEE Internet of Things Journal, 6*(3), 4377–4387. https://doi.org/10.1109/JIOT.2018.2876298
10. Zhang, K., Leng, S., He, Y., Maharjan, S., & Zhang, Y. (2018). Cooperative content caching in 5G networks with mobile edge computing. *IEEE Wireless Communications, 25*(3), 80–87.
11. Dai, Y., Xu, D., Maharjan, S., Qiao, G., & Zhang, Y. (2019). Artificial intelligence empowered edge computing and caching for internet of vehicles. *IEEE Wireless Communications, 26*(3), 12–18 (2019)
12. Wang, X., Ning, Z., Guo, S., & Wang, L. (2020). Imitation learning enabled task scheduling for online vehicular edge computing. *IEEE Transactions on Mobile Computing*. https://doi.org/10.1109/TMC.2020.3012509
13. Dai, Y., Xu, D., Lu, Y., Maharjan, S., & Zhang, Y. (2019). Deep reinforcement learning for edge caching and content delivery in internet of vehicles. In *2019 IEEE/CIC International Conference on Communications in China (ICCC), Changchun, 2019* (pp. 134–139).
14. Li, B., Fei, Z., & Zhang, Y. (2019). UAV communications for 5G and beyond: Recent advances and future trends. *IEEE Internet of Things Journal, 6*(2), 2241–2263.
15. Zhang, S., Zeng, Y., & Zhang, R. (2019). Cellular-enabled UAV communication: A connectivity-constrained trajectory optimization perspective. *IEEE Transactions on Communications, 67*(3), 2580–2604.
16. Abdulla, A. E. A. A., Fadlullah, Z. M., Nishiyama, H., Kato, N., Ono, F., & Miura, R. (2014). An optimal data collection technique for improved utility in UAS-aided networks. In *IEEE INFOCOM 2014 - IEEE Conference on Computer Communications, Toronto, ON, Canada, 2014* (pp. 736–744). https://doi.org/10.1109/INFOCOM.2014.6848000
17. Chen, M., Mozaffari, M., Saad, W., Yin, C., Debbah, M., & Hong, C. S. (2017). Caching in the sky: Proactive deployment of cache enabled unmanned aerial vehicles for optimized quality-of-experience. *IEEE Journal on Selected Areas in Communications, 35*(5), 1046–1061.
18. Hu, J., Zhang, H., Song, L., Schober, R., & Poor, H. V. (2020). Cooperative internet of UAVs: Distributed trajectory design by multi-agent deep reinforcement learning. *IEEE Transactions on Communications, 68*(11), 6807–6821.
19. https://www.iihs.org/topics/fatality-statistics/detail/state-by-state
20. https://www.romania-insider.com/romania-road-fatality-rate-2019

21. Singh, S. (2015). Critical reasons for crashes investigated in the national motor vehicle crash causation survey. Nature Highway Traffic Safety Admin., U.S. Nat. Center Statist. Anal., Tech. Rep. DOT HS 812 115, Feb. 2015 (pp. 1–2).
22. Kuutti, S., Bowden, R., Jin, Y., Barber, P., & Fallah, S. (2021). A survey of deep learning applications to autonomous vehicle control. *IEEE Transactions on Intelligent Transportation Systems, 22*(2), 712–733.
23. Eraqi, H. M., Moustafa, M. N., & Honer, J. (2017). End-to-end deep learning for steering autonomous vehicles considering temporal dependencies. arXiv:1710.03804. [Online]. Available: https://arxiv.org/abs/1710.03804
24. Rausch, V., Hansen, A., Solowjow, E., Liu, C., Kreuzer, E., & Hedrick, J. K. (2017). Learning a deep neural net policy for end-to-end control of autonomous vehicles. in *2017 American Control Conference (ACC)* (pp. 4914–4919).
25. Wang, D., Devin, C., Cai, Q.-Z., Yu, F., & Darrell, T. (2019). Deep object-centric policies for autonomous driving. In *2019 International Conference on Robotics and Automation (ICRA), 2019* (pp. 8853–8859).
26. Lv, Y., Duan, Y., Kang, W., Li, Z., & Wang, F. Y. (2014). Traffic flow prediction with big data: A deep learning approach. *IEEE Transactions on Intelligent Transportation Systems, 16*(2), 865–873.
27. Huang, W., Song, G., Hong, H., & Xie, K. (2014). Deep architecture for traffic flow prediction: deep belief networks with multitask learning. *IEEE Transactions on Intelligent Transportation Systems, 15*(5), 2191–2201.
28. Nallaperuma, D., Nawaratne, R., Bandaragoda, T., Adikari, A., Nguyen, S., Kempitiya, T., De Silva, D., Alahakoon, D., & Pothuhera, D. (2019). Online incremental machine learning platform for big data-driven smart traffic management. *IEEE Transactions on Intelligent Transportation Systems, 20*(12), 4679–4690.
29. Lin, Y., Dai, X., Li, L., & Wang, F. Y. (2018). An efficient deep reinforcement learning model for urban traffic control. *Preprint arXiv:1808.01876*.
30. Jang, K., Vinitsky, E., Chalaki, B., Remer, B., Beaver, L., Malikopoulos, A. A., & Bayen, A. (2019). Simulation to scaled city: zero-shot policy transfer for traffic control via autonomous vehicles. In *Proceedings of the 10th ACM/IEEE International Conference on Cyber-Physical Systems* (pp. 291–300).
31. Wei, H., Zheng, G., Gayah, V., & Li, Z. (2021). Recent advances in reinforcement learning for traffic signal control: A survey of models and evaluation. *ACM SIGKDD Explorations Newsletter, 22*(2), 12–18.
32. Bhattacharya, P., & Gavrilova, M. L. (2007). Voronoi diagram in optimal path planning. In *Proceedings of the 4th International Symposium on Voronoi Diagrams in Science and Engineering (ISVD 2007)* (pp. 38–47). Glamorgan, UK: IEEE.
33. Roberge, V., Tarbouchi, M., & Labonté, G. (2013). Comparison of parallel genetic algorithm and particle swarm optimization for real-time UAV path planning. *IEEE Transactions on Industrial Informatics, 9*(1), 132–141. https://doi.org/10.1109/TII.2012.2198665
34. Yijing, Z., Zheng, Z., Xiaoyi, Z., & Yang, L. (2017). Q learning algorithm based UAV path learning and obstacle avoidance approach. In *2017 36th Chinese Control Conference (CCC), 2017* (pp. 3397–3402). https://doi.org/10.23919/ChiCC.2017.8027884
35. Yan, C., Xiang, X. & Wang, C. (2020). Towards real-time path planning through deep reinforcement learning for a UAV in dynamic environments. *Journal of Intelligent and Robotic Systems, 98*, 297–309.
36. Liu, Q., Shi, L., Sun, L., Li, J., Ding, M., & Shu, F. (2020). Path planning for UAV-mounted mobile edge computing with deep reinforcement learning. *IEEE Transactions on Vehicular Technology, 69*(5), 5723–5728.

Chapter 6
Artificial Intelligence Deployment in Transportation Systems

Zhiwei Guo and Keping Yu

6.1 Review for AI-Deployment Transportation Systems

6.1.1 Overview

Since the development of intelligent transportation in 1973, the development of intelligent transportation was relatively slow in the early stage due to the limitation of communication means. From 1995 to 2000, with the rapid growth of data transmission speed by leaps and bounds, and the breakthrough of location service and communication technology, the development speed of intelligent transportation was significantly accelerated. At this time, the development of AI-Deployment transportation system (AID-TS) was mainly limited by computing capacity. From 2000 to 2010, intelligent transportation technology has been fully promoted, and high-definition video and intelligent analysis and judgment have been fully applied in the field of urban transportation. From 2010 to now, with the continuous development of big data, machine learning, and other technologies [1], AI-based vehicle–road cooperation, automatic driving, intelligent travel, and other technologies will become the key direction of the next stage of technology development of intelligent transportation system.

In recent years, the intelligent transportation industry has developed rapidly, more traffic bayonet networking and collected more vehicle traffic record information. The relevant departments can use artificial intelligence technology to analyze

Z. Guo
School of Artificial Intelligence, Chongqing Technology and Business University, Chongqing, China
e-mail: zwguo@ctbu.edu.cn

K. Yu (✉)
Global Information and Telecommunication Institute, Waseda University, Shinjuku, Tokyo, Japan
e-mail: keping.yu@aoni.waseda.jp

S. Garg et al. (eds.), *Intelligent Cyber-Physical Systems for Autonomous Transportation*, Internet of Things, https://doi.org/10.1007/978-3-030-92054-8_6

the urban traffic flow in real time, adjust the interval of traffic lights effectively, shorten the waiting time of vehicles, and effectively improve the efficiency of urban road traffic [2]. The artificial intelligence algorithm is based on the travel preference, lifestyle, consumption habits, and other factors of urban people [3] and effectively analyzes the data of urban passenger flow and traffic flow migration, urban construction, public resources, and so on. The analysis results can assist the urban planning decision and guide the infrastructure construction of public transportation facilities.

AI for transportation is equivalent to installing an AI brain on the transportation system of the whole city. Through automatic perception, the real-time search of traffic big data can be realized on the basis of not interfering with travelers. It can control the traffic information on the city road, the vehicle information of the parking lot, and so on in real time [4] and forecast the change of the traffic flow and the number of parking spaces effectively in advance [5]. In this way, resources can be reasonably allocated, traffic can be effectively channelized, and large-scale traffic linkage scheduling can be realized. It can also improve the efficiency of traffic operation in the whole city, relieve traffic congestion, and ensure the smooth travel of residents.

The application of artificial intelligence in transportation system can be regarded as a process from Internet technology (IT) to operation technology (OT) to evolution technology (ET). Originally, the transportation industry needs to invest a lot of resources to achieve informationization and digitalization. In order to excavate the value of data and output products and services, OT should be carried out to form a standardized operation process and model. Finally, to ET, that is to realize intelligence.

With the support of artificial intelligence, transportation system can realize the real-time and efficient flow of data [6]. The sensor monitoring and positioning system is responsible for data acquisition, and the data communication is carried out by means of mobile communication, satellite communication, and so on. The data is processed by the central processor of the subsystem above and then published on the Internet and other channels using information feedback and automatic control to carry out route planning, traffic signal control, ramp control, etc.

In transportation management, "human" has played an important role in various application scenarios in the past: whether it is driving, order maintenance, illegal investigation or information judgment, all need human participation to be able to complete smoothly. However, with the rapid development of the Internet of Things (IOT), big data, and artificial intelligence technology, "artificial intelligence" will replace "human brain" in the field of traffic management [7].

Advanced physical perception technology, precise positioning technology, accurate map technology and wireless communication technology, and rapid increase of computer processing and storage performance, transportation systems, and infrastructure information level unceasing enhancement. All of these have laid a good foundation for the application of artificial intelligence in traffic management, so that artificial intelligence can be developed in the transportation industry.

The urbanization process is increasingly accelerating, and the travel needs of people are becoming more diversified and personalized. How to solve the traffic problems such as traffic congestion and frequent accidents has become an important topic of urban construction. The AI-Deployment transportation systems integrating a variety of advanced technologies provide new ideas and means to promote the quality and efficiency of transportation. Its great potential value in solving traffic congestion, inefficient resource utilization, and blocked information circulation is worth looking forward. As a new type of real-time, efficient, and accurate transportation system, AID-TS are currently being widely used in developed countries such as Europe and the United States. It is predicted that the application of AID-TS can effectively improve transportation efficiency, reduce traffic congestion by 20%, reduce delay losses by 10–25%, reduce car accidents by 50–80%, reduce fuel consumption by 30%, and reduce exhaust emissions.

There are three stages in the development of AI-deployment transportation systems:

- **Eye Era: Electronic eye monitoring and auxiliary governance:** In the past two decades, the AI-Deployment transportation systems (AID-TS) have been developing rapidly in China, and the application of AID-TS technology in road traffic management is increasing. Road video surveillance, electronic police, signal machine, and so on are all part of the AID-TS. The application of these systems has played a huge role in the construction and application of the urban traffic command center. However, the AID-TS at this stage are relatively scattered, and no information network has been formed.
- **Combination of Eyes and Hands: The centralized command platform appears:** Around 2010, the development of intelligent transportation in China entered the second stage. The transportation department began to build a centralized command platform for traffic information, tried to combine the traffic information system with the traffic command system, and made a comprehensive analysis and judgment of the traffic data, the intelligent transportation ushered in a leap forward.
- **Brain era: Artificial Intelligence Involves Governance:** Artificial intelligence, big data, cloud computing, and other technologies develop at an extremely fast speed and have been applied to various industries, including the traffic governance field of the transportation industry. With the advent of these technologies and the power of the Internet, intelligent transportation has also entered a new era. The 5G technology, the Internet of Vehicles, and the coordination between vehicles and roads will also be used in daily traffic management.

Nowadays, the increase of vehicles in China and the application of a large number of sensing devices have led to the exponential growth of traffic data. Data is the foundation of artificial intelligence. Only with enough data can artificial intelligence fully learn and become a traffic brain, can better arrange and utilize traffic resources, and maximize the utilization rate of traffic resources. The new mode and technology bring new development and competition. The entry of artificial intelligence, Internet of Vehicles, Internet of Things, and Internet technologies, as well as driverless technologies, has become the hot spot of competition among countries. It has promoted the speed of high-quality development of transportation industry in various countries, accelerated the construction of AID-TS, and at the same time can better meet the travel needs of people.

6.1.2 Prevalence

In the world, the most extensive application of AI-Deployment transportation systems (AID-TS) is Japan, and the AID-TS in Japan are quite complete and mature, followed by the United States, Europe, and other regions. The AI-Deployment transportation systems are developing rapidly in China, and advanced AI-Deployment transportation systems have been built in major cities such as Beijing, Shanghai, and Guangzhou. Among them, Beijing has established four AI-Deployment transportation systems of road traffic control, public traffic command and dispatch, expressway management, and emergency management. Guangzhou has established three, including the main traffic information sharing platform, the logistics information platform, and the static traffic management system. With the development of related technology, AI-Deployment transportation systems will be more and more widely used in transportation industry.

Japan began research on AI-Deployment transportation systems as early as 1973. The AID-TS planning system in Japan includes advanced navigation systems, safety assistance systems, traffic management optimization systems, efficient road traffic management systems, public transportation support systems, vehicle operation management systems, pedestrian guidance systems, and emergency vehicle support systems. AID-TS are mainly used in traffic information provision, electronic toll collection, public transportation, commercial vehicle management, and emergency vehicle priority. There are currently more than 18 million car navigation system users in Japan.

The United States is one of the more successful countries in the application of AID-TS. In March 1995, the U.S. Department of Transportation published the "National Intelligent Transportation System Project Plan." It clearly stipulated the 7 major areas and 29 user service functions of the AI-Deployment transportation systems and determined the annual development plan to 2005. The 7 major areas include travel and traffic management systems, travel demand management systems, public transportation operation systems, commercial vehicle operation systems, electronic toll collection systems, emergency management systems, and advanced vehicle control and safety systems. According to reports, the application of AID-TS in the United States has reached more than 80%, and related products are also more advanced. The U.S. AID-TS are used in vehicle safety systems (51%), electronic toll collection (37%), highway and vehicle management systems (28%), navigation and positioning systems (20%), and commercial vehicle management systems (14%) that have been developed rapidly.

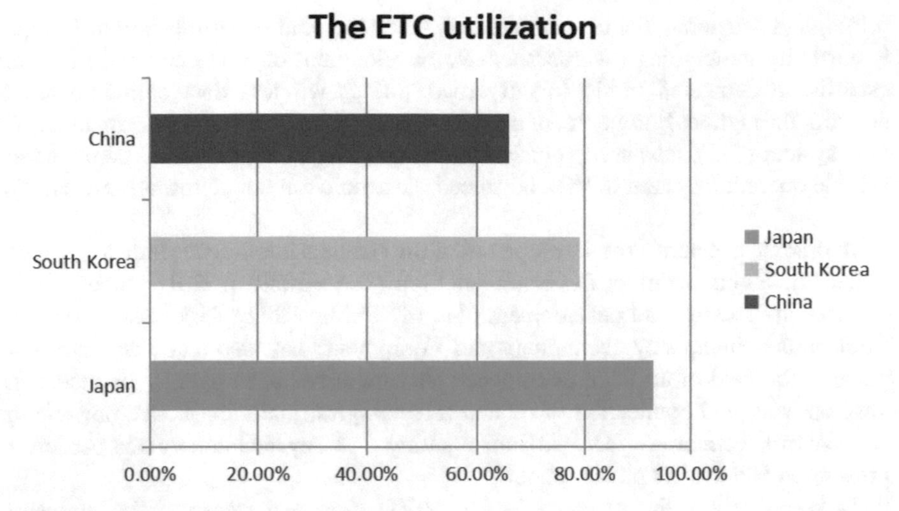

Fig. 6.1 The ETC utilization in some countries

Taking the utilization rate of electronic toll collection (ETC) in some countries as an example, as shown in Fig. 6.1, the current utilization rate of highway ETC in Japan is 92.9, in Korea 79.1, and in China 65.98.

In the future, artificial intelligence will become an infrastructure, and the transportation industry itself is a field with extensive public participation, so the two have a high compatibility. The construction of intelligent transportation will serve the daily travel of the general public and make the social resources circulate efficiently.

6.1.3 Development Status

Japan has a modern highway intelligent management system, unified consideration of people, vehicles, road traffic, and other factors. Japan believes that the highway intelligent management system is an effective way to solve the traffic problems at the present stage, and it is a way to promote the traffic reform and also promote the road traffic to be more convenient, safe, high quality, and efficient. According to this concept, the AI-Deployment transportation systems are relatively developed in Japan. In 2003, a number of highways fully implemented electronic toll collection (ETC) charges and closed the semi-automatic hybrid charging unit.

The European Union has launched its Horizon 2020 research, development, and innovation programme to support public investment in the development of the AI industry. Through the public–private partnership plan, 2.5 billion euros will be invested to strengthen the construction of advanced research centers, support the development and application of AI technology for small- and medium-sized enterprises, and accelerate the development of AI testing and experimentation. Europe is currently undergoing a comprehensive development of Telematic and plans to establish a dedicated traffic (mainly road traffic) wireless data communication network throughout Europe. It attempts to develop advanced travel information service system (TIS), advanced vehicle control system (VCS), advanced commercial vehicle operation system (CVO), advanced electronic toll collection system (ETC), etc.

In order to accelerate the development of the artificial intelligence industry, China issued the Robot Industry Development Plan (2016–2020) in 2016. Since 2017, the special research and development plan of "Public Safety Risk Prevention and Control and Emergency Technology and Equipment" has also made development plans in the field of artificial intelligence. Attention has been paid to the research, development, and application of related technologies, including active prevention and control technology for road traffic safety and key technology for proactive prevention and control police robots.

In recent years, the market of urban AI-Deployment transportation industry has maintained a rapid growth trend in China. The market demand of urban AI-Deployment transportation systems mainly focuses on traffic signal control system, traffic video surveillance system, electronic police system, and others. Specifically, the AI-Deployment transportation systems are integrated technologies such as advanced information technology, data communication and transmission technology, electronic sensing technology, and computer software processing technology. It can be effectively applied to the whole ground traffic management system, thus establishing an efficient, safe, environmental-friendly, real-time, and accurate large-scale integrated traffic management system.

AID-TS construction in Singapore focuses on advanced urban traffic management systems. In addition to traditional functions, such as signal control, traffic detection, and traffic guidance, the system also includes the use of electronic billing cards to control traffic flow. During peak hours and congested roads, tolls can also be automatically increased, and the efficiency of road use can be controlled as reasonably as possible.

The AID-TS demonstration project is selected in Gwangju City in South Korea and is expected to cost 10 billion won. It selects 9 items including traffic sensing signal system, bus passenger information system, dynamic route guidance system, automated management system, timely broadcast system, electronic toll collection system, parking forecast system, dynamic weighing system, and AID-TS center.

Australia has advanced AI-Deployment transportation systems and intelligent traffic control systems, such as Sydney coordinated adaptive traffic system (SCATS). It can actively adapt to changes in traffic conditions by randomly acquiring traffic information from sensors and cameras on the road. In addition, the Australian transportation management department has developed an active signal system that can change the speed limit on the road according to different conditions. It can detect the speed of cars and continuously monitor them. If they are going too fast, it will send a signal to the driver.

6.2 Architecture for AI-Deployment Transportation Systems

With the development of social economy and improvement of people's living standard, the pressure of social transportation is increasing. The roads are crowded with many kinds of vehicles, and the air pollution is becoming more and more serious. In order to relieve the traffic pressure, it is of great significance to implement the architecture of AI-Deployment transportation systems. With the progress of modern science and technology, taking AI technology as the representative, it has brought many aspects of influence to people's daily life and social development, including the structure of AI-Deployment transportation systems. Therefore, improving the application level of AI technology in intelligent transportation cannot only promote the development of urban traffic on the basis of intelligence, but also facilitate the overall development of the society and promote the overall progress.

The architecture of AI-Deployment transportation systems is relatively complex, which can be divided into the sensing layer, the networking layer, and the application layer as shown in Fig. 6.2. The first layer is the sensing layer, which is responsible for collecting information and where all data in the AI-Deployment transportation systems come from. The second layer is the networking layer, which is responsible for the transmission of information. The data in the AI-Deployment transportation systems can be obtained anytime and anywhere depending on it. The third layer is the application layer, which realizes the identification and perception between objects and objects, people, and objects. Thus, it can effectively manage

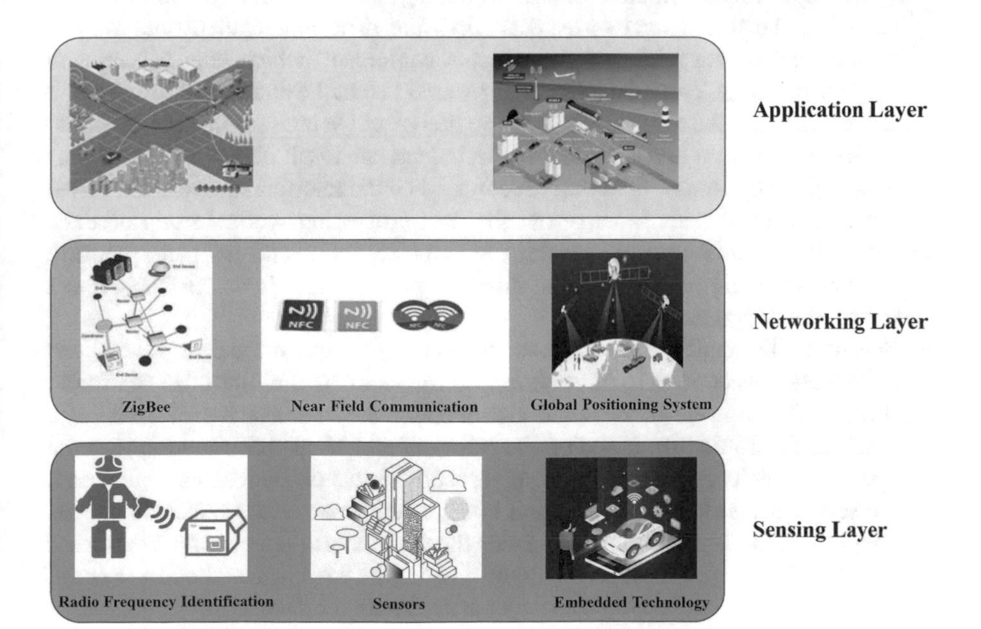

Fig. 6.2 The architecture for AI-deployment transportation systems

all kinds of data information in the traffic system and play the intelligent role of the system.

6.2.1 AI-Deployment Sensing Layer

The sensing layer is the basis for AI-Deployment transportation systems architecture and is an important link in the integration of the physical world and information world [8]. The key technologies of the sensing layer mainly include radio frequency identification (RFID), sensors, and intelligent embedded devices. In the sensing layer, the information of the object itself and its surroundings can be sensed through facilities such as vision sensors, sound sensors, radar, satellite navigation system, etc., so as to realize data collection and acquisition.

- **Radio Frequency Identification:** RFID is a simple wireless system that consists of an interrogator (reader) and a number of transponders (tags). The tag is composed of a coupling element and a chip. Each tag has the unique electronic code of the extension entry and is attached to the object to identify the target object. It transmits the radio frequency information to the reader through the antenna, and the reader is the device that reads the information. RFID technology gives objects a characteristic of traceability, that is, it can grasp the exact position of the object and its surrounding environment at any time.

  The application principle of this technology in the traffic system is mainly shown in the following two aspects: first, the data generated by the vehicle on the road is transmitted to the relevant equipment, which needs to rely on RFID technology. The principle in this process is that the antenna of information reader will send the electronic tag to the designated vehicle. When the specified vehicle travels to a specific area, the data generated will activate the tag, thus realizing the transmission of data. Second, when the relevant equipment receives the data, it also needs to carry out the data processing work. In this process, the RFID technology can interpret the received data, send it to the platform for corresponding processing, and give full play to the role of AI-Deployment transportation systems.
- **Sensors:** The application of sensor technology in transportation system can effectively collect vehicle information on road surface and optimize the traffic flow[9]. At the same time, it can also monitor the intersections of each road, optimize the traffic flow based on the calculation of all aspects, and improve the operating efficiency. Based on the aggregation points on both sides of the road, the sensor network is organized and constructed, and various collected signals are put into the signal equipment under the function of the network. The sensor terminal is deployed through the landfill under the road or installed in the road planning place to realize the data acquisition function effectively.
- **Embedded Technology:** Embedded technology is a kind of technology that takes computer as an information processing component and embeds it into the

application system. It integrates the software solidification into the hardware system and integrates the hardware system with the software system. Embedded technology has the characteristics of small code, high automation, and fast response, so it is used more and more widely. Many systems in AID-TS, such as vehicle monitoring systems, on-board GPS navigation products, set-top box electronic maps, etc., use embedded technology.

6.2.2 AI-Deployment Networking Layer

The networking layer mainly realizes the transmission, routing, and control of information, that is, the data collected from the sensing layer is transmitted through the communication network [10]. The networking layer generally uses wireless network, Internet, and wired network for data transmission. The networking layer delivers data safely, completely, and quickly through the integration of various telecommunication networks and the Internet. The networking layer is used to select different subsystems and dynamic operation, so that the system information can be fully utilized and shared. In addition, this layer can also realize the collection, storage, and processing of information, make the corresponding processing according to the requirements of different departments, and provide the corresponding information. The main technologies include: ZigBee, NFC, Wi-Fi, Bluetooth, Global Positioning System (GPS), etc.:

- **ZigBee:** ZigBee is a new wireless communication technology, which is suitable for a series of electronic components and devices with short transmission range and low data transmission rate. ZigBee wireless communication technology is a network technology applied to Internet communication that is developed and generated based on the way bees contact each other. It has the advantages of low complexity, short distance, low cost, and power consumption. Compared with the traditional network communication technologies, ZigBee wireless communication technology is more efficient and convenient. As a short-range, low-cost, and low-power consumption wireless communication technology, ZigBee is based on 802 15.4 wireless standard approved by IEEE on networking, security, and application software. This technology is particularly suitable for small data traffic services and can be easily installed in a series of fixed and portable mobile terminals.
- **Near-Field Communication (NFC):** Near-field communication is an emerging technology that is developed from the integration of non-contact RFID and interconnection technology. The "near field" in the NFC refers to radio waves near an electromagnetic field, and radio waves are essentially electromagnetic waves. The electric and magnetic fields alternate for energy conversion during the propagation from the transmit antenna to the receive antenna and enhance each other during the conversion. While within 10 wavelengths of the electromagnetic wave, the electric field and the magnetic field are independent of each other, and

the magnetic field can be used for short-distance communication, which is called near-field communication.

The near-field communication service combines the near-field communication technology and the mobile communication technology. It can realize many functions such as identity authentication, ticketing, data exchange, and anti-counterfeiting. It is a new type of service in the mobile communication field. Devices using NFC technology, such as mobile phones, can exchange data while in close proximity to each other. By integrating the functions of inductive card reader, inductive card, and point-to-point communication on a single chip, mobile terminal is used to realize mobile identity recognition and other applications.

- **Global Positioning System (GPS):** Global Positioning System is a high precision radio navigation positioning system based on artificial Earth satellites. It can provide accurate geographic position, vehicle speed, and accurate time information anywhere in the world and near Earth space. With the expansion of urban construction scale, the number of vehicles is increasing day by day. And the operation management, reasonable dispatch of transportation, and the command and safety management of police vehicles have become important problems in the public security and traffic system. The emergence of GPS provides specific real-time positioning ability for vehicles, ships, and other vehicles.

  The main systems used in traffic systems include the command management system combining vehicle GPS positioning and radio communication system, and the command management system using GPS differential technology. GPS vehicle navigation has a wide range of applications, such as transportation route navigation, emergency vehicle navigation, vehicle dispatch, etc. Intelligent vehicles and highway systems can be divided into four types according to their functions: automatic system, fleet management system, consultative navigation system, and general survey system. The fleet management system can operate multiple vehicles and has a center. The consultative navigation system integrates the automatic system and fleet management system. The general survey system usually consists of an automated vehicle-mounted camera or a digital camera, which can be used to obtain road information on time and point characteristics.

6.2.3 AI-Deployment Application Layer

The application layer can compute, process, and mine the data transmitted by the networking layer reasonably and efficiently and extract the effective information from it. The application layer can design and make different management platforms according to the different needs of users. For example, the vehicle management platform of "intelligent transportation" collects the information on the road through sensors such as coils and cameras and transmits data to the management platform through the networking layer. Then, statistics and management of traffic flow, vehicle speed, and vehicle occupation ratio on the road can be visualized.

The application layer mainly contains two important components: feedback control and service application. For feedback control, on the one hand, it can be applied to realize information processing in different services and provide the traffic management functions. On the other hand, it can continuously expand the function according to the service demand. Among them, service application mainly includes expert systems based on statistical principles, such as bus scheduling; auxiliary decisions, such as automated route planning and automated guidance; adaptive processing, that is, traffic lights are adjusted on the basis of actual traffic flow information, and also users are enable to dynamically query the corresponding traffic information.

6.3 Business Situations for AI-Deployment Transportation Systems

6.3.1 Autonomous Transportation Management

Some of the key verticals around autonomous transportation management are shown in Fig. 6.3. They are discussed below:

- **Automation of Urban Rail Transit:** The development process of urban rail transit is constantly accelerating, and the corresponding supporting technologies are also rapidly updated [11]. Many first-tier cities with a large population bear the greater pressure of passenger flow. While speeding up the construction of new lines, it is necessary to increase the number and departure frequency of shuttle buses on important lines, which is undoubtedly increase the cost investment. Automated urban rail transit can significantly reduce human and material power

Fig. 6.3 Autonomous transportation management

costs. Although there are higher costs required to build such an urban rail transit system, its positive role in improving the operation efficiency and quality of urban rail transit and reducing the operating costs is obvious. Urban rail transit automation can automatically complete train wake-up, parking, switch and doors, etc.; it can set up differentiated operation modes such as routine operation, degraded operation, and operation terminal, meet the personalized needs of different scenarios, and improve operation safety and ride experience. And the train starting and braking are more stable, which can bring a high-quality travel experience for passengers, and the speed is faster, which can reduce people's travel costs.

- **Intelligent Highway System (IHS):** IHS is a system that conducts real-time automatic safety detection of vehicles, releases relevant information, and implements real-time automatic operations. It provides safer, more economical, comfortable, and faster basic services for the realization of intelligent highway transportation. It is a system for managing multiple self-driving vehicles, which can realize the automatic and orderly operation of the fleet. All vehicles in the system can be driven automatically. Real-time communication between vehicles and roads can achieve mutual cooperation to achieve the goal of efficient and safe operation.

  The characteristic of IHS is to combine highway with cloud computing, big data, IOT, and AI to realize comprehensive real-time perception of highway information. And grasp the status quo of each section of road, each vehicle and each structure, and accurately predict the development trend. IHS includes a smart management system and a public information service system. The smart management system receives the perception data transmitted through the stable, large broadband expressway private network. Then it conducts fusion analysis on the massive sensing data to realize the visualization, mobility, intelligence and precision of business management, emergency response, and charging management. The public information service system can provide the public with various information including road conditions to assist decision-making.
- **Automation of Traffic Signal Systems:** The application of artificial intelligence can effectively alleviate the problem of urban traffic congestion [12]. Authoritative statistics show that: In terms of economy, the economic loss caused by traffic congestion in the United States every year is as high as 121 billion dollars. In terms of the environment, traffic jams cause up to 25 billion kilograms of carbon dioxide emissions every year. When the cars are running in urban areas, about 40% of the engines are idling because the city has not installed intelligent traffic signal systems. The application of intelligent traffic signal system can improve the afford ability of urban traffic roads and avoid the continuous expansion and reconstruction of relevant departments, thus reducing the cost consumption. It can accurately monitor and perceive the current road conditions and automatically adjust the light color conversion time according to the advanced artificial intelligence algorithm. The system uses a decentralized approach to accurately control the operation of the traffic network, which

can better adapt to the actual traffic conditions, and coordinate with adjacent intersections to optimize the overall road traffic management.

- **Intelligent Road Traffic Management System:** In the process of operation, the intelligent road traffic management system gives the road traffic information to the traffic management control department for in-depth excavation and analysis. The analysis results will be provided to users including residents, car owners, parking lots, logistics enterprises, etc., so as to provide effective reference data for their travel planning. At the same time, the traffic management department can use the traffic system to deal with emergency situations and traffic accidents in time. Transportation departments can quickly understand the status of vehicles and optimize their own management and service systems accordingly. With the widespread application of artificial intelligence in the field of road traffic management, the work of the traffic police will be undertaken by the police robot. This robot can patrol the road 24 hours and implement a full range of supervision, so as to improve the work efficiency of the public security traffic management department.
- **Highway Traffic Safety Prevention and Control System:** By using the highway traffic safety prevention and control system, the relevant departments can keep abreast of the traffic situation of each section of the road, supervise the traffic violations, and deal with them quickly to restore normal traffic order. Also, improve their own control ability and the efficiency of service management as a whole, correct traffic violations in time, keep the whole urban road system unblocked, and reduce the probability of major traffic accidents, so that improve the safety of urban traffic.

  Traffic behavior monitoring, traffic safety research and judgment, traffic risk early warning, and traffic law enforcement technology are the main components of road traffic safety prevention and control system. At present, these technologies have been integrated with artificial intelligence, so that the traffic management department can clearly see the running state of highway traffic, find the path of vehicles, catch the key illegal behaviors, eliminate safety risks, quickly respond to the road cooperation, and improve the level of traffic information application service.

6.3.2 Vehicular Control

When a vehicle is driving, the driving environment is complex and changeable. Only relying on the driver to process and make decisions on complex information, and to control the vehicle, cannot ensure the safe operation of the vehicle. The advanced vehicle control system (VCS) can assist fully autonomous driving, thereby improving driving safety [13]. VCS uses advanced sensor technology, computer technology, vehicle control technology, and positioning to detect vehicle peripheral information [14]. Through information fusion and processing, the system can automatically recognize dangerous states and assist drivers in safe assisted driving or automatic driving to improve driving safety and increase road capacity. It can greatly improve the driver's safety factor and driving efficiency. The key technologies of VCS include sensor technology, machine vision, communication technology control technology, driving condition monitoring technology, and weather detection technology. According to the degree of intelligence, VCS can be divided into Intelligent Connected Car, Intelligent Vehicle–Road Collaboration System, and Driverless.

6.3.2.1 Intelligent Connected Car (ICV)

ICV refers to the organic combination of the Internet of Vehicles and smart cars [15]. It is equipped with advanced in-vehicle sensors, controllers, actuators, and other devices and integrates modern communication and network technologies. It can realize intelligent information exchange and sharing between cars and people, cars, roads, and backgrounds, as well as safe, comfortable, energy-saving, and efficient driving, and finally realize a new generation of cars that replace human operations.

ICV is emerging industries under the background of a new round of technological revolution. It can significantly improve traffic safety, realize energy-saving and emission reduction, eliminate congestion, and improve social efficiency. And to promote the coordinated development of automobiles, electronics, communications, services, and social management. It is of great strategic significance to promote the transformation and upgrading of my country's industry.

6.3.2.2 Intelligent Vehicle–Road Collaboration System (IVRCS)

The IVRCS is based on artificial intelligence, information processing, positioning and navigation, wireless communication, electronic sensing, and many other technologies to obtain vehicle and road information and achieve seamless docking between vehicle and vehicle and vehicle and road [16]. Through the collection and analysis of real-time traffic data, the active safety control of vehicles and the cooperative management of roads are realized, and the cooperative linkage between people, vehicles, and roads is brought into full exertion. Finally, achieve the goal of improving traffic efficiency and safety of road traffic system. In the construction of intelligent transportation, vehicle–road cooperative system is an important part.

IVRCS is one of the hot research directions in the field of ITS, which has high dependence on many technologies such as information security, state awareness, multi-mode communication, data fusion, and collaborative processing. These technologies cross and interact with each other and become the key technologies of IVRCS, such as intelligent communication, intelligent vehicle technologies, intelligent system collaborative control, and intelligent road measurement system:

- **Intelligent Communication:** In the field of intelligent transportation and vehicles, mobile communication is a basic technology, which is the core of real-time interaction between people, vehicles, and roads [17–19]. Based on vehicle-to-vehicle and vehicle-to-road communication technology, the intelligent communication technology realizes real-time and efficient information interaction between vehicle-to-vehicle and vehicle-to-road in the process of high-speed vehicle driving.

- **Intelligent Vehicle Technology:** The key technologies of intelligent vehicles mainly include: real-time collection of vehicle position, operating status, driving environment, and other information by means of sensing devices such as positioning system, gyroscope, electronic compass, and lidar on-board unit installed on the vehicle. With the help of the electric hydraulic braking system installed on the vehicle, industrial control computer and other control equipment can realize intelligent control of the vehicle unit, and in order to avoid rear-end collision and other traffic accidents in time [20].
- **Collaborative Control of Intelligent System:** Collaborative control of intelligent systems can be divided into two categories, one for efficiency and the other for safety. Among them, the efficiency oriented intelligent system collaborative control includes: accurate parking control technology, dynamic collaborative dedicated lane technology, intersection intelligent control technology, cluster guidance technology, traffic control and traffic guidance collaborative optimization technology, etc. Another kind of intelligent system cooperative control for safety includes: Collaborative Driver Assistance System (C-DAS), vehicular cooperative safety early warning system.
- **Intelligent Road Measurement System:** The key technologies of the intelligent road measurement system can realize a variety of functions, such as multi-channel road condition information acquisition, multi-mode wireless data transmission, multi-channel traffic information acquisition, information fusion, and rapid identification and location of unexpected and abnormal events [21].

6.3.2.3 Driverless

Driverless vehicles rely on environment sensing technology, use on-board sensors to perceive the surrounding environment of the vehicle, and process the captured environmental information [22]. Using computer information technology to control, adjust the driving speed and driving direction of the car, to ensure that the vehicle can drive normally on the road and reach the destination according to the original plan. In the process of driving, the driverless car can use the safety control function equipped with the system to deal with various emergencies and ensure driving safety. In a broad sense, driverless vehicles can be seen as a kind of Internet-based vehicles that realize autonomous driving through the integrated application of computer technology, network communication technology, and intelligent control technology.

- **Driverless System in Highway Environment:** The highway has perfect road signs, and the driverless system applied in this structured environment needs to undertake the function of tracking road signs and automatically identifying vehicles. For high-speed autonomous driving in this highly standardized environment, the driverless system aims to achieve the fully autonomous driving. Although it is difficult for such applications to be applied in other environments, motorway driving is quite dangerous, and if fully automated driving can be achieved, the application value of driverless technology can be fully demonstrated.

- **Driverless System in Urban Environment:** Compared with highway driving, driverless driving in cities has lower requirements for car speed but higher requirements for safety, so it has a broad development prospect. At present, driverless system can share the pressure of urban mass public transport and alleviate the traffic tension in urban areas. The specific application scenarios include airports, industrial parks, parks, campuses, and other public places.
- **Driverless System in Special Environment:** Some countries have mastered relatively advanced driverless vehicle technology and have been committed to applying driverless technology to military and other special environments. In these fields, the applications of driverless technology have similarities with the highway and urban environment, but there are differences in the pertinence of technical performance.

6.3.3 Public Transportation Scheduling

The public transportation scheduling (PTS) system has the functions of collecting, transmitting, and processing bus operation data and is the core subsystem of the public transportation system. PTS realizes real-time monitoring and visual scheduling of bus operating vehicles. It can increase the full load rate of vehicles and the transportation capacity of the public transportation system, improve the efficiency of public transportation enterprises, and further enhance the information and intelligence of the entire city.

PTS optimizes the company's vehicle operation organization through information technology and rationally arranges the working hours of drivers and passengers, so as to maximize the transportation efficiency of the company and create good corporate and social benefits. PTS is mainly composed of bus scheduling center, sub-scheduling center, vehicle-mounted mobile station, and electronic stop sign.

6.3.3.1 Public Transportation Scheduling Center

The public transportation scheduling center is mainly composed of an information service system, geographic information system (GIS), a large-screen display system, a coordinated scheduling system, and an emergency handling system. The information service system is responsible for providing users with public transportation information (pre-trip information, transfer information, travel schedule information, fare information, etc.). GIS receives positioning data and completes the map mapping of vehicle information. Its functions include input/output geographic information and data information, display and edit maps, query vehicle roads and other information, maintain databases, receive and process GPS data, map matching GPS data, process and display vehicle roads and other information, save and manage vehicle operating data, etc.

6.3.3.2 Sub-scheduling Center

The sub-scheduling center is composed of two parts: the vehicle positioning system and GIS. The vehicle positioning system is responsible for completing the positioning and monitoring of vehicles under the jurisdiction of the scheduling center, two-way communication between vehicles, sending scheduling instructions to vehicles, and sending data to electronic station signs.

6.3.3.3 Vehicle Mobile Station

The vehicle-mounted mobile station uses differential GPS technology for positioning. Vehicle-mounted dedicated terminal (GPS receiver), single-chip microcomputer (SCM), wireless MDOEM, data, voice communication radio, and other equipment are installed on mobile public transport vehicles, as shown in Fig. 6.4. They can automatically complete the positioning of moving vehicles and the return of positioning information without human intervention. When necessary, they can provide short messages to the sub-scheduling center, and at the same time, they can set aside an interface for external on-board display equipment.

6.3.3.4 Electronic Stop Sign

The electronic stop sign is responsible for receiving and displaying the next bus arrival information and service information. It is composed of a set of MODEM, radio, SCM, and electronic display stop sign, as shown in Fig. 6.5. The function of SCM is to receive information, process it, and send it to the electronic stop sign

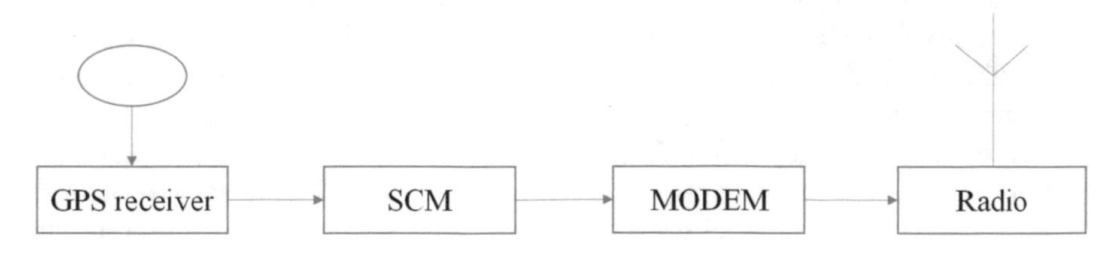

Fig. 6.4 Autonomous transportation management

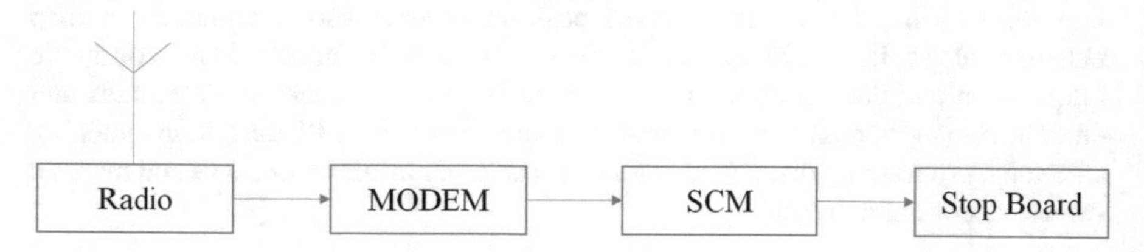

Fig. 6.5 Electronic stop sign composition diagram

for display. The electronic stop sign adopts a rolling information work method. In addition to displaying vehicle operation information, it can also display other information (date and time, weather forecast, and advertisement).

6.3.4 Transportation Information Service

Advanced traffic information service system (TIS) is an important part of ITS. TIS can be defined as: make full use of advanced information technology, information transmission technology, collect static and dynamic service information of various transportation facilities, and provide travelers with current public transportation and road conditions. It helps travelers to choose travel mode, travel time, and travel route and provides high-quality traffic information services for traffic participants, making the travel experience more comfortable.

Its function is to disseminate information to users, requiring good information transmission channels, which is a necessary condition and direction for contemporary intelligent construction and management. TIS is mainly divided into the following four aspects.

6.3.4.1 Traveler Information Service Needs

The travel needs of the public can provide TIS with information needs, including information content, structure, and form, thereby improving the level of traffic information service and adapting to the needs of economic and social development:

- According to the public's demand for relevant information: weather conditions (weather information); degree of road congestion and slippery road surface (road condition information); traffic accident information (route information); location of gas stations and public toilets (information on supporting facilities).
- Before traveling, the public paid more attention to information. Road condition information, weather information, and route selection information have all been highly valued. In the travel process, people are almost entirely focused on the knowledge of road conditions.

- Before traveling, traffic radio, newspapers and television, text messages, and mobile phones were the top three ways for people to obtain information. Among them, traffic broadcasting is the most widely used channel.
- In the process of travel, traffic broadcasts, mobile phone maps, and variable electronic display static signs are ranked in the top three ways for people to obtain information.

6.3.4.2 The Content of Travel Information Services

The information service content in TIS mainly includes: pre-travel information service, driving driver information service, public transportation information service on the way, personalized information service, etc.

- **Information Service Before Travel:** Before the start of the trip, travelers can log into the pre-trip information service system at any starting point to provide a reference for the planning and decision-making of the trip. This is achieved by browsing and understanding travel routes, travel modes, travel costs, current road network systems, and bus operations. This service is mainly for the mass travelers.
- **Driver Information Service While Driving:** The multimedia terminal is used to release accurate information about travel decisions and vehicle operating status, as well as road information and warning information, and provide guidance for drivers who are not familiar with the terrain and road conditions. The user subject is the driver.
- **Public Transportation Information Service on the Way:** Using advanced electronics, communication, multimedia, and network technologies, bus travelers in the journey can obtain information about bus travel services in real time on the side of the road, at the bus station, or on the bus. So that people can make further decisions and adjustments on their travel routes, methods, and time. The target user is the passenger.
- **Personalized Information Service:** Obtain personalized information and log into the personalized information service system through a varicty of media and personal devices to obtain information about comprehensive social services and facilities related to travel. It includes the address, business or office hours of places such as restaurants, parking lots, auto repair shops, hospitals, and police stations.

6.3.4.3 Construction Content of Public Travel Information Service Platform

The public travel information service platform includes “1 public travel information center, 7 application systems, and 7 release channels.”

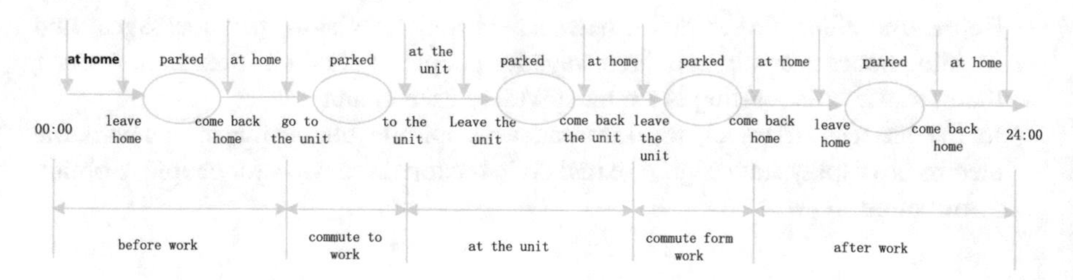

Fig. 6.6 Typical commuter travel chain diagram

- **Public Travel Information Center:** To build a public travel platform information center, a traffic public travel database framework should be built, and a basic traffic travel database and a special database should be built. And moderately connect with the business data of the Meteorological Bureau, the Tourism Bureau, and other related industries. Then, realize the unified collection, cleaning, transmission, mining, and analysis of comprehensive public travel information to form a unified transportation travel data resource platform.
- **7 Application Systems for Public Travel:** Build and develop 7 public travel service systems, including a traffic electronic map service system, a travel route planning service system, a comprehensive traffic real-time status system, a traffic dynamic guidance system, a traffic weather information service system, a road rescue information service system, and a tourism information service system. It meets the needs of public travel comprehensive traffic information services and traffic dynamic guidance services and realizes the in-depth mining and application of public travel information.
- **7 Release Channels for Public Travel:** Public travel mainly includes 7 information service sub-platforms, including public travel service websites, public travel service hotlines, traffic information short message services, public travel smart terminal services, traffic guidance services, traffic radio and television, public travel Weibo, and WeChat.

6.3.4.4 Travel Chain Analysis

The travel chain refers to the round trip itinerary composed of people's travel purposes sorted in time in order to complete one or more activities. Figure 6.6 shows a typical commuter travel chain diagram. The travel chain includes commuting mode, commuting time, and activity information seen at different times. It is a product of the travel decision-making mechanism and an important forecast target for travel demand and transportation mode management at the beginning of the month. Mastering the travel chain of travelers is an important guarantee for improving the traveler information service system and formulating the best solutions for travelers.

References

1. Zhu, L., Yu, F. R., Wang, Y., Ning, B., & Tang, T. (2019). Big data analytics in intelligent transportation systems: A survey. *IEEE Transactions on Intelligent Transportation Systems, 20*(1), 383–398.
2. Soares, E. F. de S., Quintella, C. A. de M. S., & Campos, C. A. V. (2021). Smartphone-based real-time travel mode detection for intelligent transportation systems. *IEEE Transactions on Vehicular Technology, 70*(2), 1179–1189.
3. Veres, M., & Moussa, M. (2020). Deep learning for intelligent transportation systems: A survey of emerging trends. *IEEE Transactions on Intelligent Transportation Systems, 21*(8), 3152–3168.
4. Zhou, F., Yang, Q., Zhong, T., Chen, D., & Zhang, N. (2021). Variational graph neural networks for road traffic prediction in intelligent transportation systems. *IEEE Transactions on Industrial Informatics, 17*(4), 2802–2812.
5. Du, B., Peng, H., Wang, S., Bhuiyan, Md. Z. A., Wang, L., Gong, Q., Liu, L., & Li, J. (2020). Deep irregular convolutional residual LSTM for urban traffic passenger flows prediction. *IEEE Transactions on Intelligent Transportation Systems, 21*(3), 972–985.
6. Zeroual, A., Harrou, F., Sun, Y., & Messai, N. (2018). Integrating model-based observer and Kullback–Leibler metric for estimating and detecting road traffic congestion. *IEEE Sensors Journal, 18*(20), 8605–8616.
7. Shengdong, M., Zhengxian, X., & Yixiang, T. (2019). Intelligent traffic control system based on cloud computing and big data mining. *IEEE Transactions on Industrial Informatics, 15*(12), 6583–6592.
8. Liang, X., Zhang, Y., Wang, G., & Xu, S. (2020). A deep learning model for transportation mode detection based on smartphone sensing data. *IEEE Transactions on Intelligent Transportation Systems, 21*(12), 5223–5235.
9. Wang, Q., Zheng, J., Xu, H., Xu, B., & Chen, R. (2018). Roadside magnetic sensor system for vehicle detection in urban environments. *IEEE Transactions on Intelligent Transportation Systems, 19*(5), 1365–1374.
10. Lin, J., Yu, W., Zhang, N., Yang, X., & Ge, L. (2018). Data integrity attacks against dynamic route guidance in transportation-based cyber-physical systems: Modeling, analysis, and defense. *IEEE Transactions on Vehicular Technology, 67*(9), 8738–8753.
11. Zhou, Y., Liu, L., Shao, L., & Mellor, M. (2018). Fast automatic vehicle annotation for urban traffic surveillance. *IEEE Transactions on Intelligent Transportation Systems, 19*(6), 1973–1984.
12. Zhang, R., Ishikawa, A., Wang, W., Striner, B., & Tonguz, O. K. (2021). Using reinforcement learning with partial vehicle detection for intelligent traffic signal control. *IEEE Transactions on Intelligent Transportation Systems, 22*(1), 404–415.
13. Flores, C., Merdrignac, P., de Charette, R., Navas, F., Milanés, V., & Nashashibi, F. (2019). A cooperative car-following/emergency braking system with prediction-based pedestrian avoidance capabilities. *IEEE Transactions on Intelligent Transportation Systems, 20*(5), 1837–1846.
14. Barmpounakis, E. N., Vlahogianni, E. I., & Golias, J. C. (2018). Identifying predictable patterns in the unconventional overtaking decisions of PTW for cooperative ITS. *IEEE Transactions on Intelligent Vehicles, 3*(1), 102–111.
15. Chakeri, A., Wang, X., Goss, Q., Ilhan Akbas, M., & Jaimes, L. G. (2021). A platform-based incentive mechanism for autonomous vehicle crowdsensing. *IEEE Open Journal of Intelligent Transportation Systems, 2*, 13–23.
16. Rosenstatter, T., & Englund, C. (2018). Modelling the level of trust in a cooperative automated vehicle control system. *IEEE Transactions on Intelligent Transportation Systems, 19*(4), 1237–1247.

17. Song, X., Guo, Y., Li, N., & Zhang, L. (2021). Online traffic flow prediction for edge computing-enhanced autonomous and connected vehicles. *IEEE Transactions on Vehicular Technology, 70*(3), 2101–2111.
18. Aujla, G. S., Singh, A., Singh, M., Sharma, S., Kumar, N., & Choo, K.-K. R. (2020). BloCkEd: Blockchain-based secure data processing framework in edge envisioned V2X environment. *IEEE Transactions on Vehicular Technology, 69*(6), 5850–5863.
19. Singh, A., Aujla, G. S., & Bali, R. S. (2021). Intent-Based Network for Data Dissemination in Software-Defined Vehicular Edge Computing. *IEEE Transactions on Intelligent Transportation Systems, 22*(8), 5310–5318.
20. Yang, P., Duan, D., Chen, C., Cheng, X., & Yang, L. (2020). Multi-sensor multi-vehicle (MSMV) localization and mobility tracking for autonomous driving. *IEEE Transactions on Vehicular Technology, 69*(12), 14355–14364.
21. Zhu, B., Tao, X., Zhao, J., Ke, M., Wang, H., & Deng, W. (2020). An integrated GNSS/UWB/DR/VMM positioning strategy for intelligent vehicles. *IEEE Transactions on Vehicular Technology, 69*(10), 10842–10853.
22. Erkent, Ö., & Laugier, C. (2020). Semantic segmentation with unsupervised domain adaptation under varying weather conditions for autonomous vehicles. *IEEE Robotics and Automation Letters, 5*(2), 3580–3587.

Part III
Cyber-Physical Systems

Chapter 7
Cyber-Physical Systems: Historical Evolution and Role in Future Autonomous Transportation

Bhawna Rudra and S. Thanmayee

7.1 Introduction

The Internet of Things has emerged in all the industries for various purposes along into the automobile industry [1, 2]. The best example for the IoT in the automobile industry is the connected car [3, 4], which is connected with the onboard sensor that enhances the in-car experience with the help of Internet connectivity. Connected car will establish a connection between the cars for the communication to surf not only the Internet but also between other cars present in the vicinity. There were less number of Internet-enabled cars by 2017, but by now it has increased and allows many innovations with the help of sensors and was estimated to increase to 38 billion by 2020 [5]. Many companies have adopted and still adopting for the development of the devices for communication in automobile industry. Many companies around the world have adopted renewable energy sources and come up with the idea of smart cities for the generation of the electricity and other appliances to control with the help of sensors. The integration of the conventional grid with telecommunication and information technologies with IoT will allow the resource utilization to optimize power consumption and exchange the generated power among the distributed sources. Many companies started to save energy by using renewable resources or Green computing.

B. Rudra (✉) · S. Thanmayee
National Institute of Technology Karnataka, Mangaluru, India

S. Garg et al. (eds.), *Intelligent Cyber-Physical Systems for Autonomous Transportation*, Internet of Things, https://doi.org/10.1007/978-3-030-92054-8_7

7.2 Internet of Things: Communication

The usage of automated methods for retaining the data based on the content, consolidation of IoT data securely without considering where it has appeared and where to store and finally offer new services to access the information in a productive manner are the ways to manage IoT. Vehicles communicating with the outside world were started in 1996. So many sensors are used in connected vehicles for sensing the physical properties. The main aim of the connected vehicles is to reduce traffic congestion and fuel consumption, avoid accidents and to provide comfortable drive and road safety. For communication purposes, short-range signals are used on highways. If the connection is between vehicle and its internal parts, then local area network connection can be used. If the communication is between vehicle to vehicle, vehicle to Internet, or vehicle to road, then Bluetooth that covers 100mts, Wi-Fi that covers a range of 32mts or Zigbee that covers a range of 50mts communication channels can be used [6, 7].

Wi-Max Wi-MAX is an advanced technology still not used in in-vehicle communication. The information like speed and position of vehicle data can be transferred between vehicles on an ad hoc mesh network.

Radiofrequency Identification (RFID) It uses radio waves in the form of numbers for the identification of the objects. The RFID tags are used to detect the objects with the help of RFID readers. These tags contain electronically stored information that is transformed using radio-frequency electromagnetic fields from an object. The RFID device will scan the RFID object to retrieve the identifying information. The reader will gather information from the RFID tag to track the individual objects. The range will cover 3–300 feet [3, 8].

Near Field Communication NFC is customer oriented if RFID is integrated with mobile phones. It covers a short range of 20 m and a low power wireless link that is used to send a small amount of data between two devices within that specific range. It uses the unlicensed radio frequency band of 13.56 MHz. NFC plays a major role in connecting smart objects [9].

Seamless, Optimized Connectivity, Mobility It uses multiple wireless WAN interfaces like Wi-Fi, 3G/4G(LTE), which provide secure mobility and seamless user experience, reliable connection and application-aware and flow-based connectivity. Wi-Fi allows the systems to communicate with each other in far ranges. Wi-Fi technology includes personal computers, video-game consoles, digital audio players and modern printers. Wi-Fi devices can unite to the Internet via a WLAN network and a wireless access point [5].

Deterministic Networking This can be a wired or wireless network and consists of the key elements that include network synchronization, timely transmission and centralized scheduling. Wired network requires time-triggered Ethernet, and wireless will use 802.15.4 for low power low rate (4 Hz), and Wi-Fi (802.11ac) for a higher rate (100 Hz).

Machine-to-Machine Communication (M2M) It is a communication between embedded systems, sensors, actuators and the devices attached to them wirelessly. This communication is used in various applications like health care, robots, smart home technologies and so on. This involves sensing, heterogeneous access and the application. M2M includes PAN technologies like ultra-wideband and Bluetooth or local networks. This provides connectivity between the devices and the gateways. M2M provides an interconnection between the devices that communicate with each other. These applications contain a middleware, where the data are communicated between various applications and used for specific business engines [10].

7.3 Vehicle Communication

The automobile industry has undergone and is still undergoing many changes in terms of technology and transformation from display modules to driverless cars. The introduction of IoT has created more innovative opportunities in this industry like power connected cars. The cars are connected through the Internet and wireless LAN. This system will allow the vehicles to communicate with each other and take decisions about how much traffic is nearby and whether to opt for a particular route for reaching the destination, etc., thus avoiding traffic congestion, saving fuel and so on. The rise of mobile devices and their connectivity is increasing the use of connected cars. The mobile apps are allowing more functionality for the users. The manufacturers are taking care of vehicle connections with IoT and infotainment systems. The allied car services and applications along with further improvements on IoT in the automotive sector are discussed below [6, 11].

Vehicle-To-Vehicle Communications (V2V) This comprises of wireless network that communicates about the speed and location information to the nearby vehicles for the prevention of accidents, thus improving the safety of commuters [6] as shown in Fig. 7.1. The communicated vehicles will communicate by broadcasting the messages continuously about the speed and position of the vehicles over a mesh network. Dedicated Short-Range Communications (DSRC) technology was designed specifically for automotives. Vehicles communicate with others within a range of 100 m. The communication will be between the vehicles and the vehicle to the road infrastructure [7]. Messages can be shared safely by vehicle communication. It is an ad hoc network and contains a large number of nodes. By using DSRC, protocols can be fit in RFID and WAP for medium access control [12]. Providing security for VANET is a very essential task. Various challenges of VANET like mobility, scalability and various attacks and threats and their solutions are discussed in [13–16]. For better results, VANETs are combined with IoT. Various applications of IoV and its security issues are discussed in [17]. To overcome the security issues, authors proposed a lightweight protocol for RFID, which increases the performance with less CPU and memory utilization. To alert the vehicles, LCD displays are placed on the backside of the vehicle. With this

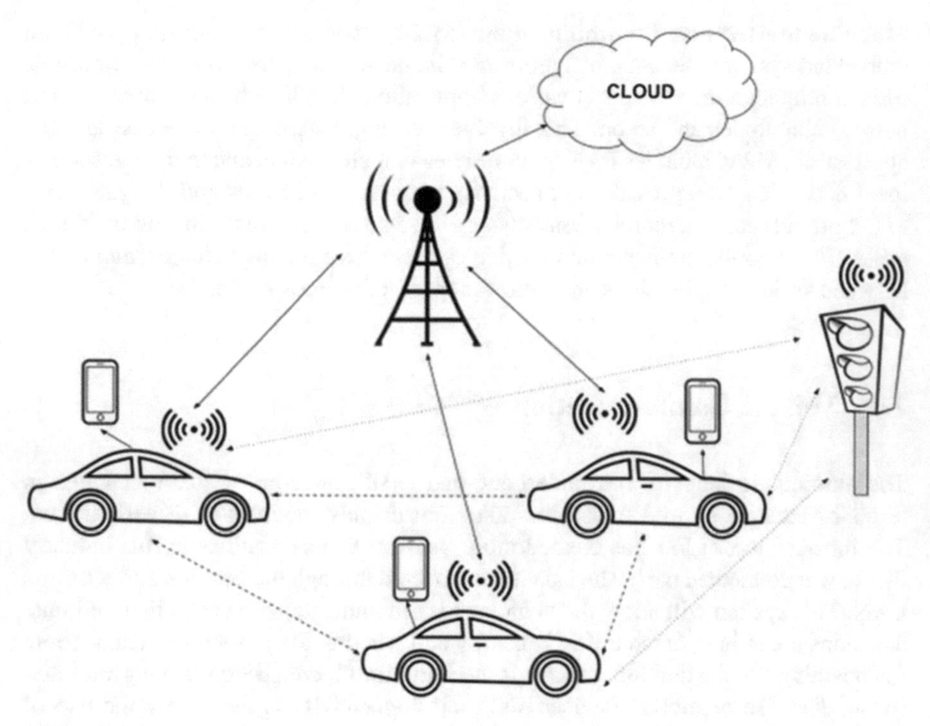

Fig. 7.1 V2V and V2X communication

driver, one can easily find which vehicles are coming behind and he or she can alert the driver of that vehicle with a wireless transceiver that is placed in his or her vehicle. This method of communicating one system to another is very simple and cost effective [18].

Vehicle to Infrastructure (V2X) V2X communication is a wireless transmission of information between vehicles and roadside infrastructure to avoid or mitigate accidents and also to provide a wide range of other safety, mobility and environmental benefits. Figure 7.1 shows the communication with traffic light. The vehicles sense the signal lights and can automatically take decisions on stopping and moving at traffic junctions. The vehicles on the roads will communicate for the digital signage, traffic lights and safety and control systems to avoid crashes and traffic congestion through intelligent safety applications. This will help in the reduction of accidents and the casualties but also help in tracing and monitoring the vehicles on the move. At the time of emergency, vehicles will be able to quickly intimate details to roadside assistance, as well the emergency services like ambulance, insurance providers as well as family members [19, 20].

Vehicle-to-Retail (V2R) This presents the drivers with location-based advertising and issues the discount coupons for the shopping mall nearby. Some applications allow for booking the restaurants and easy order through voice interface. OEMs are exploring for vehicle-to-retail (V2R) applications like the Volvo has set up its Roam Delivery trial and locate and unlock a connected vehicle to deposit an item purchased online [21].

Vehicles and Smartphone Integration People prefer to be online for most of the time even when it comes to being on board a care or even being a driver. Using the On-Board Diagnostics OBD/OBD-II port, information regarding the engine and other crucial vehicle parameters can be displayed on the driver's smartphone. The analysis of the obtained parameters can be performed to know the status of the vehicle. Alerts like open doors, lights ON and hand brake ON and performing actions on certain vehicle parts such as lock/unlock vehicle doors, roll windows up/down and AC temperature $+/-$ are becoming seamless [22]. Some of the services provided by the connected vehicles are as follows:

Infotainment This refers to a system in vehicles, which is a combination of entertainment and information services. The features of an In-Vehicle-Infotainment (IVI) system are managing audio/visual entertainment content, delivering rear-seat entertainment, and connectivity with smartphones for hands-free experience with the help of voice commands along with navigation. It creates a safer in-car experience and has the options like the apps for accessing the features and avoids the driver's distraction. Audio is the primary way for the acceptance of the commands from the users. Advanced infotainment features integrate the user behaviour and the cloud-based infotainment systems like CarPlay, Google Projected Mode, and Mirror Link, etc. [19].

Driving Insights—Analytics In order to find the driving model, the smart sensors present in smart phone like GPS, gyroscope or orientation sensor and accelerometer can be used. Once the smart phone is mounted in the vehicle, data collected from these sensors can be used for the detection of the driving patterns like sharp turns, sudden acceleration, etc. This data can be used for the analysis of the driver's behaviour with respect to the driving. Pay As You Drive (PAYD) and Pay How You Drive (PHYD) are the upcoming use-based insurance packages provided by auto insurance companies that reward safe drivers and penalize rash ones with different premiums [7, 23, 24].

On-Board Diagnostics for Predictive Maintenance The On-Board Diagnostics (OBD/OBD-II) is commonly used for self-diagnosis and reporting of any issues that may happen in near future or might have occurred [25]. The information such as emissions, mileage, faults, vehicle and engine speed, etc. can be monitored. The connector is connected to the check engine light or MIL-malfunction indicator light, which illuminates when a problem is found. The pairing of vehicles with smart phones will help the owners to know the condition of the vehicle as well as can provide the details of nearby service stations for the better performance of the car. This will help in generation of proactive service alerts for the early diagnosis of the faults [24].

Safety: Real-Time Driver Monitor For the safe driving, drivers are screened and evaluated based on their driving habits. The sensor technologies will help to monitor the behaviour and know the fatigue level of the drivers by allowing the vehicle to be more intelligent by which accidents can be avoided. Real-time monitoring will help the drivers to control the speed of the vehicles and prevent the damages that may happen due to the speed. The sensor components involved in the vehicle will able to detect the eye blink, gas, impact sensors, alcohol detection sensors and so on as well as allow to track the location, which will help to send the information to the predefined numbers [7, 26].

Advanced Driver Assistance System (ADAS) The main objective of these systems is to help drivers in driving safely. They provide non-distractible warnings in the form of audio or visuals. There are several scenarios when the ADAS proves to be a life saver system. Most of the accidents happen due to mistakes that a driver makes. The primary task of a driver is to drive safely by focusing on the traffic. Most of the drivers involve themselves in secondary tasks such as answering phone calls. ADAS keeps track of these distractions and provides alerts to the drivers and thereby helping them in preventing accidents [27].

Geofencing and Speed Monitoring These applications can be used to inform the owner about the vehicle movement, whether the drive was fast or slow measured by speed sensors. This is very useful for the monitoring of remote company cars [28].

Stolen Vehicle Tracking The tracking device attached to the vehicle will be able to trace the stolen vehicle with the help of GSM and GPS modules present in it. The satellite signal will be sent remotely along with latitude and longitude to determine the position of the vehicle. Once the vehicle is traced, the owner will be notified [20, 26, 29].

Biometrics Information for Driver Identification The biometric information can be used for the identification and to authenticate a driver. Biometrics include face recognition, fingerprints or voice recognition. Voice can be used to provide hands-free experience for the navigation through the connected car environment. The data obtained can be used as an anti-theft protection tool and increase safety. Cameras along with sensors within the car can be used for the biometric authentication and will allow the quick car setting to change like seat position, mirror settings, etc. to accommodate different driving patterns of the people. Even the health parameters can be checked before the driver as well as can be monitored continuously with the help of sensors for the prevention of accidents [6, 23, 24].

Context-Aware Data-Centric Misbehaviour Detection Scheme To improve road safety, [30] proposed a context-aware data-centric misbehaviour detection scheme (CA-DC-MDS). It analyses the spatial and temporal correlation between vehicles. It uses a dynamic context reference model to receive the information from neighbouring vehicles.

A Geographical Segment Architecture For integrating various vehicles, we need better protocols that can take the data from various sensors, which are placed in different locations in cities. To do this task and to increase the performance, authors proposed GSA. It contains GSA-SR, GSA-GR and IGSAR protocols. With these GSA protocols, the performance is increased compared to the existing protocols.

7.4 Applications

7.4.1 Intelligent Transportation

Smart Toll System This application is used to collect toll charges without resulting in traffic jams or delay. It is the most convenient way to collect toll charges. The framework consists of sensors such as RIFD tags and scanner. It may also contain cameras to automatically detect the number plates. There are many other methods such as the one which uses GPS that is based on telematics [19, 29] and also cellular networks to facilitate the toll collection process.. The recent technology, which is smart card/tag based—FASTag, is known to eliminate several toll collection issues and provide a smooth functioning of toll collection system [7, 11].

Smart Parking The most common problem in the developed cities is lack of sufficient parking space. Finding parking space is a biggest challenge for drivers [9, 31]. Parking problems also lead to traffic jams. Another reason being an increase in the number of vehicles. The goal of smart parking systems is to intelligently manage the parking space usage. This is expected to provide efficient parking and eliminate traffic jams. The system is built by integrating sensors, microcontrollers, cloud and edge devices. The system will use the real-time data to provide information about the parking space. It can also be integrated with an automated parking fee payment module. This system will be of great help to the drivers, which is hassle free as they can reserve the parking space well ahead of time. This can reduce traffic jams and air pollution as well.

Energy Conservation This application has gained a lot of importance due to the resulting benefits, mainly significant energy saving. Basically the Intelligent Street Lighting system saves energy by switching the lights on and off by sensing the movement of pedestrians or vehicles. Such an adaptive lighting system can be used in smart buildings as well. This can save a large unit of power. This results in a good power management strategy. The overall design includes many sensors. IR sensors are deployed to detect the presence of people or objects. It can be integrated with camera sensors to provide more efficient service. This design can be clubbed with energy-efficient LED-based street lights. This can be one of the good solutions for power saving. IoT-based smart lighting systems can be integrated with IP connectivity module and energy-efficient LED-based street lights. This system can be used for traffic control. Consider a scenario in which emergency vehicles like ambulances need to be given priority to move ahead. Such vehicles can be provided with a high priority lane. RFID technology can be used to sense the presence of such vehicles. The traffic signal can be controlled in such a way that the vehicles in the high priority lane can be allowed to pass the signal without any delay. Also the lane with a higher number of vehicles can be detected and managed accordingly. This system results in good energy conservation and increased road safety [11, 32].

Post-Event Diagnostics This technology is used to record vehicle data just before or during or soon after the accident. It records important information like the vehicle speed, acceleration level, seatbelt status, airbag condition, etc. that gives a very good insight about the reason for the vehicle crash. This system incorporates multiple sensors like temperature sensor, pressure sensor, proximity and ultrasonic sensors. In order to retrieve data from the Event Data Recorder (EDR), some additional tools can be integrated to the vehicle with a good visualization tool. This can represent data in a graphical form [25].

Traffic Control with Priority This system is considered to be intelligent traffic control systems. In this system the vehicles are prioritized based on the type of users. There can be different categories of users like emergency services, VIP users like Politicians, public transport. Vehicles can be tagged with RFID chips. The system deploys RFID scanners on the intersection points on the road. The system can then analyse the traffic density. The system makes use of Round Robin method for controlling the traffic lights for normal traffic like public transport. In case of emergency vehicles detected on the high priority lane, it comes out of the Round Robin sequence and turns on the green light for high priority traffic [11].

Warrant of Fitness—Feedback to Road Transport Authorities Fitness test of vehicles particularly the Light Motor Vehicles (LMVs) is very much important. Based on the fitness of the vehicles, a Warrant of Fitness (WoF) certification mark will be issued to the LMVs. This mark indicates that the vehicle is fit to be on road. It examines the different parts of a vehicle like tyres, brakes, lights, wipers, seatbelts, fuel system, airbags, speedometer, etc. The WoF sticker on the windscreen will contain the date on which the test was performed and the due date for next test. In case the WoF sticker is not found on the vehicle, the police will take appropriate actions on the vehicle owner. RFID technology can be used to automate the process of scanning the WoF stickers and retrieving the information about the vehicle's fitness test [33]. This results in an automated WoF feedback system.

Environmental Control—Emissions and Air Quality Monitoring Air pollution is increasing due to the emission of gases from the vehicles [34]. One of the reasons for this is incomplete combustion of fuel. IoT smoke and temperature sensors can be used to monitor and control the pollution. With the help of sensors, the system will generate an alert message when the pollution level crosses the predefined threshold value using the GSM module and the system will be switched off automatically. To find the location of the vehicle, a GPS module is used. This can effectively control the pollution emitted by vehicles and save the environment [7, 26, 29].

Traffic Flow Prediction Usage of vehicles has increased exponentially. Traffic flow detection is very essential for vehicles to make decisions based on surroundings. To achieve this, we are using machine learning strategies. But with existing models, it is difficult to collect the data from surroundings due to complex relations between various models. So for better results we can make use of deep learning models that are very useful in smart transportation [35].

Traffic-Aware Electric Vehicle Charging Management System Air pollution can be reduced with electrical vehicles. But charging electrical vehicles will be a very big task. If the vehicle is not charged, then it will not work and we need many charging stations. Otherwise vehicles need to wait for a long period of time and due to this trip time of an electric vehicle will be reduced. To overcome this problem, reservation for charging slots can be performed with VANET. Geographical routing protocols are used for charging slot reservation [7, 36].

7.4.2 Supply Chain Management

Automotive supply chain management (SCM) has been majorly impacted by IoT. The SCM is an organized chain that connects different suppliers, customers, service providers, manufacturers, dealers and distributors who are geographically distributed. Automotive SCM is now smart due to the integration with IoT. This smart system can solve most of the challenges faced by different stakeholders of the system. The challenges such as dynamic customer demands, dynamic cost, risk and globalization are now tackled smoothly by the IoT-based automotive SCM [37, 38].

Raw Material Procurement The use of sensors and RFID technology can help in providing good visibility in procuring raw materials. Smart devices can track the entire assembly line, which involves getting materials from the suppliers to the production shop floor. These smart tags can monitor and track specific automobile components in order to give customized and optimized orders.

Manufacturing—Process Optimization IoT can be used to monitor the production process. This includes monitoring each step in the manufacturing process that in turn increases visibility of the entire process. IoT can couple the shop floor and top floor. This can eliminate any human interventions required to monitor the process for any deviations. Sensors are used to measure certain operating parameters like temperature, pressure, alignment, etc. These readings are useful for automatically maintaining the present values of those operation parameters at the desired value. This automation increases the productivity and efficiency of the manufacturing process and thereby reduces the overall process cost [20, 39].

Quality Assurance Both product and process must meet certain specified requirements. Quality assurance is most important for any industry. It audits and inspects the process and product for assuring that it meets the desired requirements. The demand for connected smart cars is rising due to which the automotive industry is facing a paradigm shift in production of cars that need to be smarter. Software

Quality Assurance skill will have a major impact on manufacturing and delivery of great connected cars that are intelligent [40, 41].

Sales Order Management IoT and its products have changed the trend of sales and marketing specifically in automotive industries. It has completely renovated the old sales and marketing process. With IoT-integrated sales order management system, the OEMs will be continually connected to the customers throughout the production process. This results in high cross-selling and upselling. After the product is delivered to the customers, the companies can still be in touch with the customers in the connected environment. They can gather the usage data of the product and analyse them to check on its performance. This will build a good customer relationship by knowing the features being used by different customer categories. This can help the industry in providing personalized post-sale services [42]. Also it can help in improving future marketing and sales.

Delivery Management—Dispatch and Logistics The main goal of IoT-integrated delivery management system is to improve the cost effectiveness in transportation and warehouse. The data gathered in the system can be analysed using Data Analytics, which in turn optimizes the distribution networks. Supply chain managers can analyse customer locations, production and delivery cost, order quantities and required production time. This analysis can help them plan the exact number of distribution centres and also where they could be located. This ensures improved standard of service [41].

7.5 Case Study

Self-Driving Car The self-driving car senses the environment and navigates without any human intervention. It provides a new experience by connecting the car with the environment using the network connection of the Internet of Things. Once the destination is predetermined by the sensors, the whole system will communicate with the environment and the on-going traffic to avoid the disturbances and reach the destination smoothly. With the help of this, drivers can enjoy the driving freedom, thereby allowing them to use mobiles, laptops and other devices without any fear of accidents on the roads. Tesla vehicles are the best examples of self-driving cars [4, 38].

Cars with IoT-installed devices will communicate with the vehicles around them. They also communicate with central infrastructure that is responsible for broadcasting live traffic information and information about the drivers if required, to prevent accidents or any danger that will happen due to weather conditions as illustrated in Fig. 7.2. The vehicles will be able to detect each other's location using sensors and will be able to predict the risks on the road. Authorities expect the number of accidents to reduce by a staggering 90% once self-driving technology is implemented across the globe. Telematics is a subset of IoT, which collects the

Fig. 7.2 Self-driving cars with RSU and GPS tracking

information about the vehicle like its location and speed, how much time it was on the road and fuel consumption to name a few [5, 43, 44].

Electric Cars Electric cars are replacing petrol and diesel cars due to their sustainability and because they are cheap and easy to maintain and more efficient. It is powered by an electric motor. It makes use of rechargeable batteries to store energy. The usage is increasing day by day due to their plug-in technology that allows the owners to charge at home; the powerful lithium-ion batteries help to drive for longer distances. In order to make long distance travel economical, many countries are providing incentives for public charging stations. There are many automotive companies that have started developing wireless charging processes for electric cars. The main intention is to provide ease in the charging process. Some of the IoT applications for electric cars are home charging solutions, scheduling a charging slot, nearest charging station with tariff rates, battery charge status with estimated driving time [9, 45, 46].

7.6 Connected Vehicle Challenges

Vehicles are now an integral part of IoT. They are no longer an independent entity. Connected vehicles are the end result of integrating various technologies such as cloud computing, fog computing, smartphones, Bluetooth connectivity and data analytics. The benefit of such integration leads to security challenges. The first and foremost challenge is the integration of heterogeneous elements on IoT in the connecting vehicle. To provide intelligent transportation is the other challenge, and data has to be synchronized uniformly for the proper function of the device. A separate cloud platform is required for the proper functioning of the overall system. Cloud services are allocated for the communication between infrastructure and the vehicles and these services fall under three categories [23, 26, 29, 47].

Infrastructure as a Service (IaaS) The IoV and the traffic-related services computed are based on the cloud framework that involves the traffic status data storage, vehicle monitored status, safety status, and real-time traffic analysis. Platform as a Service (PaaS): This service includes GPS data processing, cloud storage, information mining and analysis, and information security. Software as a Service (SaaS): This allows the third party to develop any kind of application and support the IoV and allows access from various terminals.

The data could be hacked [48]. The result would be disastrous as the vehicle can be controlled remotely by a hacker. The other problems that are associated with the connected vehicles are loss of data at unconnected huge vehicles that come between the connected vehicles, and at building intersections. The vehicular network is basically ad hoc and mobile in nature and hence the communication is affected by multipath fading, Doppler effects and shadow. Due to high mobility, the network topology is highly dynamic. This results in frequent disconnection of communication. There will be obstacles that will result in interruption of data link between the communicating vehicles. The stored data about the vehicles will raise the issue of security [47]. The attacks like STRIDE of information security, authenticity attacks, data availability and data authenticity attacks are to name a few. The demand for infotainment services is increasing. But the major challenge posed by infotainment services is the bandwidth requirement. The connected vehicles have many Electronic Control Units. The sensors produce a large amount of data that needs to be analysed in real time. This is challenging because little latency may lead to a massive disaster [49]. Fog computing plays an important role in preventing latency issues.

7.7 Summary

The Internet of Things has evolved in such a way that it has resulted in innovative research and business developments. A lot of applications come under the umbrella of IoT such as smart agriculture, smart city, smart water management, and IoT-based healthcare applications. IoT has become a major trend in automotive sectors. IoT-based vehicle communications that lead to connected cars applications and intelligent transportation are recent research trends in the area of Internet of Things. Adding a flavour of IoT into transportation leads to intelligent transportation and new generation cars. IoT-based automotive supply chain management has also gained a lot of focus due to its immense benefits in the overall business development. Connected cars include smart cars that communicate with other cars and edge devices like smartphones. The communication generates data that can be used for analytical processing. The results of the analytical processing can be used by drivers to know the traffic and car's condition. It can also be used by the road safety authorities to know the overall traffic situation. So that they can have good decision making and timely actions. The future of vehicle network is to get networked with IoT. The framework can include gateways, sensors, hubs and pocket switches integrated into the vehicles. It will eliminate the dense cabling that otherwise leads to design and deployment complications.

References

1. Ashton, K., et al. (2009). That 'internet of things' thing. *RFID Journal, 22*(7), 97–114.
2. Aujla, G. S., & Jindal, A. (2020). A decoupled blockchain approach for edge-envisioned IoT-based healthcare monitoring. *IEEE Journal on Selected Areas in Communications, 39*(2), 491–499.
3. Wollshlaeger, D., Foden, M., Cave, R., & Stent, M. (2015). Digital disruption and the future of the automotive industry. *IBM Corporation*.
4. Nukala, M. R., Bhargave, S., & Patwardhan, B. (2014). Transforming the automotive industry with connected cars. *CSI Communications*, 31.
5. DST Bhosale. (2018). IoT based cars: A paradigm shift in automobile industry. *International Journal of Trend in Scientific Research and Development*, (Special Issue-ICDEBI2018), 110–113.
6. Hossain, M. K., & Haq, S. S. (2013). Detection of car pre-crash with human, avoidance system & localizing through GSM. *International Journal of Scientific and Research Publications, 3*(7), 1–4.
7. Bautista, P. B., Cárdenas, L. L., Aguiar, L. U., & Igartua, M. A. (2019). A traffic-aware electric vehicle charging management system for smart cities. *Vehicular Communications-ScienceDirect, 20*, 100188.
8. Sharma, S., Pithora, A., Gupta, G., Goel, M., & Sinha, M. (2013). Traffic light priority control for emergency vehicle using RFID. *International Journal of Innovations in Engineering and Technology, 2*(2), 363–366.
9. Kasznar, A. P. P. (2018). *Analysis of the Applicability of Internet of Things Projects from China in the Brazilian Scenario*. PhD Thesis, Universidade Federal Do Rio De Janeiro.
10. Liu, T., Yuan, R., & Chang, H. (2012). Research on the internet of things in the automotive industry. In *2012 International Conference on Management of e-Commerce and e-Government* (pp. 230–233). IEEE.
11. Mlinarić, M. (2016). *Intelligent Traffic Control with Priority for Emergency Vehicles*. PhD Thesis, University of Zagreb. Faculty of Transport and Traffic Sciences. Division of ..., 2016.
12. Katherine, A, V., Muthumeenakshi, R., & Vallilekha, N. (2014). Vehicle to vehicle communication using RFID along with GPS and WAP. *International Journal for Computer science and mobile computing, 3*, 21–27.
13. Samara, G., Al-Salihy, W. A. H., & Sures, R. (2010). Security issues and challenges of vehicular ad hoc networks (VANET). In *4th International Conference on New Trends in Information Science and Service Science* (pp. 393–398). IEEE.
14. Mishra, R., Singh, A., & Kumar, R. (2016). VANET security: Issues, challenges and solutions. In *2016 International Conference on Electrical, Electronics, and Optimization Techniques (ICEEOT)* (pp. 1050–1055). IEEE.

15. Lv, F., Zhu, H., Chang, S., & Dong, M. (2017). Synthesizing vehicle-to-vehicle communication trace for VANET research. In *2017 IEEE International Conference on Smart Computing (SMARTCOMP)* (pp. 1–3). IEEE.
16. Arena, F., & Pau, G. (2019). An overview of vehicular communications. *Future Internet, 11*(2), 27.
17. Sharma, S., & Kaushik, B. (2019). A survey on internet of vehicles: Applications, security issues & solutions. *Vehicular Communications, 20*, 100182.
18. Alanezi, M. A. (2018). A proposed system for vehicle-to-vehicle communication: low cost and network free approach. *Indian Journal of Science and Technology, 11*, 12.
19. White paper connected vehicles - from building cars to selling personal travel time well- spent. *Cisco*.
20. Singh, P., Sethi, T., Biswal, B. B., & Pattanayak, S. K. (2015). A smart anti-theft system for vehicle security. *International Journal of Materials, Mechanics and Manufacturing, 3*(4), 249–254.
21. Autonomous-vehicles-and-the-internet-of-things.
22. White paper- the in-car app experience: Convergence and integration. *UIEvolution*.
23. Lee, B.-G., & Chung, W.-Y. (2012). A smartphone-based driver safety monitoring system using data fusion. *Sensors, 12*(12), 17536–17552.
24. Sowmya, D., Suneetha, I., & Pushpalatha, N. (2014). Driver behavior monitoring through sensors and tracking the accident using wireless technology. *International Journal of Computer Applications, 102*(2), 2014.
25. Reddy, P. A. K., & Kumar, P. D. . K. Bhaskar Reddy, E. Venkataramana, M. Chandra Sekhar Reddy (2012). Black box for vehicles. *International Journal of Engineering Inventions, 1*, 06–12.
26. Kamel, M. B. M. (2015). Real-time GPS/GPRS based vehicle tracking system. *International Journal of Engineering and Computer Science, 4*(8), 648–652.
27. Biondi, F., Strayer, D. L., Rossi, R., Gastaldi, M., & Mulatti, C. (2017). Advanced driver assistance systems: Using multimodal redundant warnings to enhance road safety. *Applied Ergonomics, 58*, 238–244.
28. OKane, T., & Ringwood, J. V. (2013). *Vehicle Speed Estimation*.
29. Mehta, V., Punetha, D., & Bijalwan, V. (2016). A real time approach to theft prevention in the field of transportation system. *IJIMAI, 3*(7), 77–80.
30. Ghaleb, F. A., Maarof, M. A., Zainal, A., Rassam, M. A., Saeed, F., & Alsaedi, M. (2019). Context-aware data-centric misbehaviour detection scheme for vehicular ad hoc networks using sequential analysis of the temporal and spatial correlation of the consistency between the cooperative awareness messages. *Vehicular Communications, 20*, 100186.
31. Jog, Y., Singhal, T. K., Barot, F., Cardoza, M., & Dave, D. (2017). Need & gap analysis of converting a city into smart city. *International Journal of Smart Home, 11*(3), 9–26.
32. Ashton, K. (2009). That 'internet of things' thing. *RFID journal, 22*(7), 97–114.
33. Warrant of fitness warrant of fitness.
34. Tiwari, D., Shekhar, S., Joshi, A., & Deep, A. Automated system for air pollution detection and control in vehicles.
35. Miglani, A., & Kumar, N. (2019). Deep learning models for traffic flow prediction in autonomous vehicles: A review, solutions, and challenges. *Vehicular Communications, 20*, 100184.
36. Aujla, G. S., Jindal, A., & Kumar, N. (2018). EVaaS: Electric vehicle-as-a-service for energy trading in SDN-enabled smart transportation system. *Computer Networks, 143*, 247–262.
37. Berman, S. J. (2012). Digital transformation: Opportunities to create new business models. *Strategy & Leadership, 40*(2), 16–24.
38. Bhattacharya, S., Mukhopadhyay, D., & Giri, S. (2014). Supply chain management in Indian automotive industry: complexities, challenges and way ahead. *International Journal of Managing Value and Supply Chains (IJMVSC), 5*, 8.
39. Whitepaper-designing for manufacturing's internet of things. (2014). *Cognizant*.

40. Schnabel, R. B., Haeusler, K. G., Healey, J. S., Freedman, B., Boriani, G., Brachmann, J., Brandes, A., Bustamante, A., Casadei, B., Crijns, H. J. G. M., Doehner, W., Engström, G., Fauchier, L., Friberg, L., Gladstone, D. J., Glotzer, T. V., Goto, S., Hankey, G. J., Harbison, J. A., ... &Yan, B. (2019). Searching for atrial fibrillation poststroke: a white paper of the AF-SCREEN International Collaboration. *Circulation, 140*(22), 1834–1850.
41. Balasubramaniyan, M. C. & Manivannan, D. (2016). IoT enabled air quality monitoring system (AQMS) using Raspberry Pi. *Indian Journal of Science and Technology, 9*(39), 32–39.
42. How-the-iot-is-changing-sales-andmarketing.
43. Lacity, M. C., & Willcocks, L. P. (1998). An empirical investigation of information technology sourcing practices: Lessons from experience. *MIS Quarterly, 22*, 363–408.
44. The new auto insurance ecosystem: Telematics, mobility and the connected car (2012). *Cognizant Report*
45. Whitepaper 2025 every car connected: Forecasting the growth and opportunity. *GSMASBD* (2012).
46. Berckmans, G., Messagie, M., Smekens, J., Omar, N., Vanhaverbeke, L., & Van Mierlo, J. (2017). Cost projection of state of the art lithium-ion batteries for electric vehicles up to 2030. *Energies, 10*(9), 1314.
47. Kshetri, N. (2014). Big data s impact on privacy, security and consumer welfare. *Telecommunications Policy, 38*(11), 1134–1145.
48. Eiza, M. H., & Ni, Q. (2017). Driving with sharks: Rethinking connected vehicles with vehicle cybersecurity. *IEEE Vehicular Technology Magazine, 12*(2), 45–51.
49. Xiao, Y., & Zhu, C. (2017). Vehicular fog computing: Vision and challenges. In *2017 IEEE International Conference on Pervasive Computing and Communications Workshops (PerCom Workshops)* (pp. 6–9). IEEE.

Chapter 8
Cyber-Physical Systems in Transportation

Yi He, Alireza Jolfaei, and Xi Zheng

8.1 Introduction to Cyber-Physical Systems

Cyber-Physical Systems, also known as CPS, are derived from the term "**cybernetics**," which was introduced by Norbert Wiener [1] in 1948. The name was used to describe these automatic machines with interactive communications and consummate controls like sophisticated human nervous systems. Wiener developed this idea from the Greek word *kybernetes*, which describes the steersman, with the implication of feedback and control mechanisms. Although, at that age, the idea aroused many controversies in multi-disciplines, a book named "Cybernetics" written by Wiener was published, with the belief that one day this novel ideal world would be achieved.

Soon after the publication of *Cybernetics*, Edmund C. Berkeley and S. P. Frankel [2] proposed the concept of "**mechanical brains**" in his book *Giant Brains or Machines That Think*. The book described a type of machine that was capable of communicating with other devices without human interference. Different components of the "*brains separately handled storage, computation, and control functions*". Information was delivered among devices by the cooperation of these components. In other words, these machines were able to process data, extract information, and memorize knowledge in their brains. Their computing capability, such as calculation, reasoning, and other operations, was carried out as intelligently as human brains. The machine language (e.g., binary codes), acting as a communication protocol, was used to express meanings and manage electrical equipment.

Y. He (✉) · A. Jolfaei · X. Zheng
Macquarie University, Sydney, NSW, Australia
e-mail: yi.he9@hdr.mq.edu.au; alireza.jolfaei@mq.edu.au; james.zheng@mq.edu.au

S. Garg et al. (eds.), *Intelligent Cyber-Physical Systems for Autonomous Transportation*, Internet of Things, https://doi.org/10.1007/978-3-030-92054-8_8

In 1954, Dr. Dennis Gabor, the recipient of the Nobel Prize in Physics 1971, illustrated a learning filter case in his article *Communication Theory and Cybernetics* that cybernetics is a science of purposeful action based on information. It "*involves collating data of experience and taking decisions, with the purpose of adjustment to changing circumstances, and control over them*" [3]. He advocated the **self-adjustment capability** of machines was similar to humans' learning nature that can be trained on filtration, recognition, and prediction.

Until 2006, the terminology "**Cyber-Physical Systems** (CPS)" was proposed at the First National Science Foundation (NSF) Workshop on CPS in Austin, Texas, United States (US) [4, 5], replacing "cybernetics" as an individual research field. In the August of 2007, the CPS program was prioritized as a top tactic of the US Federal research investments by the President's Council of Advisors on Science and Technology (PCAST). From 2009, NSF funded around 30 million US dollars per year on research related to CPS. Studies included but were not limited to the employment of CPS in multi-disciplines such as transportation, healthcare, communication, energy, and manufacturing industries. The NSF CPS program further evolved as the current Cyber-Physical Systems Virtual Organization (https://cps-vo.org/). Thereupon, the professional society Association for Computing Machinery (ACM) and the international technical organization Institute of Electrical and Electronics Engineers (IEEE) jointly launched the first International Conference on Cyber-Physical Systems (ICCPS) in 2010. This event accelerated the scientific and technological research on CPS. On the other hand, academic institutions conducted numerous CPS-related projects on motors, grid, and aerospace. For instance, Carnegie Mellon University (CMU) launched an *Autonomous Transportation Systems* program, and State University of New York (SUNY) promoted a *Cyber Transportation Systems* research plan based on CPS [6]. The booming period of CPS started at that time.

With the full bloom of CPS's applying, **Transportation Cyber-Physical Systems** (TCPS) also came into being. As the name implies, TCPS refers explicitly to the application of CPS in the field of transportation systems. TCPS is divided into three types [7]: (1) infrastructure-based TCPS, (2) vehicle-infrastructure coordinated TCPS, and (3) vehicle-based TCPS. These three types of TCPS communicate through the network and interact with physical objects to achieve the transportation system's higher efficiency and reliability. Additionally, the extensive use of TCPS also reduces environmental stress. Facing continuous growing demands across air, land, and water transport of both humans and goods, TCPS can remarkably save waiting time and energy consumption due to its real-time information feedback and efficient communication.

There are even some works [8, 9] that state that a transportation-based cyber-physical system is similar to an **Intelligent Transportation System (ITS)**. It improves traffic efficiency and optimizes planning routes by analyzing the real-time data shared under vehicular networks. If TCPS is an emerging concept, then ITS, in fact, has existed for a long time. As early as 1989, Alan Chachich [10] offered ITS in his transportation research. He expressly referred ITS to the smart highway's traffic management, which belongs to one type of infrastructure-based TCPS. He

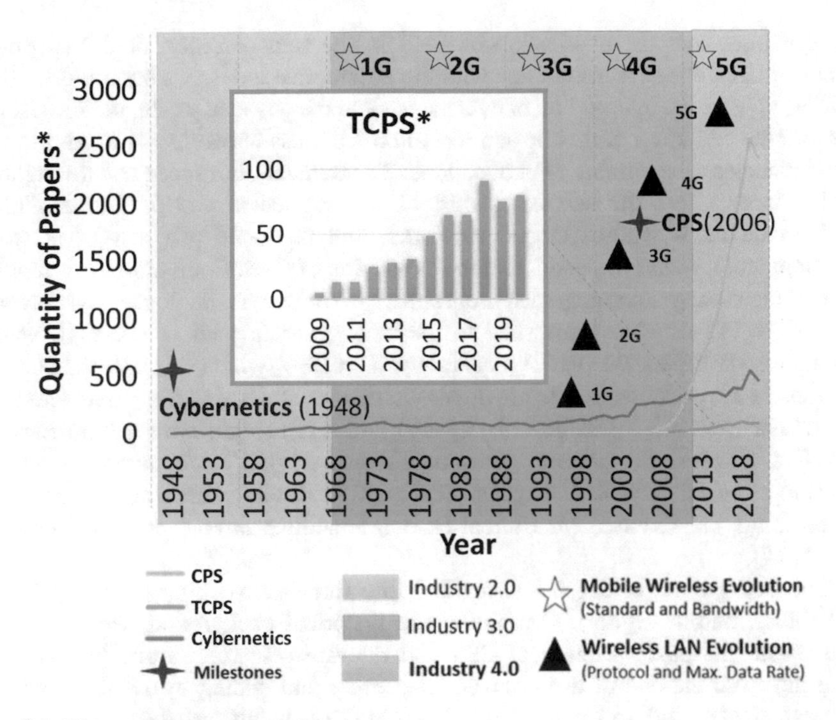

Fig. 8.1 The research trend on CPS by year

also divided vehicle sensing into two parts: real-time operation and offline analysis. For example, the control of traffic lights relies on real-time sensor data, so it pertains to real-time operation. Whether set up infrastructures or where the cameras should be installed, on the contrary, depends on the in-depth analysis of offline data.

Figure 8.1 outlines the CPS research trend from 1948 to 2020. We use three specific keywords "cybernetics," "CPS," and "TCPS" to filter our topics in *Title* and *Abstract* fields. The statistics do not include the papers in 2021 for an impartial consistent measurement on an annual basis. All sources are from the publications collected by Scopus, one of the largest worldwide abstract and citation databases operated by Elsevier.

As the orange line demonstrates, the quantity of cybernetics-related articles published each year increases to over 300 papers since 2011. Nevertheless, it shows a slow rise in the last 72 years when comparing to the CPS research. The sharp slope of the yellow line has made it evident that research related to the CPS received unprecedented attention and has been extensively developed. This surging prosperity is due to solid supports by policies, the myriad demands of consumption, and the innovation of continuous-upgraded "new generations (of devices and networks)." The green line represents the number of papers on TCPS, which exhibits a distinct rising in the past ten years. From the green histogram, we can infer that the application of CPS in the transportation field has grown exponentially, which is also in line with the automotive industry's booming.

Since CPS is an intersection of cyberspace and the physical world, in Fig. 8.1, we showed the evolution of mobile and local-area wireless networks [4] in addition to the statistics on the number of publications. The transmission speed and throughput undoubtedly affect the service quality of the connected world. The evolution of CPS demands well-developed networks, and the rapid progress of network development has, in turn, facilitated the expansion of CPS. From the 1970s, mobile communication systems upgraded progressively. The core technologies and network bandwidth [11–13] keep refreshed to the next-generation about every 10 years. Though the IEEE 802.11 (WiFi) standard arose in 1997, which was later than that of the mobile network, the update of wireless network standards and apparatus iterated faster. For example, IEEE 802.11 (legacy) and 802.11b/g/n have a frequency of 2.4 GHz, which is the same as Bluetooth. However, 802.11n/ac is 5 GHz, which leads to a higher transmission speed. The multiplied rate of transmission commits to reaching the requirement of real-time interactions among the CPS devices [5, 11, 14].

The background of Fig. 8.1 is divided into three parts: Industry 2.0 (yellow), 3.0 (blue), and 4.0 (gray). Combining the historical background, we can easily understand the past evolution of CPS and relevant research trends. *Industry 1.0* (starting from the end of the seventeenth century and ending by 1870) is outside our statistical period, so we have excluded it in Fig. 8.1. *Industry 2.0*, also known as *the Age of Electricity*, has introduced mass production by machines and therefore relieved human labor. In that era, the power grid was the main energy source, and management science was gradually applied to factories and companies. In 1969, the first programmable logic controller "Modicon" was invented, symbolizing the start of the 3rd Industrial Revolution (*Industry 3.0*)—*Information Age*. The new digital era was surrounded by heterogeneous information systems. Computing power was growing exponentially, electronics were popularized, and manufacturing became automated. Today, we are in the Fourth Industrial Revolution (***Industry 4.0***). This term was proposed at Hannover Fair 2011 and later officially recognized as a national strategic initiative by the Germany Trade and Invest (GTAI) in 2013. Its development focus lays on CPS, aiming to fulfill personalization needs and enhance network communication between machines. Consequently, it goes by the name of **the Age of CPS** [15–17].

The remainder of this chapter is organized as follows: Sect. 8.2 summarizes the theories on CPS and TCPS frameworks. Next, we discuss the influential roles of applying CPS in ITS. Section 8.4 is an outlook on the challenges of CPS in transportation and corresponding solutions. Finally, Sect. 8.5 concludes the chapter.

8.2 Design and Modeling

In order to better apply CPS in life, improve production or service quality, and monitor as well as feedback operation processes in real-time, there are many framework models for dissecting CPS to help people understand and use CPS, such as the "*3C (Communication, Computation, and Control) concept*" and "*5C (Connection, Conversion, Cyber, Cognition, Configuration) architecture*." The following paragraphs will explain these theories one by one.

8.2.1 CPS Architecture

*CPS is the intersection of the cyber and the physical world by the integration computation, communication, and control (**3C**) with feedback loops* [18].

As part of the new business model in Industry 4.0, CPS is expected to accomplish various customer needs at the cost of mass production with real-time connections among objects, systems, and humans [4]. It is a "system of systems," which is able to interact with many individually operated subsystems. It is an automated system, which is capable of dynamic self-reconfigurations. It is a coupling of the cyber and the physical, in which the computing components are connected and interactively react on the real world [6, 18].

As shown in Fig. 8.2, *sensors* are the physical devices used to perceive the *environment* (real world) [19]. The connected sensors transmit collected data (e.g., in-car temperature) to the main computing components—*processors* (e.g., Centre Console)—for information extraction and data analysis for further decision-making.

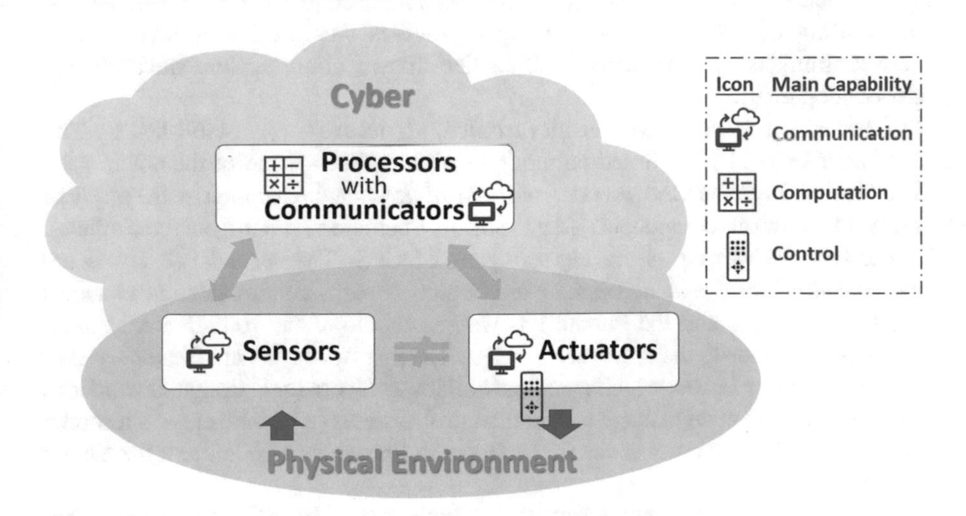

Fig. 8.2 3C concept of CPS

The computation results (e.g., the temperature is lower than the preset value) generated by the processors are shared with other inside subsystems (e.g., Climate Control Unit) or outside systems via the *communicators* that connect with the cyber, and vice versa. The processors act as a gathering place of information integration, with computing, data filtering, and action reasoning capabilities. Once decisions (e.g., lower the temperature) have been made, the *actuators* (e.g., the compressor, the refrigerant) will receive instructions from the processors and react to the environment (e.g., in-vehicle air) when the control function is achieved. The sensors keep sensing the environment after the reactions of the actuators. The new sensor data is regarded as feedback for the next-time actuators' reconfiguration and adjustments.

In addition to dividing the CPS by the capabilities of its physical components, some studies [4] have labeled its functions into different **maturity levels**. At the basic maturity level, the CPS can preset general conditions. When knowledge can be extracted from raw data, CPS makes the information transparent. As the processing goes more profound, the understanding of the environment by the CPS is deeper. Links are established in a semi-mature CPS and connect with other subsystems. The more links exist the more intelligent decisions that can be made. Lastly, the highest level of maturity is achieving self-optimization and self-adjustment and maintaining the interaction between internal and external CPS.

Another popular approach to model the CPS is the "**3-layer architecture**," which splits the CPS according to the distance between the source (closest to the environment) and destination (nearest to the user). The three layers are: physical system layer, information system layer, and user layer [20]. (1) The *physical system layer* contains sensors and sensor networks, actuators and actuator networks, smart chips, and embedded systems. It is in charge of the collection, transmission, and execution of data. (2) The *information system layer* is composed of the data center, control center, and next-generation network, responsible for the processing and transmitting of data. (3) The *user layer* provides the interaction environments between humans and computers, fulfilling decision-making functions and achieving closed-loop control.

Similarly, a "**6-layer, 3-level hierarchical structure**" was established by Zhu and Basar in 2011 [21], based on the industrial control system of the power grid. The first two layers are the *physical and control layers*, which comprise the physical plant and control components (e.g., sensors, actuators, intelligent controllers). These two are viewed as the *physical part* of CPS. The second two layers are the *communication and network layers*, which contain communication channels (wireless channels and the Internet) as well as topology and routing architecture. These two composite the *cyber part* of CPS. The last two layers are the *supervisory and management layers*, which coordinate all lower layers (e.g., design, instruction) and make decisions on management and control systems (e.g., budget plan, resource allocation). These two act as the CPS's brain and provide *the interfaces for human operations* [21].

With the continuous refinement of functions over time, CPS has been further divided. In 2015, the "**5C architecture of CPS**" [22] (shown in Fig. 8.3) was

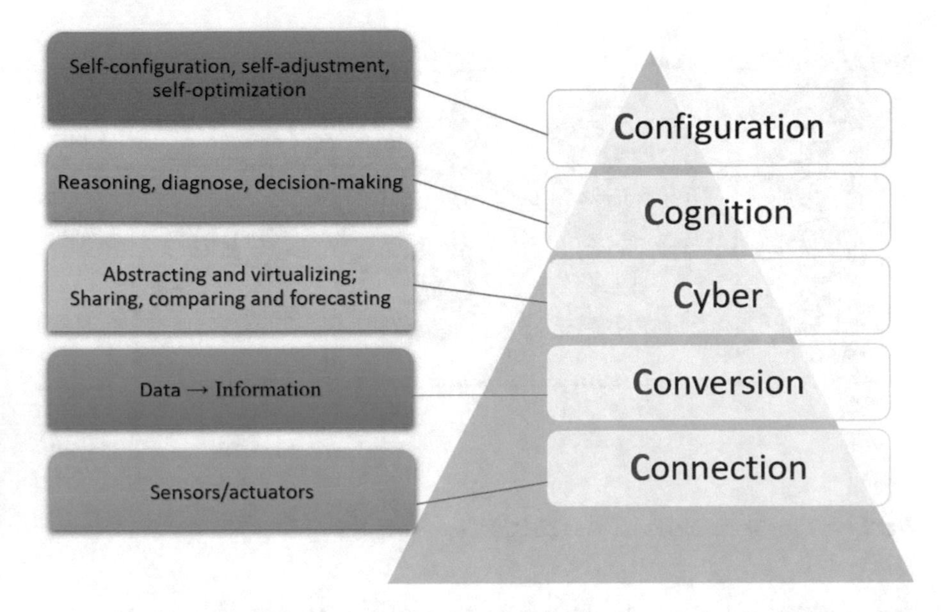

Fig. 8.3 5C architecture of CPS

suggested by J. Lee et al. and became one of the most popular structures cited by researchers. The first layer of 5C architecture is the smart "*connection*" level, which covers the sensor and actuator networks for data collection and feedback. The second layer is the data-to-information "*conversion* " level, which deals with the data conversion and information extraction. The third layer is the "*cyber*" level, where the physical world can be abstracted in the virtual world and be replicated for modelling, comparing, and forecasting. The fourth layer is the "*cognition*" layer, where diagnosing, logical reasoning, and decision-making are realized. The uppermost layer is the "*configuration*" level, in which the CPS achieves self-configuration, self-adjustment, and self-optimization.

Other architectures [23] such as "*Spatio-temporal Event Model*" [24] and "*Model-based Design Methodology*" [25] are not to be described in detail in this chapter, but they are worth reading if you have a strong interest on the model and design of CPS.

8.2.2 TCPS Architecture

In addition to the division of CPS, there are also particular models for TCPS.

A "**5-layer structure**" [26] (as shown in Fig. 8.4) has been recommended to model the ground transportation cyber-physical system. The bottom layer contains sensors and actuators. The second layer is Vehicle-to-Vehicle (V2V) communica-

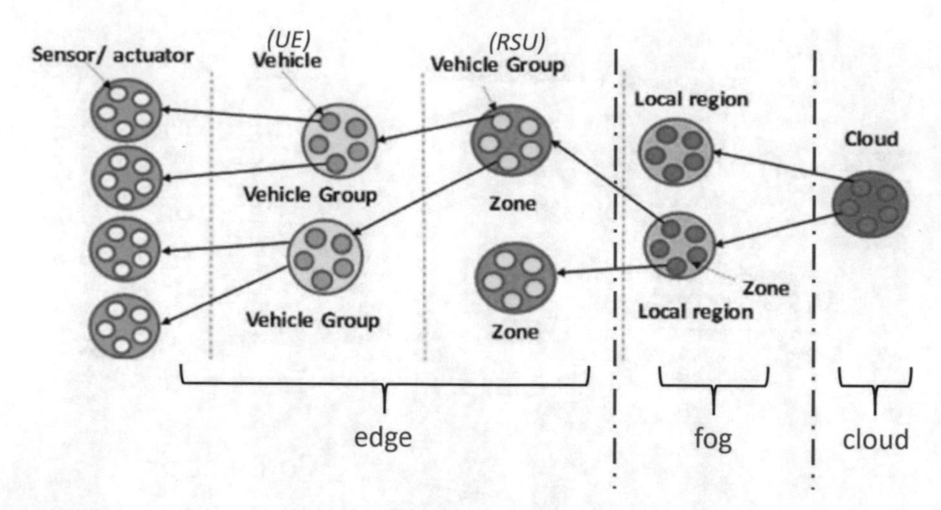

Fig. 8.4 5-layer hierarchical structure of TCPS [26]

tion, in which an individual vehicle can be regarded as User Equipment (UE). The third layer is the Vehicle-to-Infrastructure (V2I) communication, where a Road-Side Unit (RSU) is treated as an *edge node*, monitoring a zone of vehicles near to it. It will transmit road hazards and rules (e.g., speed limits) to groups of vehicles within its zone (edge, a.k.a. *fog node*). The fourth layer is the local region (fog) layer, inside which are dozens of zones. The topmost is the *cloud* layer that supervises all local regions like a city's master control center, where long-term data storage and traffic analysis perform.

Specifically speaking, in a **vehicle-based TCPS** [7], sensors such as intra-vehicular radars and cameras perceive the surrounding environment and transmit data to an Electronic Control Unit (ECU). This intelligent in-vehicle console supports making decisions and instructions (e.g., deceleration) on the TCPS actuators (e.g., brake) according to the computation results by preset algorithms. Meanwhile, the console plays a role of a communicator, connecting to the regional traffic management center (fog) directly or through road-side infrastructures for real-time road-condition feedback. When the traffic conditions in various regions are integrated into the cloud, a traffic report for the entire city is generated.

On the other hand, in an **infrastructure-based TCPS** [7], sensors are installed on the infrastructures. For instance, speed sensors will record the vehicle's speed and transmit it to the traffic management center. The traffic management center will send commands to TCPS actuators like surveillance cameras to capture the speeding vehicles' license plate numbers. In the end, the sensor and actuator data will be sent to the traffic management center for reference in consequent traffic fines and driver's license deductions on legal grounds.

Besides the above division of TCPS by its physical objects, some studies have carried out different segmentation by its *computing distribution*.

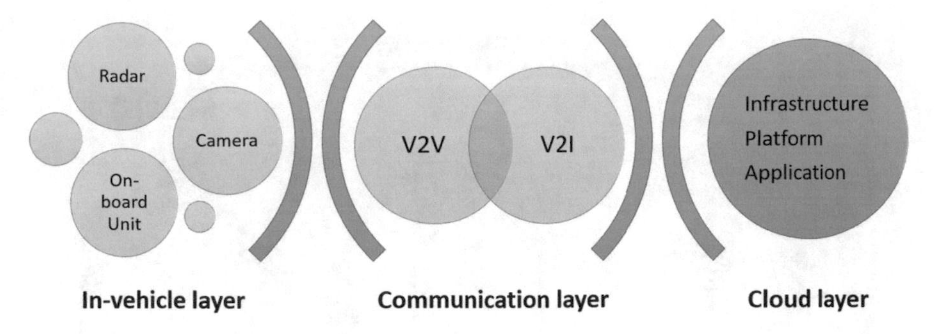

Fig. 8.5 VCC cloud computing architecture

Whaiduzzaman et al. [27] have mapped the TCPS in a **Vehicular Cloud Computing (VCC) architecture**, shown in Fig. 8.5. The *inside-vehicle layer* composes radars, cameras, and Inertial Navigation Sensors (INS) embedded On-Board Units (OBU). They have limited computation and storage functions, with Geographic Information Systems (GIS) and GPS modules. The *communication layer* achieves V2V and V2I communication, and generates reports (e.g., Wireless Roadside Inspection (WRI) and Emergency Warning Message (EWM)). The *cloud layer* is where to accomplish the main computation and storage of the data. It contains three parts: cloud infrastructures, cloud platforms, and cloud applications. Such as Network as a Service (NaaS), Information as a Service (INaaS), Cooperation as a Service (CaaS), Storage as a Service (STaaS), Entertainment as a Service (ENaaS), Pictures on a wheel as a Service (PicaaS), and Computing as a Service (CompaaS) are the cloud primary application services provided by the cloud platforms. More importantly, real-time applications, like fuel feedback, driver health recognition, human activity trace, and environmental perception are also realized in the cloud. Hence the model names Vehicular Cloud Computing (VCC) architecture.

Another approach that emphasized the distributed structure was suggested by Wang et al. [28], underlining the advantages of higher efficiency, lower latency, and less energy consumption for intermediate proximity services and flexible resource allocations over distributed computing. They pioneered a novel architecture called **Collaborative Vehicular Edge Computing (CVEC) framework**, vertically and horizontally dismembering the computation offloading [29, 30]. As shown in Fig. 8.6, *vertical collaborations* are divided into three layers (*infrastructure layer, edge computing layer, and core-computing layer*) and three types of computing (*local computing, edge computing, and remote cloud computing*) that occur between any two neighbor layers. *Horizontal collaborations* for vehicular networks are categorized into inter-domain and intra-domain classes. Inter-domain class was illustrated through 3 subclasses: *Mobile Edge Computing (MEC)*, *Fog Computing (FC)*, and *Cloudlet*.

Specifically, in a *MEC collaboration*, a BS (RSU) plays as an intermediary between remote computing centers and proximal vehicles. Through the MEC servers in the BS, Software-Defined Networking (SDN) controllers achieve better

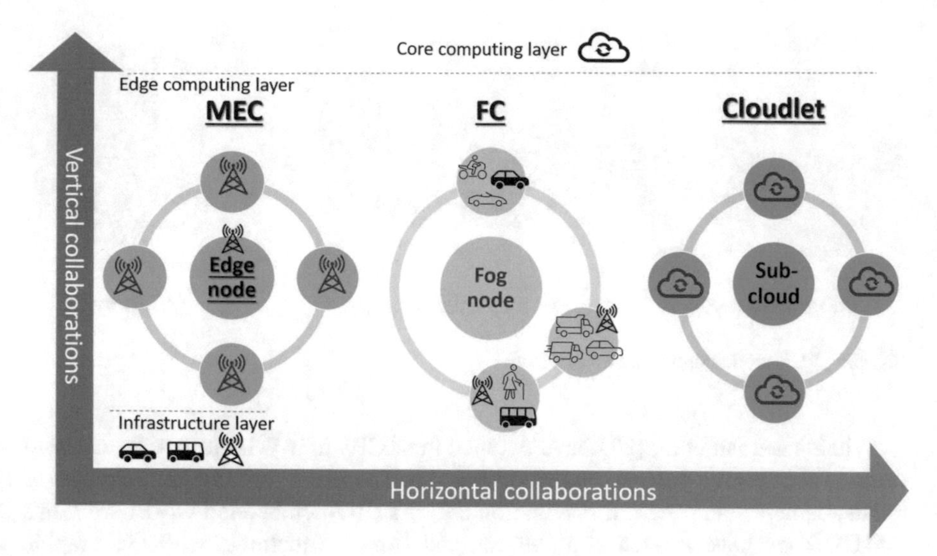

Fig. 8.6 CVEC edge computing framework

service quality. In an *FC collaboration*, computing tasks are distributed to diverse fog nodes. The fog nodes could be vehicle groups, local region BSs, and even WiFi access points. They are clustered by the distance and bandwidth availability, without a preassigned RSU as a mandatory component. Thus, an FC collaboration provides more flexibility than a MEC collaboration on node assembling and reassembling. In a *Cloudlet collaboration*, the network is treated as a whole, and collaborative computation is assigned to the divided sub-clouds, namely, Cloudlets. In this type of collaboration, if packet propagation is at the speed of light, a cloudlet's service range can support up to 300 km.

8.3 The Roles of CPS in ITS

The CPS evolution facilitates Internet-of-Vehicles (IoV) communication, which covers V2V, V2I, Vehicle-to-Sensors, and Vehicle-to-Personal-Devices. The equipment, such as display screen, sensors, camera, GPS receiver, antenna, and Central Processing Unit (CPU), communicate via IoV. By installing CPS devices to help in route planning, reducing the waiting time of traffic lights, optimizing fuel consumption, and minimizing polluted air emissions, an eco-driving plan [31] was proposed and verified in a real prototype of an on-board unit. Through V2V and V2I communications, TCPS is sharing traffic indicator information, detecting the number of passengers, GPS positioning of the location, predicting the route, and calculating the optimal driving speed to reduce sudden stops of the buses and idle time at intersections, so as to achieve a high-efficiency intelligent transportation system, with the purpose of emission reduction and energy saving. Burg et al. [5] have also presented a typical use of CPS to fulfill V2V, V2I, and in-vehicle communications. Anti-lock Braking Systems (ABS), Forward Collision Warning (FCW), temperature sensing, Adaptive Cruise Control (ACC), tire pressure monitoring, and Battery Management Systems (BMS) collaborate intelligently over the in-vehicle network. Bluetooth technique and IEEE 802.11p standard are used in Wireless Access in Vehicular Environments (WAVE) for V2V and Vehicle-to-Everything (V2X) connections. Amplitude Modulation (AM) and Frequency Modulation (FM) broadcasting from the Base Station (BS) provides traffic and entertainment information to the connected vehicles. Car location and trajectory are shared among Global Positioning System (GPS) embedded smart devices, car navigation boards, satellites, and the Internet. These advanced technologies applied to TCPS make ITS more informative and collaborative.

Figure 8.7 shows the advancement of road infrastructure in the Auckland Harbor Bridge since 1959. As shown in the figure, over the time the transportation system has become more intelligent. Nowadays, a sheer number of devices are being used in TCPS, such as intra-vehicle sensors, speed detectors, and lane signal indicators, which help in maximizing the traffic efficiency, road utilization, and eventually lead to saving commute time. The level of reliability and safety we observe on our roads has improved over the time.

With the increase in the number of vehicles, the transportation system needs to expand its services by utilizing more CPS devices in order to improve the traffic efficiency and safety. Some of the CPS devices include emergency vehicle notification systems, dynamic traffic light control, and electronic toll collection. For instance, surveillance cameras (CCTVs), infrared sensors, even mobile phones with GPS are used to collect vehicle speed, location, and throughput inside the city. Not only historical but also real-time data will be collected, integrated, and shared in the ITS via Traffic Management Centers (TMC). More comprehensive information, such as knowing the crowd density and detecting social events, is perceived to facilitate the intellectualization and urbanization of highly efficient public transportation. The reporting and referring interactions by connected devices enable a high-efficient, safe, intelligent, and comfortable urban transportation [33].

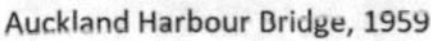

Fig. 8.7 Advancement of road infrastructure in the Auckland Harbor Bridge [32]

In addition, TCPS also contains huge commercial potential. Widespread network coverage and lower operation cost have attracted considerable commercial interest from industries, including equipment manufacturers, big data processing centers, and communication service providers. Technologies such as intelligent decision-making, split computing, and AI are extensively utilized to reach the ideal traffic condition. It is expected that the automobile industry will enlarge its sales to over $81 million by 2025 [34] and 80% of the new cars will embed connected driving technologies, bringing massive convenience and efficiency to urban transportation. The added value in the economy is predicted to range from $210 billion to $740 billion per year in the near future. The popularity of autonomous cars like Tesla, for instance, reveals an unprecedented positive attitude to the smart TCPS from the mass. It is anticipated that the collaboration of physical machines and virtual networks will accelerate future transportation progress with optimized designs and solutions. In such collaborative traffic systems, ample transportation facilities are provided to the public, and TCPS plays a vital role in achieving intelligent, safe, efficient, and environment-friendly transportation.

8.4 Challenges and Solutions

Enormous data are generated and shared in cyberspace with the rapid increase in the number of connected devices. Diverse unlicensed spectrums coexist and are used by novel devices for their easy access and low cost. This complex environment makes TCPS face unprecedented challenges, such as trustworthy authentication mechanisms and highly efficient real-time communication, which requires high *availability, efficiency, reliability, and security*. In an ITS, vehicular communication requires more dynamic capability because of vehicles' mobility. It has been long-awaited to change the conventional infrastructures by applying various sensors [10], such as magnetic field detectors, infrared devices, radars, ultrasonic equipment, and video sensors, leading to a more convenient and efficient traffic system. Though we have made great progress in the past 30 years, communication features have not been fully achieved in Commercial Vehicle Operations (CVO) and Vehicle-to-Roadside Communication (VRC). A series of detailed studies on CPS is in need. Figure 8.8 lists various challenges and the corresponding solutions, which we will explain in more detail in the following subsections.

8.4.1 Standardization and Availability

In the standardization progress of transmission rules, a variety of communications, including Bluetooth, Near-Field Communication (NFC), 5G cellular system, Wireless Local Area Network (WLAN), Wide Area Network (WAN), Ethernet, and Controller Area Network (CAN) adopt a series of non-uniform standards. For

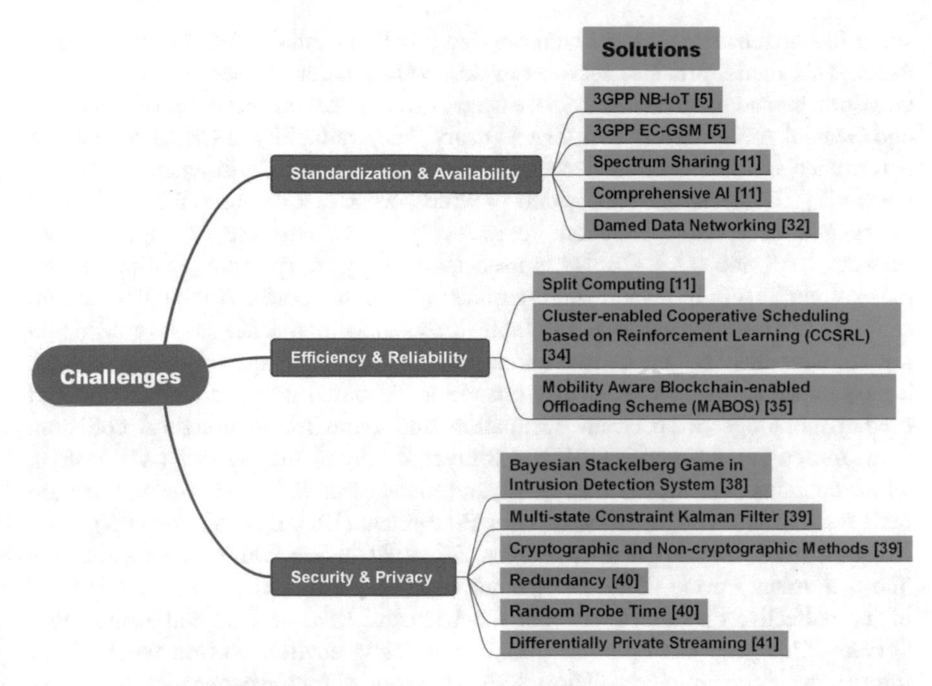

Fig. 8.8 Challenges and solutions

example, *ISO18000 Part 3* is one of the standards used in NFC, where the "readers" (RF signal senders) can identify massive "tags" (RF signal receivers)" [5]. The NFC-enabled mobiles can use the location-based interactions between devices to pay bus fees and taxi charges. *ZigBee* is a norm for wireless networks in transportation. It adds a network layer between the physical layer and the MAC layer, which enable connection to the Internet. On-board units (OBUs) gather information from sensors and mobiles under ZigBee protocol, feedbacking real-time road conditions. An electronic key with ZigBee can lock, unlock, and start a car remotely in a nearby range. Besides, Personal Area Network (PAN), Home Area Network (HAN), and WLAN also have their standards for short-range communication, where Low-Power Wide Area Network (LPWAN), WAN, and cellular networks are for long-range transmission. However, only the Third Generation Partnership Project (**3GPP**) standards (e.g., NB-IoT, EC-GSM) [5] are licensed among these diverse protocols. Other frequently used spectrums are unlicensed. Although many CPS can save much cost via unlicensed-spectrum communication, it is fraught that there is a mass of sensitive data shared in such an unbinding open network.

To solve the *availability* problem, Samsung Research [12] underlines **Spectrum Sharing** technology, which emphasizes spatial reuse. Exclusive licensees often underutilize the scarce licensed spectrum resources, especially those at low-frequency ranges. In order to achieve spectrum sharing and utility maximization, the US Federal Communications Commission (FCC) uses a three-tiered access model to share its Citizens Broadband Radio Service (CBRS) band (3.55–3.7 GHz) to other users. This model provides services in descending order of user priority: Federal Government and fixed satellite service users, Priority Access Licensees (PAL) users, and General Authorized Access (GAA) users. The availability of CBRS channels is determined by the Spectrum Access Systems (SAS) and Environmental Sensing Capability (ESC) mechanism, which provide access to a database for availability query and occupation detection, respectively. A Coexistence Manager (CxM) between SAS and GAA CBSDs is used to manage the spectrum sharing among the assigned users in a semi-static manner (database-based). Although spectrum sharing can ease the pressure of availability, dynamic approaches can be considered a good alternative option in urban areas, where traffic and spectrum demands are highly concentrated and constantly changing. AI-based network architecture can predict the usage of spectrum occupation and avoid (or minimizing) collision. **Comprehensive AI** model [12] is a 3-layer Artificial Intelligence (AI) system, which contains three layers: local, joint, and End-to-End (E2E) AI. The architecture includes four types of participants: User Equipment (UE), Base Station (BS), Core Network (CN), and Application Server. *Local AI* serves within one local entity like a terminal device (UE) for channel detection and coding. *Joint AI* is used in the collective operations between the UEs and BSs, or CNs and Application Servers. This joint can optimize complex wireless environment on the basis of future network condition prediction. *E2E AI* is applied for the connection from UEs to the applications, which facilitates the optimization of the entire communication channels. Wang et al. [35] have proposed a dynamic naming approach based on a Named Data Networking (NDN), one of the widely adopted methods under Information-Centric Network (ICN). It solves the shortage of IP address resources where numerous CPS in ITS are still based on the IP protocols. This content-based network topology ICN is one of the most promising solutions to service availability, utilizing the limited bandwidth more efficiently. A packet consists of three components: the interest packet, the data packet, and Forwarding Information Base (FIB). The interest packets correspond to Pending Interest Table (PIT) in NDN, while data packets refer to Content Store (CS). All the transmissions are based on the "interest" and corresponding "content" so that the shortage of IP addresses does not impact the information delivery.

8.4.2 Efficiency and Reliability

To solve the *efficiency* problem, **Split Computing** [5] was proposed. It is an advanced concept that transfers the computing capability among different types of equipment according to their computation capabilities (as shown in Fig. 8.9). For instance, in the 4G era, computation is completed in the cloud (by data centers) and Core Network (CN), emphasizing centralized *cloud computing* over the Internet. Current emerging 5G era focuses on *edge computing*. A Base Station (BS) is regarded as an edge device, which is part of the distributed systems with

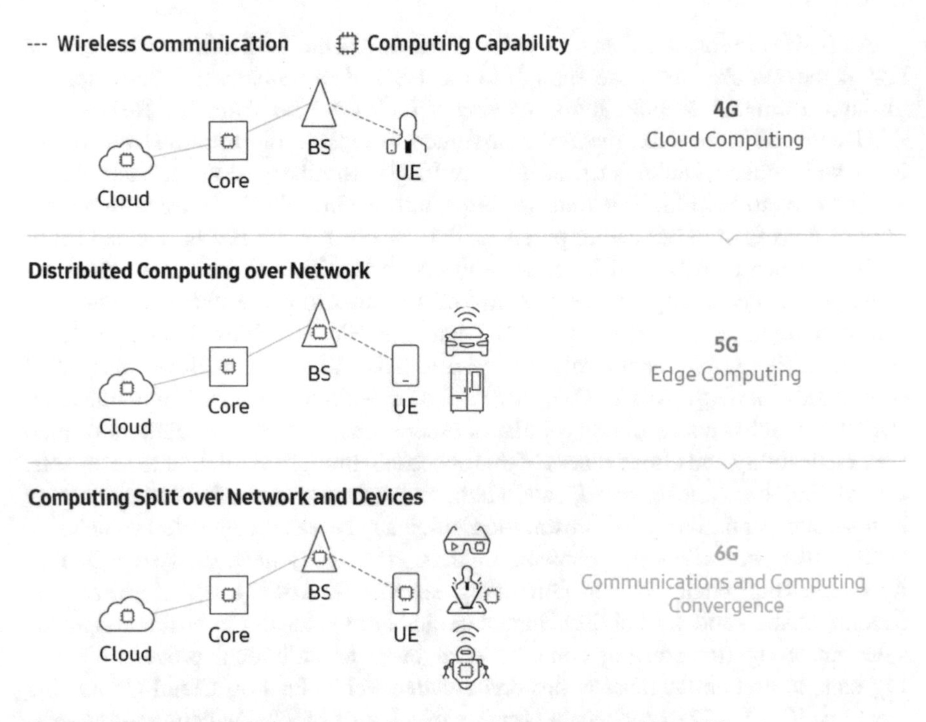

Fig. 8.9 Evolution of wireless communication and computing capability [12]

computation capability. A User Equipment (UE), namely, the terminal, is a human-operated or automatic device without (or with minimal) computing capability in the current 5G era. Such intelligent appliances and smart cars are UEs. In the future 6G era, *split computing* makes communication and computing involve everything. This computing is distributed intelligently according to the devices' computation capability to maximize the effect and utility. In other words, the computing jobs are more dynamically and accurately split among the networks and devices than distributed computing in 5G era. This technique can be used in intelligent robots, virtual reality equipment, and augmented reality glasses. How many computing jobs are assigned to a CPS device depends on its computation and communication capabilities so that we can view a UE as a special "edge node." In the "Machine as a Main User" 6G world, new computing techniques (e.g., quantum computing, edge computing) and new network types (e.g., intelligent cloud, AI-based network slicing) are combined, promising to make "the Age of CPS" come true [12, 36].

Xia et al. [37] proposed a novel approach to improve the information sharing efficiency named a Cluster-enabled Cooperative Scheduling based on Reinforcement Learning (CCSRL). It categorizes connected vehicles within one zone (cluster), where an RSU responds to a group of vehicles. These vehicles are divided into three types by clustering, namely, the cluster Head Vehicle (HV), Auxiliary Vehicle (AV), and Destination Vehicle (DV).

An RSU is to forward data to a series of moving vehicles directly or through the HV. If the vehicles are close enough to the RSU, data transmission is completed through a direct V2I link. If the moving vehicles are far from the RSU or the RSU does not store data, the HV is assigned to replace the transmission role of RSU with other vehicles such as AVs and DVs. Similarly, AVs in each cluster are the ones to help the HV transmit information. Only the DVs are without any duty of transmission but acting purely as data receivers. The HV is selected based on its distance to RSU and its bandwidth capability. The AV as an HV assistant is employed depending on the rewards of transmission and channel condition. The communication between the RSU and the HV is achieved by a cellular-subscriber-like Long-Term Evolution Vehicle (LTE-V) network. In contrast, V2V communication (e.g., AVs to DVs) uses a Dedicated Short-Range Communication (DSRC), which is not involving cellular or other infrastructures. The authors verified the practicability and effectiveness of this approach through simulation experiments, considering the distance, speed, bandwidth, and rewards for reinforcement learning into account, optimizing the transmission efficiency between connected vehicles.

To solve the *reliability* problem, Lakhan et al. [38] have devised a Mobility Aware Blockchain-enabled Offloading Scheme (MABOS), which utilized the decentralization and verifiability characteristics of blockchain, to enable the multi-side offloading (transferring computational tasks to individual processors) in a fog computing communication prototype called Vehicular Fog Cloud Computing Network(VFCN). The blockchain consists of a list of blocks, which cryptographically record the previous blocks' transaction and timeframe data. The *efficiency* is evaluated by the total cost of time, which includes communication and computation costs. The design idea of the model can be summarized as

$$F_a^{com} = \Phi_{com} \times \tau_u \times B_a, \tag{8.1}$$

$$F_a^{comp} = \sum_{i=1}^{N} T_{a,i}^{e} \times \Phi_{comp}, \tag{8.2}$$

$$F_a = \sum_{a=1}^{A} F_a^{com} + F_a^{comp}, \tag{8.3}$$

where the total cost F_a of application $a = \{1, \ldots, A\}$ is the sum of communication cost F_a^{com} and computation cost F_a^{comp}. Φ_{comp} is the computing cost of each node distributed by the scheme during the execution, and Φ_{com} is the utilization rate by the bandwidth B_a of application a. $\sum_{i=1}^{N} T_{a,i}^{e}$ is the average execution time of all the N tasks shared on each application. τ_u is the offloading time on data sending.

8.4.3 Security and Privacy

In addition to communication availability, efficiency, and reliability, CPS also faces security challenges. In [39], Alguliyev et al. used a tree diagram (shown in Fig. 8.10) to depict the threats and attacks that may cause CPS failures. For example, eavesdropping, remote spying, and corruption of data can occur during communication. Tampering with hardware or software can impact the actuation functions. Equipment malfunction and failure may interrupt the sensing process. Apart from the abstract concepts, Burg et al. [5] have analyzed a few representative security incidents in the CPS implementation domain, such as automotive,

Fig. 8.10 Overview of attacks and threats on CPS [39]

transportation, and traffic domains. Attacks were found on the CAN bus, ECU, speed sensor, GPS receiver, and broadcasting network. Sensors were subject to signal collision and identity spoofing threats, while airgap (used to connect outside networks without a network interface) attacks and brute-force hacks (using trial-and-error to guess login information or encryption key) on wireless transmissions. Malware (malicious software to cause damage or gain unauthorized access) and side-channel attacks (utilizing sensitive information from a system and reducing cryptography security) befall in controllers, while DoS attacks and IP thefts occur in intelligent applications.

An effective and safe TCPS relies on comprehensive and thoughtful design and requires establishing an extensible security framework. Dibaei et al. [40] have conducted a detailed investigation of potential attacks and countermeasure defenses. For example, Denial-of-Service (DoS) attacks occur on RSU. A large number of requests flood into RSU, where the authentication, management, and transmission functions are completed in a vehicular network. The traditional solution to combat DoS is to block the attacker's IP address. However, if the attacker uses a crafty strategy with distributed IP addresses, the defenses become vain. Similarly, attacks such as replay, Sybil, and impersonation are targeting connected vehicles. A replay attack intends to repeat or delay a transmitted packet between an RSU and a car. Although information encryption can reduce the damage caused by this attack, in most cases, the attacker is highly mobile and difficult to be detected. In order to solve this problem, Halabi et al. [41] proposed a novel approach to improve the effectivity of the Intrusion Dection System (IDS). The adoption of Bayesian Stackelberg Game can help to identify false data injection, reply attack, wormhole and Sybil attack. Bian et al. [42] adopted a multi-state constraint Kalman filter to evaluate the vehicles' deviation from the platoon and their trajectory when crowd sensing (a.k.a. collective perception). The authors have proposed cryptographic and non-cryptographic security mechanisms. After encryption, any modification of messages can be found via the hash value and error correction technique. However, jamming attacks, for instance, cannot be identified by cryptographic method because the noise cannot be distinguished from the communication channel. These kinds of assaults are analyzed by non-cryptographic methods, such as comparison with other information provided from other sources, estimating whether the information is consistent, or establishing a channel hopping process. Deng et al. [43] have also indicated that physical sensor attacks (e.g., jamming attacks) can be defended via redundancy and random probe time approaches. Redundancy is a way of using multiple sensors to collect the same type of data in order to detect anomalies after comparison, and random probe time for sensor data (e.g., LiDAR signals) can effectively recognize attacks like spoofing. To address the privacy concerns, Ezabadi et al. [44] have proposed a Differentially Private Streaming (DPS) mechanism to protect sensitive data from untrusted ITS servers. The proposed mechanism consists of three steps including an encoding, a perturbation, and a dynamic compression step. These steps can effectively preserve privacy, distinguish noises, and improve perception data accuracy.

8.5 Concluding Remarks

This chapter systematically introduced the Cyber-Physical Systems from the origin of "cybernetics" to the vision of "the Age of CPS." From the statistics, research enthusiasm has advanced rapidly along with megatrends. Many theories were established to understand and utilize CPS better, such as "3C concept" (Computation, Communication, and Control) and "5C architecture" (Connection, Conversion, Cyber, Cognition, and Configuration). The technologies used in Transportation Cyber-Physical Systems (TCPS) comprise "5-layer structure" (sensor/actuator, user equipment, road-side unit, the local region, and cloud processing), "Vehicular Cloud Computing" (inside-vehicle, communication, and cloud), and "Collaborative Vehicular Edge Computing" (Mobile Edge Computing, Fog Computing, and Cloudlet). Even though science delivers tremendous changes, both physically and virtually, many challenges remain, such as availability, efficiency, reliability, and security issues. The solutions to mitigate network availability challenges include Spectrum Sharing and Comprehensive AI (local, joint, and end-to-end AI). Communication efficiency can be improved via Split Computing and CCSRL. Blockchain-based MABOS is used to improve transmission reliability, and both cryptographic and non-cryptographic methods can be utilized to ensure CPS security. Future research is expected to focus on: (1) standardized protocols, (2) large-scale availability, (3) communication efficiency and reliability, and (4) cybersecurity and privacy. The ideal TCPS or ITS requires the new network topologies to coexist with current heterogeneous networks harmoniously. Key technologies such as machine intelligence and cybersecurity mechanisms also need to be optimized. Only by realizing these breakthroughs can people enjoy fully autonomous vehicles and intelligent transportation in their "movie-like" dream lives.

References

1. Wiener, N. (1948). Cybernetics. *Scientific American, 179*(5), 14–19.
2. Berkeley, E. C., & Frankel, S. P. (1950). Giant brains or machines that think. *Physics Today, 3*(5), 39–39.
3. Gabor, D. (1954). Communication theory and cybernetics. *Transactions of the IRE Professional Group on Circuit Theory, 1*(4), 19–31.
4. Monostori, L., Kádár, B., Bauernhansl, T., Kondoh, S., Kumara, S., Reinhart, G., Sauer, O., Schuh, G., Sihn, W., & Ueda, K. (2016). Cyber physical systems in manufacturing. *Cirp Annals, 65*(2), 621–641.
5. Burg, A., Chattopadhyay, A., & Lam, K. (2017). Wireless communication and security issues for cyber physical systems and the internet-of-things. *Proceedings of the IEEE, 106*(1), 38–60.
6. Park, K.-J., Zheng, R., & Liu, X. (2012). Cyber physical systems: Milestones and research challenges. *Computer Communications, 36*(1), 1–7.
7. Deka, L., Khan, S. M., Chowdhury, M., & Ayres, N. (2018). Transportation cyber-physical system and its importance for future mobility. In L. Deka, & M. Chowdhury, (Eds.), *Transportation cyber-physical systems* (pp. 1–20). Elsevier.

8. Lin, J., Yu, W., Yang, X., Yang, Q., Fu, X., & Zhao, W. (2017). A real-time en-route route guidance decision scheme for transportation-based cyber physical systems. *IEEE Transactions on Vehicular Technology, 66*(3), 2551–2566.
9. Jindal, A., Aujla, G. S., Kumar, N., Chaudhary, R., Obaidat, M. S., & You, I. (2018). SeDaTiVe: SDN-enabled deep learning architecture for network traffic control in vehicular cyber-physical systems. *IEEE Network, 32*(6), 66–73.
10. Chachich, A., Klocek, P., & Sigel, G. H. (1989). Highway-based vehicle sensors. In *Infrared fiber optics*. Society of Photo-optical Instrumentation Engineers (Vol 10318, pp. 175–201). ISSN: 0277-786X Journal Abbreviation: Proceedings of SPIE–the international society for optical engineering.
11. Surantha, N., Sutisna, N., Nagao, Y., & Ochi, H. (2017). SoC design with HW/SW co-design methodology for wireless communication system. In *2017 17th International Symposium on Communications and Information Technologies, ISCIT 2017* (Vol. 2018-January, pp. 1–6).
12. Samsung Research. (2020). 6G Vision: The next hyper connected experience for all. *Samsung's 6G white paper*, 1–39.
13. Adedoyin, M. A., & Falowo, O. E. (2020). Combination of ultra-dense networks and other 5G enabling technologies: A survey. *IEEE Access, 8*, 22893–22932.
14. IEEE Standards Association. (2021). IEEE Guide for EMF Exposure Assessment of Internet-of-Things (IoT) Technologies and Devices. *IEEE Std 1528.7-2020*, 1–90. ISSN: 978-1-5044-7078-0.
15. Demir, K. A., & Cicibas, H., Industry 5.0 and a critique of industry 4.0. In *4th International Management Information Systems Conference*, Istanbul, Turkey, October, 2017, pp. 17–20.
16. Hozdić, E. (2015). Smart factory for industry 4.0: A review. *International Journal of Modern Manufacturing Technologies, 7*(1), 28–35.
17. Xu, L. D., Xu, E. L., & Li, L. (2018). Industry 4.0: state of the art and future trends. *International Journal of Production Research, 56*(8), 2941–2962. Taylor & Francis
18. Lee, E. A. (2010). CPS foundations. In *Proceedings of the 47th Design Automation Conference*, DAC '10 (pp. 737–742). Association for Computing Machinery. Event-Place: Anaheim, California.
19. Gronau, N., Grum, M., & Bender, B. (2016). Determining the optimal level of autonomy in cyber-physical production systems. In *2016 IEEE 14th International Conference on Industrial Informatics (INDIN)* (pp. 1293–1299). ISSN: 2378-363X.
20. Liu, Y., Peng, Y., Wang, B., Yao, S., & Liu, Z. (2017). Review on cyber physical systems. *IEEE/CAA Journal of Automatica Sinica, 4*(1), 27–40.
21. Zhu, Q., & Başar, T. (2011). Robust and resilient control design for cyber-physical systems with an application to power systems. In *2011 50th IEEE Conference on Decision and Control and European Control Conference* (pp. 4066–4071). ISSN: 0743–1546.
22. Lee, J., Bagheri, B., & Kao, H.-A. (2015). A cyber-physical systems architecture for industry 4.0-based manufacturing systems. *Manufacturing Letters, 3*, 18–23.
23. Khaitan, S. K., & McCalley, J. D. (2015). Design techniques and applications of cyber physical systems: A survey. *IEEE Systems Journal, 9*(2), 350–365.
24. Tan, Y., Vuran, M. C., & Goddard, S. (2009). Spatio-temporal event model for cyber-physical systems. In *2009 29th IEEE International Conference on Distributed Computing Systems Workshops* (pp. 44–50). ISSN: 1545–0678.
25. Jensen, J. C., Chang, D. H., & Lee, E. A. (2011). A model-based design methodology for cyber-physical systems. In *2011 7th International Wireless Communications and Mobile Computing Conference* (pp. 1666–1671). ISSN: 2376–6506.
26. Jolfaei, A., & Kant, K. (2019). Privacy and security of connected vehicles in intelligent transportation system. In *2019 49th Annual IEEE/IFIP International Conference on Dependable Systems and Networks – Supplemental Volume (DSN-S)* (pp. 9–10).
27. Whaiduzzaman, M., Sookhak, M., Gani, A., & Buyya, R. (2014). A survey on vehicular cloud computing. *Journal of Network and Computer applications, 40*, 325–344.
28. Wang, K., Yin, H., Quan, W., & Min, G. (2018). Enabling collaborative edge computing for software defined vehicular networks. *IEEE Network , 32*(5), 112–117.

29. Aujla, G. S., Singh, A., Singh, M., Sharma, S., Kumar, N., & Choo, K.-K. R. (2020). BloCkEd: Blockchain-based secure data processing framework in edge envisioned V2X environment. *IEEE Transactions on Vehicular Technology, 69*(6), 5850–5863.
30. Singh, A., Aujla, G. S., & Bali, R. S. (2021). Intent-based network for data dissemination in software-defined vehicular edge computing. *IEEE Transactions on Intelligent Transportation Systems, 22*(8), 5310–5318.
31. Zhang, L., Liang, W., & Zheng, X. (2018). Eco-driving for public transit in cyber-physical systems using V2I communication. *International Journal of Intelligent Transportation Systems Research, 16*(2), 79–89. Springer.
32. *The 2019 international workshop on safety, security, and trust in intelligent transportation system (SST-ITS)* (5–8 August 2019).
33. Rajendar, S., Rathinasamy, D., & Kaliappan, V. K. (2019). Intelligent transportation cyber-physical system: On crowd through event detection for public urban transportation. *International Journal of Recent Technology and Engineering, 8*(2), 2375–2379.
34. Kaiwartya, O., Abdullah, A. H., Cao, Y., Altameem, A., Prasad, M., Lin, C., & Liu, X. (2016). Internet of vehicles: Motivation, layered architecture, network model, challenges, and future aspects. *IEEE Access, 4*, 5356–5373.
35. Wang, C., Wu, J., Zheng, X., Pei, B., Zhang, X., Yu, D., & Tang, J. (2021). Leveraging ICN with network sensing for intelligent transportation systems: A dynamic naming approach. *IEEE Sensors Journal, 21*(14), 15875–15884.
36. Dogra, A., Jha, R. K., & Jain, S. (2020). A survey on beyond 5G network with the advent of 6G: Architecture and emerging technologies. *IEEE Access, 9*, 1–1.
37. Xia, Y., Wu, L., Wang, Z., Zheng, X., & Jin, J. (2020). Cluster-enabled cooperative scheduling based on reinforcement learning for high-mobility vehicular networks. *IEEE Transactions on Vehicular Technology, 69*(11), 12664–12678.
38. Lakhan, A., Ahmad, M., Bilal, M., Jolfaei, A., & Mehmood, R. M. (2021). Mobility aware blockchain enabled offloading and scheduling in vehicular fog cloud computing. *IEEE Transactions on Intelligent Transportation Systems, 22*, 4212–4223.
39. Alguliyev, R., Imamverdiyev, Y., & Sukhostat, L. (2018). Cyber-physical systems and their security issues. *Computers in Industry, 100*, 212–223.
40. Dibaei, M., Zheng, X., Jiang, K., Abbas, R., Liu, S., Zhang, Y., Xiang, Y., & Yu, S. (2020). Attacks and defences on intelligent connected vehicles: A survey. *Digital Communications and Networks, 6*(4), 399–421.
41. Halabi, T., Wahab, O. A., Al Mallah, R., & Zulkernine, M. (2021). Protecting the internet of vehicles against advanced persistent threats: A Bayesian Stackelberg game. *IEEE Transactions on Reliability, 70*(3), 970–985.
42. Bian, K., Zhang, G., & Song, L. (2017). Toward secure crowd sensing in vehicle-to-everything networks. *IEEE Network, 32*(2), 126–131. Institute of Electrical and Electronics Engineers Inc.
43. Deng, Y., Zhang, T., Lou, G., Zheng, X., Jin, J., & Han, Q.-L. (2021). *Deep learning-based autonomous driving systems: A survey of attacks and defenses*. _eprint: 2104.01789.
44. Ezabadi, S. G., Jolfaei, A., Kulik, L., & Kotagiri, R. (2019). Differentially private streaming to untrusted edge servers in intelligent transportation system. In *2019 18th IEEE International Conference On Trust, Security And Privacy In Computing And Communications (TrustCom)* (pp. 781–786).

Part IV
Application Use Cases of Autonomous Transportation Systems

Chapter 9
Correlation Between Traffic Lights and Emergency Vehicles in Intelligent Transportation System

Gurpreet Kaur and Sumit Sharma

9.1 Introduction

Urbanisation and movement of masses towards cities have led towards concentration of population in smaller areas. This has put forward various challenges in front of planning commissions of states and countries. The continuously increasing vehicular traffic on the road is one of the major issues in urban areas. The increase in the number of vehicles on the road congestion is the main concern in urban areas, especially in rush hours. The major demerits related to road congestion include wastage of fuel, degradation of vehicular performance, anomalous driving behaviour, etc. Intelligent Transportation Systems (ITS) aim at utilising the sensing, analysing, disseminating, and computing abilities to provide solutions to transportation-related issues [1]. ITS are also being used for supporting other causes like data offloading, energy trading, etc. The mobility of the vehicles is a major difficulty in data offloading, which can be tackled using various approaches like Software-defined network-based controller [2]. Moreover, electric vehicles are also being put to use for energy trading in smart cities, which can be secured by using concepts like block chain [3]. Highly advanced deep learning networks are also being put to use to optimise the network traffic for enhancing the quality of service in smart vehicles [4–6]. There are a plethora of other highly advanced support services as well that contribute towards improving the quality of life in society, but the most haunting side effect of ITS is the road congestion. The root cause of road congestion is observed to be inefficient traffic control at traffic lights [7]. To manage the traffic, we use static traffic light method on single and multiple intersection. The static traffic light has a fixed time of lights. There are three lights in a traffic light, which are red, yellow, and green. The red light means stop, the yellow light means wait,

G. Kaur · S. Sharma (✉)
Chandigarh University, Mohali, India

S. Garg et al. (eds.), *Intelligent Cyber-Physical Systems for Autonomous Transportation*, Internet of Things, https://doi.org/10.1007/978-3-030-92054-8_9

and the green light means go. The static traffic light approach has fixed time slots for each side of the intersection irrespective of the density of each side. The magnitude of the losses incurred increases manifold in country like India where in January 2019, more than a million-and-half (1,607,315) new vehicles were registered. The bifurcation of this new registration includes 74% of the vehicles being two wheelers (1,187,998) and more than 80% of the total vehicles consuming petrol [8]. This method is not efficient in terms of cost, performance, and maintenance of traffic light [8].

The average waiting time of vehicle increases at the intersection due to static traffic light approach. To overcome this problem, intelligent traffic light system (ITLS) needs to be deployed to manage the traffic automatically on the intersection of roads. ITLS has varying time period based on analysis performed instead of fixed time slots. This method manages the traffic and aids in minimising the fuel consumption of vehicles running on the road. ITLS minimises the average waiting time and congestion of vehicles on the intersection as well. The ITS consist of surveillance system, communication system, traffic light control system, and energy efficiency system [9]. The focus of the existing ITLS has been mainly on these factors only, but lesser consideration is given on adding parameter of high-priority vehicles and their early clearance from intersections. Thus, there is a need of an architecture of ITS working in light of ITLS for better management with respect to emergency vehicles. The different types of emergency vehicles are fire brigades, police cars, and ambulances. A scheme also needs to be established for multiple lane clearance only in case of conflicting emergency vehicle lane sides based on prior declared non-overlapping of the paths chosen. The whole paper consists of 4 parts. Section 9.2 discusses the proposed architecture of ITS in light of intelligent traffic light. Section 9.3 explains the proposed scheme to give priority to emergency vehicles. Section 9.4 discusses the conclusion and future enhancement of this work.

9.2 Architecture

As shown in Fig. 9.1, the architecture of ITLS is in the form of top-down stack where the flow of information starts from the top layer and reaches the bottom layer. The decisions based on the analysis are propagated back to the top layer. The proposed architecture of ITLS contains four layers, i.e., hardware layer, network layer, analytical layer, and application layer. The hardware layer represents the different types of sensors that perform specific functions such as count of vehicles, motion, weight, location and type of vehicle, and an interface manager that helps in maintaining the state of different sensors.

Hardware Layer

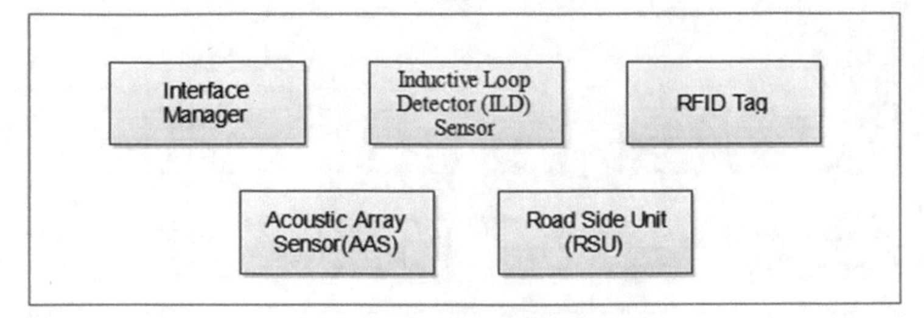

Network Layer

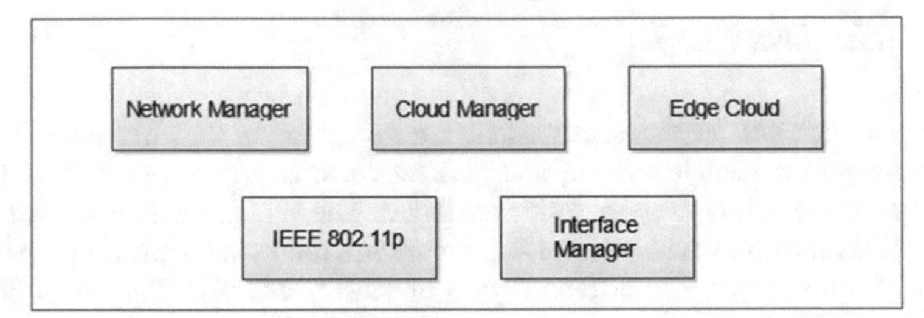

Analytical Layer

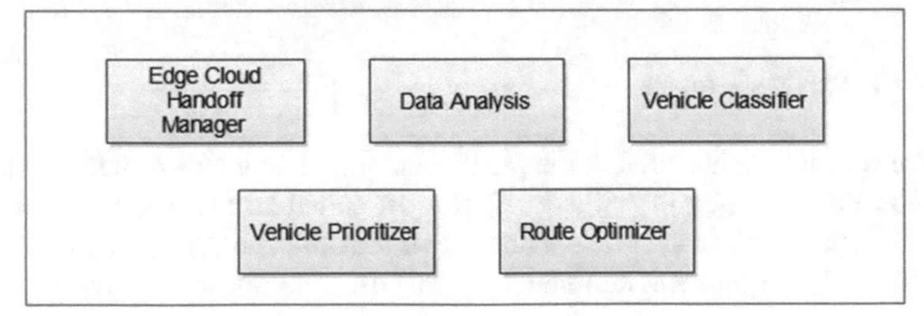

Application Layer

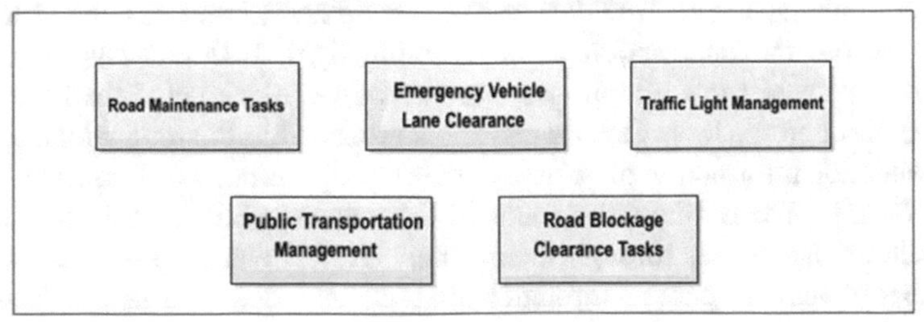

Fig. 9.1 Architecture of intelligent traffic lighting system

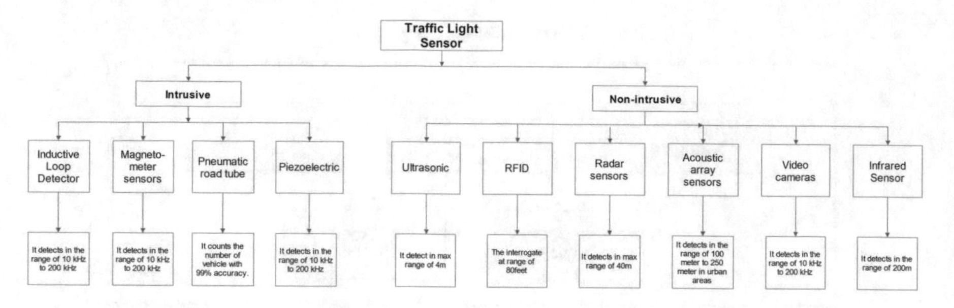

Fig. 9.2 Traffic light sensors classification

9.2.1 Hardware Layer

This layer includes all the sensors that are going to be the first point of contact/exchange for vehicles as the first and foremost requirement of ITLS implementation is the ability to sense traffic attributes. The vehicles are also a part of the hardware layer as they will be receiving the information via backward propagation from the bottom layer. The different types of sensors available that could be used to implement ITLS are intrusive sensors and non-intrusive sensors and are shown in Fig. 9.2.

9.2.1.1 Intrusive Sensors

Intrusive sensors are installed in the pipeline so that they intersect with the process flow of the entities under investigation. These are called intrusive sensors because of the fact that they intrude the process flow either via direct contact or indirect contact. These have high equipment and maintenance cost. The different types of intrusive sensors are:

- Inductive Loop Detector (ILD) Sensor: The inductive loop detector sensor is used in traffic management. ILD detects the vehicle arrival and passing at a certain point before the intersection point of traffic light. ILD consists of inductive metallic wire to form a loop under the surface of the road. When the vehicle passes through the loop, electric current is produced in the loop which transmits the detection information of vehicle to the control station. The frequency range is 10–200 kHz. The different variations of ILD can even help in detecting the type of vehicle that passes through them. Thus, an ILD can help to count the total number of vehicles passing through a road segment as well as help in bifurcating the type of vehicles passed.
- Magnetometer Sensor: The magnetometer sensor is used with the secondary sensor to detect the large loaded weighted vehicles like trucks, rail cars on the road before the traffic light. It uses a passive approach of sensing where change in the ambient magnetic field corresponds to the detection of vehicle. Therefore,

during installation of magnetometer sensor, it must be ensured that no vehicle is present when the magnetometer is being installed and taught about the ambient magnetic conditions.

- Pneumatic road tube sensors: The pneumatic road tube sensor is detecting the number of vehicle count passes through a particular range on the road. When the vehicle passes over the road tube, it sends the burst pressure, which produces the electric signal that is transmitted to analysis software. If the requirement of the scenario is to evaluate the wrong direction vehicle movement, then a pair of tubes can be drawn on various lanes. Hence, on the basis of which tube is pressed first can help to decide the direction from which the vehicle came.
- Piezoelectric sensors: The piezoelectric sensor can also be used in traffic management. The basic principle of piezoelectric sensors is the conversion of mechanical energy into the electric energy. The installation of piezoelectric sensors includes making a groove cut in the road and fitting the sensor in it. When a car crosses the groove cut, it applies pressure on the sensor that further transforms it into electrical energy, and hence, this data may be transmitted through RS232 connection or an Ethernet connection to the server.

9.2.1.2 Non-intrusive Sensors

The non-intrusive sensors are installed at different places on the roads. It does not interfere with the flow profile. It gives the information of the number of vehicles in a lane at the intersection, weather conditions, and traffic conditions on the road [10]. It has high maintenance cost. The different types of non-intrusive sensors are:

- Ultrasonic Sensor: Ultrasonic sensor is used for automation purposes in traffic management on the road. The ultrasonic sensor is placed at one side of the road near the intersection of traffic light. It covers some particular area where the vehicle is restricted to pass during the red light on ITL [11]. The max range of ultrasonic sensor is 4 m.
- Radio Frequency Identification (RFID) Sensor: The RFID sensor is placed at the vehicles. The RFID reader is installed at the road-side units (RSU). When a vehicle approaches the RSU, the RFID reader installed on the RSU reads the active RFID tag of the car. The data stored in the RFID tag is mostly a unique number that corresponds to all the information related to the car. It is used to count the number of vehicles in a specific duration [12].
- Radar Sensor: These radar sensors detect the motion of vehicle either the vehicle is moving or stationary. It is also not affected by the weather condition. The max range of radar sensor is 40 m.
- Acoustic Array Sensors (AASs): AAS is used for traffic monitoring. The acoustic sensors are deployed in a geometrical pattern where the sound sensed by an array of sensors is collaborated and analysed to evaluate the number of vehicles passed as well as the type of vehicle passed. The major challenge in the acoustic sensors

in the country like India is unavailability of real-life data pertaining to sound produced by electric vehicles when passing through AAS [13].
- Video Camera Sensor: The video camera sensor is used to capture the vehicles near the intersection of traffic lights. It checks the speed of vehicles. The video camera sensor has a large amount of variants available with different capabilities like motion detection. The amount of data generated by video camera sensors is large and, hence, requires a high amount of computational costs to process and analyse the data [14].
- Infrared Sensor: The infrared sensor detects the length of vehicle and also checks the speed of vehicles. The infrared sensors could be either active or passive. The active infrared sensor works on spectral range, whereas passive infrared sensor works on thermal range. The max range of infrared sensor is 200 m [15].

The types mentioned above are not exhaustive of all the available sensors but cover most of the sensors being used in traffic management scenarios. For the purpose of implementing ITLS, either single sensor or a combination of sensors could be used. The decision related to placement of the sensors also plays a big role in the overall performance of the system. The sensors are not confined to be placed only on the intersection points but are required to be placed on selected RSUs for predictive analytics to be fed into the ITLS.

9.2.2 Network Layer

The second layer in the architecture is the network layer that is responsible for all the communication between all the entities of the ITS structure. The various entities involved are: vehicles, sensors, road-side units, processing units, and application units. The different types of communication that are possible amongst entities are wired communication and wireless communication. The various intrusive sensors can be connected to Fog and/or Edge devices installed on the RSU via wired channels like Ethernet/RS232 connections. The fact that the sensors deployed are immobile makes the selection of wired channels ideal for the case as it will provide dedicated and congestion-free channels with fastest possible transfer speeds.

The network layer is also responsible for transmitting various information broadcasts and control messages to the vehicles for ensuring optimised traffic flow. The wireless communication system is embedded on the surface of the road near traffic light at the intersection and at RSU. The wireless communication has low maintenance cost as compared to the wired communication system. The wireless communication system helps to communicate between the vehicle to vehicle and vehicle to infrastructure. ITS use wireless communication system amongst the vehicles and traffic light/RSU. The wireless communication is of two types: short-range communication system and long-range communication system. The main communication system of using IEEE 802.11 protocols is Wireless Access in Vehicular Environments (WAVE), the Dedicated Short-Range Communications standard (DSRC), Wifi, WiMax, etc. [16]. The control messages to normal vehicles would require to be sent via wireless channels when an emergency vehicle is about to approach the road segment so that the normal vehicles provide passage clearance for fast passing through emergency vehicles. Thus wireless communication channels are used for data exchange between vehicle to vehicle and between vehicles and the RSU or traffic light.

9.2.3 Analytical Layer

The third layer of the architecture is the analytical layer. This layer is responsible for performing all the analysis on the data saved on the Edge/Cloud device based upon the request of the application layer. Replacing road-side units with Edge devices has been demonstrated to increase efficiency in the smart vehicle environments [17]. The functionalities provided by analytical layer are directly mapped with the application layer services. Thus both the layers work hand in hand, but there is a level of encapsulation maintained between these layers just like MVC architecture for easier maintenance of the system. Analytical layer would be executing various algorithms for different application layer requirements. For Example: application layer needs to propagate the route information to the emergency vehicles for fastest possible commute to the location of emergency. This further requires analysis of traffic densities at different road segments, real-time information about any possible road blockages due to accidents/herds of cattle, ongoing road maintenance work, religious processions/parades, etc. The raw information sensed from the physical layer would be transmitted via network layer to the edge/cloud devices for storage. Analytical layer would trigger functional units based upon the request received from the application layer.

9.2.4 Application Layer

The final layer of the ITS architecture is the application layer. This layer represents the different applications of ITS that need to be acted upon. The various types of domains are Road Maintenance Tasks, Road Blockage Clearance Tasks, Traveller Safety Tasks, Emergency Vehicle Lane Clearance, Traffic Light Management, Public Transportation Management, etc., under which a plethora of applications exists. The application layer tasks are shown with respect to the frequency of data collection, permissible latency, and processing unit being used for both emergency vehicles and normal vehicles as shown in Table 9.1. The processing unit could either be Fog device, Edge device, or Cloud device. The permissible latency for different applications could be near real time, very low, low, medium, or high. The frequency of data collection from the sensors could be per second, per minute, per hour, per day, and so on as per the need of the analysis to be performed. The different types of

Table 9.1 Major applications of ITS

Application	Vehicle type	Data type: Processing unit	Frequency: Processing unit	Permissible latency: Processing unit
Route selection	(i) For emergency vehicle	Text: Fog	Per second: Fog	Near real time: Fog
		Image: Edge	Per minute: Fog	Very low: Fog
			Per hour: Edge	Low: Edge
		Video: Cloud	Per day: Cloud	Medium: Edge
				High: Cloud
	(ii) For normal vehicle	Text: Fog	Per second: Fog	Near real time: Edge
		Image: Edge	Per minute: Edge	Very low: Edge
			Per hour: Edge	Low: Cloud
		Video: Cloud	Per day: Cloud	Medium: Cloud
				High: Cloud
Density	(i) For emergency vehicle	Text: Fog	Per second: Fog	Near real time: Fog
		Image: Edge	Per minute: Fog	Very low: Fog
			Per hour: Edge	Low: Edge
		Video: Cloud	Per day: Cloud	Medium: Edge
				High: Cloud
	(ii) For normal vehicle	Text: Fog	Per second: Fog	Near real time: Edge
		w2Image: Edge	Per minute: Edge	Very low: Edge
			Per hour: Edge	Low: Cloud
		Video: Cloud	Per day: Cloud	Medium: Cloud
				High: Cloud
Weather forecasting alerts	(i) For emergency vehicle	Text: Fog	Per second: Fog	Near real time: Fog
		Image: Edge	Per minute: Fog	Very low: Fog
			Per hour: Edge	Low: Edge
		Video: Cloud	Per day: Cloud	Medium: Edge
				High: Cloud
	(ii) For normal vehicle	Text: Fog	Per second: Fog	Near real time: Edge
		Image: Edge	Per minute: Edge	Very low: Edge
			Per hour: Edge	Low: Cloud
		Video: Cloud	Per day: Cloud	Medium: Cloud
				High: Cloud

(continued)

Table 9.1 (continued)

Application	Vehicle type	Data type: Processing unit	Frequency: Processing unit	Permissible latency: Processing unit
Accident details	(i) For emergency vehicle	Text: Fog	Per second: Fog	Near real time: Fog
		Image: Edge	Per minute: Fog	Very low: Fog
			Per hour: Edge	Low: Edge
		Video: Cloud	Per day: Cloud	Medium: Edge
				High: Cloud
	(ii) For normal vehicle	Text: Fog	Per second: Fog	Near real time: Edge
		Image: Edge	Per minute: Edge	Very low: Edge
			Per hour: Edge	Low: Cloud
		Video: Cloud	Per day: Cloud	Medium: Cloud
				High: Cloud

data being produced by sensors that would be required to be saved on the temporary or persistent storage devices could be binary, text-based, image-based, video-based, audio-based, and binary large object-based. All the unit values depend on whether the application is required by an emergency vehicle or a normal vehicle. Hence, the values of all the attributes are being explained for both.

9.3 Proposed Scheme

This chapter proposes a scheme for emergency vehicles lane clearance and multiple emergency vehicles route intersection avoidance. The proposed scheme consists of different phases.

9.3.1 Centralised Light-Weight Reporting

The proposed scheme focuses on providing priority to the emergency vehicles in ITS, i.e., ambulance, fire brigade, and police as compared to public vehicles and private vehicles [18]. The reason for prioritising emergency vehicle over other vehicles is the reason that saving lives always have higher priority as compared to poorer public/private transportation performance. The first step for achieving this objective is to ensure easier and effective reporting. The data type of the report should also be as light as possible. The scheme proposes the use of PON data type as it is evaluated to be a very light data interchange format [19]. Moreover, the reports

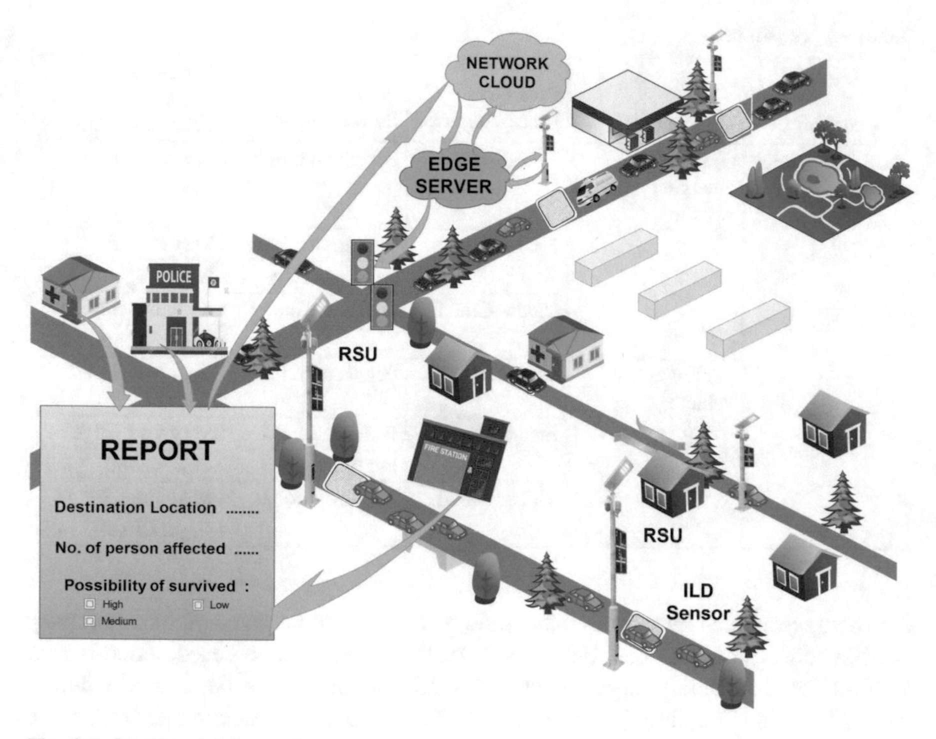

Fig. 9.3 Light-weight reporting

are also proposed to keep a minimum number of attributes. The attributes selected in the approach are destination coordinates, a number of persons affected, and severity of incident. This report is filled by the emergency stations where actual incident is reported. The hands-on information is provided to immediate response team following which the data will be entered by data entry operator at the emergency stations and the updated information will be transferred to immediate response teams as soon as it is synced to cloud device. The application layer executes different required algorithms and dispatches the information to response team for actions.

Figure 9.3 shows how a light-weight report is transferred to a centralised repository.

9.3.2 Real-Time Route Information Dissemination to Emergency Vehicles

The information shared at time instance t with response team is based on the predictive analytics performed on the date received till t-x where x is the minimum taken for processing the data received and generating results. The selection of route for transit of emergency vehicles is made by feeding the above-mentioned data in the Dijkstra's algorithm. The predictions made using models with high accuracy can falter due to the fact that the traffic is highly volatile in nature. Thus information received from RSU lying on all the candidate routes also needs to be transmitted continuously to cloud. In case connectivity of RSU to the cloud fails, the RFID sensors installed at the RSU will detect the near emergency vehicle in lane through RFID that is embedded in vehicles and pass on the information to all the neighbouring RSS for emergency lane clearance as the information related to the updated route is not amongst the RSUs. Similarly if the connectivity of emergency vehicle with the cloud fails, then the RSU will feed the recent most information received from the cloud to them. This information will continuously be processed for generation of routes from emergency vehicle's road segment's starting point/ending point to destination. The ILD detects the number of vehicles passing through it and submits the information to the RSU. The RSU lying on the selected route will continuously update the road density to cloud and broadcast information to the normal vehicles about clearing the emergency lane for the incoming emergency vehicle. The RSU nearest to the traffic light will keep on updating the recent most information about the emergency vehicle position to the traffic light controller. As soon as the emergency vehicle enters the last road segment before the traffic lights; the traffic lights for the lane would be turned to green and will be kept green till the vehicle passes the traffic light. The intelligent traffic light system that works on the density parameter of the different lanes will continue to work as proposed in the literature but will be overridden in the case of an approaching emergency vehicle [20].

9.3.3 Multiple Emergency Vehicles Intersection Crossing

Emergency vehicles include fire brigades, ambulances, and police cars. For single emergency vehicle, network cloud gives priority on the intersection of the road to turn on the green light on ITLS till the emergency vehicle crosses the traffic light. The third phase of the proposed scheme highlights the fact that it is not always possible that multiple emergency vehicles would not need to cross the intersection in a short difference of time span. In such cases, a mechanism needs to be developed for taking the decision about how to ensure that the maximum number of people in emergency situation could be saved. For multiple emergency vehicle approaching traffic signal from multiple directions at same time, then the selection of the traffic lane to be given prioritised green traffic signal, the below equation is to be used, where x is the intersection crossing impact,

$$f(x) = (a * .3) * (b * .7), \tag{9.1}$$

a = the number of persons affected and b = incident severity. The weightage assigned to the incident severity is high, and this is due to the below mentioned reasons. In case the number of persons involved is very large but the incident severity is very low, the overall criticality of the event is considerably low. Similarly even if the number of persons involved is low but incident severity is very high, then the criticality of the event is considerably high as the untimely reaching of emergency vehicle may lead to loss of lives.

Another case of multiple emergency vehicles reaching the intersection point could be that the emergency vehicles are coming from opposite directions and need to go straight ahead. Thus, in such case strict straight green traffic light signal could be turned on for both the sides.

Final case could be that multiple emergency vehicles are coming from all the sides of the intersection. In such scenario, collective intersection crossing impact of different combinations needs to be evaluated such that intersection crossing impact is maximised and the emergency vehicles crossing the intersection do not collide with each other.

9.4 Conclusion and Future Scope

This chapter discusses a four-layered architecture for ITS as well as proposes a scheme for implementation of ITLS while considering various parameters related to the emergency vehicles and emergency situations. The proposed scheme suggests the usage of PON data notation for light-weight communication of reporting data and explains the various factors related to the need of real-time route evaluation and dissemination for emergency vehicles. The scheme takes into consideration that either emergency vehicle may lose connection to cloud/edge or RSU may lose connection to the cloud/edge and proposes a solution when only one of them fails. The proposed scheme needs to be evaluated by using real-time data or simulated data for the validation of the benefits proposed. Furthermore, research may be carry-forwarded for offline–online sync-based information dissemination mechanisms in the smart cities.

References

1. Chaudhary, R., Jindal, A., Aujla, G. S., Aggarwal, S., Kumar, N., & Raymond Choo, K.-K. (2019). BEST: Blockchain-based secure energy trading in SDN-enabled intelligent transportation system. *Computers & Security, 85*, 288–299.
2. Aujla, G. S., Chaudhary, R., Kumar, N., Rodrigues, J. J. P. C., & Vinel, A. (2017). Data offloading in 5G-enabled software-defined vehicular networks: A Stackelberg-game-based approach. *IEEE Communications Magazine, 55*(8), 100–108.
3. Jindal, A., Aujla, G. S., & Kumar, N. (2019). Survivor: A blockchain based edge-as-a-service framework for secure energy trading in SDN-enabled vehicle-to-grid environment. *Computer Networks, 153*, 36–48.
4. Jindal, A., Aujla, G. S., Kumar, N., Chaudhary, R., Obaidat, M. S., & You, I. (2018). Sedative: SDN-enabled deep learning architecture for network traffic control in vehicular cyber-physical systems. *IEEE Network, 32*(6), 66–73.

5. Aujla, G. S., Singh, A., Singh, M., Sharma, S., Kumar, N., & Choo, K.-K. R. (2020). BloCkEd: Blockchain-based secure data processing framework in edge envisioned V2X environment. *IEEE Transactions on Vehicular Technology, 69*(6), 5850–5863.
6. Singh, A., Aujla, G. S., & Bali, R. S. (2020). Intent-based network for data dissemination in software-defined vehicular edge computing. *IEEE Transactions on Intelligent Transportation Systems, 22*, 5310–5318.
7. Wen, W. (2008). A dynamic and automatic traffic light control expert system for solving the road congestion problem. *Expert Systems with Applications, 34*(4), 2370–2381.
8. Yousef, K. M., Al-Karaki, M. N., & Shatnawi, A. M. (2010). Intelligent traffic light flow control system using wireless sensors networks. *Journal of Information Science and Engineering, 26*(3), 753–768.
9. Zhou, B., Cao, J., Zeng, X., & Wu, H. (2010). Adaptive traffic light control in wireless sensor network-based intelligent transportation system. In *2010 IEEE 72nd Vehicular Technology Conference-Fall* (pp. 1–5). IEEE.
10. Guerrero-Ibáñez, J., Zeadally, S., & Contreras-Castillo, J. (2018). Sensor technologies for intelligent transportation systems. *Sensors, 18*(4), 1212.
11. Dhole, R. N., Undre, V. S., Solanki, C. R., & Pawale, S. R. (2014). Smart traffic signal using ultrasonic sensor. In *2014 International Conference on Green Computing Communication and Electrical Engineering (ICGCCEE)* (pp. 1–4). IEEE.
12. Sundar, R., Hebbar, S., & Golla, V. (2014). Implementing intelligent traffic control system for congestion control, ambulance clearance, and stolen vehicle detection. *IEEE Sensors Journal, 15*(2), 1109–1113.
13. Na, Y., Guo, Y., Fu, Q., & Yan, Y. (2015). An acoustic traffic monitoring system: Design and implementation. In *2015 IEEE 12th Intl Conf on Ubiquitous Intelligence and Computing and 2015 IEEE 12th Intl Conf on Autonomic and Trusted Computing and 2015 IEEE 15th Intl Conf on Scalable Computing and Communications and Its Associated Workshops (UIC-ATC-ScalCom)* (pp. 119–126). IEEE.
14. Pothirasan, N., & Rajasekaran, M. P. (2016). Automatic vehicle to vehicle communication and vehicle to infrastructure communication using NRF24L01 module. In *2016 International Conference on Control, Instrumentation, Communication and Computational Technologies (ICCICCT)* (pp. 400–405). IEEE.
15. Ahmed, S. A., Hussain, T. M., & Saadawi, T. N. (1994). Active and passive infrared sensors for vehicular traffic control. In *Proceedings of IEEE Vehicular Technology Conference (VTC)* (vol. 2, pp. 1393–1397).
16. Morgan, Y. L. (2010). Notes on DSRC & WAVE standards suite: Its architecture, design, and characteristics. *IEEE Communications Surveys & Tutorials, 12*(4), 504–518.
17. Garg, S., Singh, A., Kaur, K., Aujla, G. S., Batra, S., Kumar, N., & Obaidat, M. S. (2019). Edge computing-based security framework for big data analytics in VANETs. *IEEE Network, 33*(2), 72–81.
18. Tonguz, O. K., & Viriyasitavat, W. (2016). A self-organizing network approach to priority management at intersections. *IEEE Communications Magazine, 54*(6), 119–127.
19. Nordahl, M., & Magnusson, B. (2016). A lightweight data interchange format for internet of things with applications in the PalCom middleware framework. *Journal of Ambient Intelligence and Humanized Computing, 7*(4), 523–532.
20. Dangi, K., Kushwaha, M. S., & Bakthula, R. (2020). An intelligent traffic light control system based on density of traffic. In *Emerging Technology in Modelling and Graphics* (pp. 741–752). Springer.

Chapter 10
Use Case for Underwater Transportation

Muhammad Waqas, Abdul Wahid, and Muazzam A. Khan Khattak

10.1 Introduction

A major portion of the earth is covered with water and a big number of important natural resources are hidden under the water. Underwater Sensor Networks are being used widely to explore those resources and to achieve ocean monitoring, underwater surveillance, pollution monitoring and disaster in time information and prevention, oil leakage detection, and underwater military applications. The International Seabed Authority issued more than 30 exploration contracts in 2018 to 104 cover almost a million square kilometers of the Deep Ocean. Underwater sensor networks are made of non-stationary sensors, and many researchers have proposed different nodes deployment strategies to cover maximum areas using less number of sensors. Optical waves are prone to scattering and not suitable for underwater sensor networks, but we can make use of optical waves for short-range communication to save energy. The underwater environment is dynamic, so nodes are considered mobile and floating with the water current. Speed and direction of nodes may vary from node to node. Keeping the mobility of nodes in mind, routing paths in UWSNs are to be updated regularly and control messages exchanged get familiar with neighbor nodes. Because of all these factors, sensors in underwater communication consumed 100 times more energy than the sensors used in territorial networks [1, 2].

M. Waqas (✉) · A. Wahid
School of Electrical Engineering and Computer Science, National University of Sciences and Technology (NUST), Islamabad, Pakistan
e-mail: mqureshi.msit18seecs@seecs.edu.pk; abdul.wahid@seecs.edu.pk

M. A. K. Khattak
Quaid-e-Azam University, Islamabad, Pakistan
e-mail: muazzam.khattak@qau.edu.pk

S. Garg et al. (eds.), *Intelligent Cyber-Physical Systems for Autonomous Transportation*, Internet of Things, https://doi.org/10.1007/978-3-030-92054-8_10

Dynamic environment in the water makes it difficult for any protocol to remain efficient; we have to make modifications according to deployment area. We have assumed 3D deployment of nodes to cover maximum area using less number of nodes. In 3D model sensors have an additional freedom for deployment to easily move in any direction. As in underwater environment sensors move horizontally with the current and may be slightly in vertical direction too [3]. A number of supplementary techniques (i.e., data aggregation, data fusion, and data dissemination, etc.) also exist in underwater communication to support the routing protocols. Among all the supplementary techniques, data aggregation is very useful in improving network performance by reducing the network overhead. We are using round based clustering (RBC) technique to re-cluster after a successful transmission because change in sensors residual energy and position may occur. Again we will go through all the procedure to form a cluster and then select a cluster head based on energy level of each node.

Aggregation is being performed on CH level after receiving data from source nodes and before transmitting it to sink node or any forwarding node. Some of the key aggregation techniques are shown in Fig. 10.1. Similarity function is very commonly used aggregation, which remove data redundancy and reduce data size while maintaining high accuracy. Using this supplementary technique with any routing protocol results in reducing size of data before transmission, which actually saves energy. We have divided our network into two layers, bottom layer is where nodes form clusters and upper layer is where nodes are called forwarding nodes. All the nodes in bottom layer sense data and transmit to their respective CHs [4]. Aggregation functions are performed at CH level and then CH finds the next suitable forwarder from upper layer. Forwarding nodes are selected on their distance from the sink node and on their residual energy level. If two nodes are at the distance

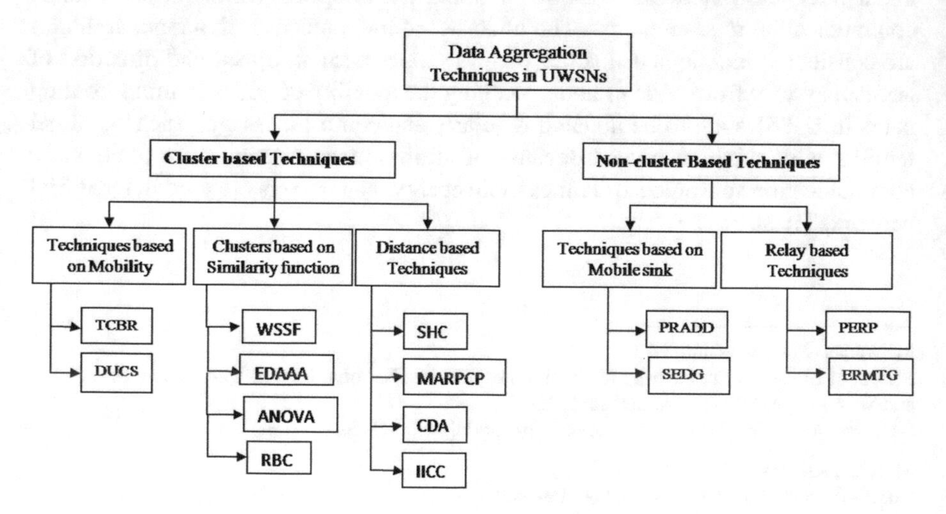

Fig. 10.1 Aggregation techniques

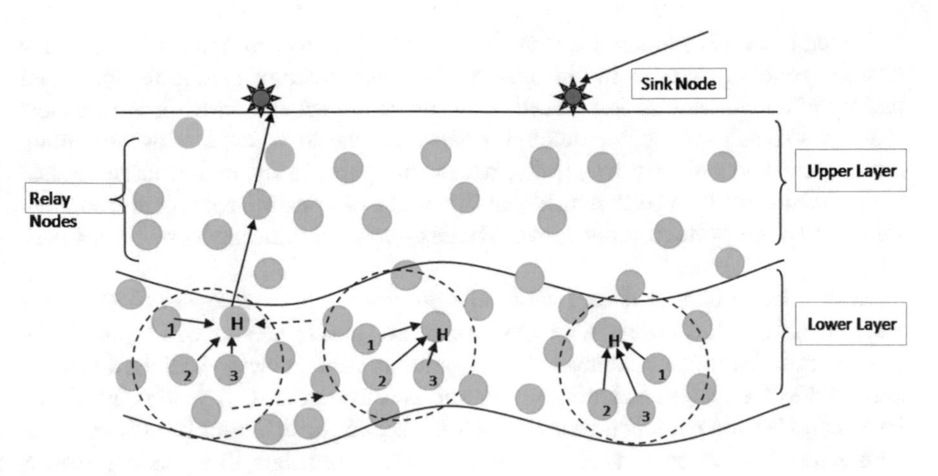

Fig. 10.2 Network structure

from sink node, then the node with high residual energy level will be selected to pass the data to sink node.

We are using Round Based Clustering (RBC) [5] and similarity based clustering technique with a little improvement. We skip neighbor and fast moving nodes at the time of cluster formation. Neighbor nodes are the nodes that came closer to each other, which may collect similar information. Hence, they are required to be skipped to minimize the similarity and to save the energy of n/2 nodes of each cluster. Fast moving nodes are the nodes moving with the water current; their movement is fast enough that they cannot stay for enough period of time to form cluster, collect, and transmit data to CH. We are using last five positions of a sensor to find its direction and to measure speed. We calculate the distance between nodes in order to find neighbor nodes to skip the nodes if distance is less than the defined threshold. Our second modification is to use different modes of communication for inter-cluster and intra-cluster for communication over long and short ranges. Network structure is shown in Fig. 10.2.

10.2 Related Work

In this section, we have reviewed a number of protocols with respect to their efficiency and suitability in underwater dynamic environment. In [6], the authors proposed an algorithm for UWSNs named as Energy Efficient Routing Protocol (EEDBR), which is specifically designed to deal with energy related issues. In this protocol a criteria is defined to select relay node or forwarding node. Relay node must be a node with minimum depth and maximum residual energy. Consequently, EEDBR has been proven energy efficient routing protocol.

Zhang et al. [7] proposed a new link state based protocol named as feedback adaptive routing protocol. In this protocol the mechanism of getting feedback and finding efficient route is introduced. Link detection scheme is to detect shortest route as well as the adaptive routing feedback method. In old scenarios the routing tables were updated very frequently, but in this paper, a dynamic routing update mechanism is introduced to reduce that frequency of updating routing information. Resultantly, this protocol reduces overall energy consumption and increases network life.

In [8], the authors explained about this proposed scheme known as RBS. RBS works in rounds. It executes in various phases, i.e., cluster head selection, create the cluster, initialization of scheme, and aggregation of data. In phase-I, all the nodes are deployed in the system including sensor and sink nodes. Every round is initialized by Sink node for a particular time interval. In second phase cluster head is selected. Cluster head is selected based on some specific parameters like residual energy, location coordinates, and distance from the sink node. After the selection of cluster head, clusters are formed based on geographical region in third phase. Third phase is data aggregation phase where data is aggregated and aggregated data is passed to the sink node. After data transfer, Re-clustering phase occurs. After each data transfer that is end of each round, clusters are reconstructed. New cluster head will do re-clustering whenever any fluctuations occur in system circumstances due to energy depletion, movement in network, etc. Re-clustering prolongs the lifetime of network.

Zonglin et al. [9] proposed a routing protocol "Relative Distance Based Forwarding (RDBF)," which is efficiently saving energy and minimizing delay. Its focus is on forwarding nodes to find a way to minimize the forwarding process. Hence fitness factor is used to find the degree of appropriateness of a forwarder node. Overall the proposed methodology reduced the energy consumption by involving fewer fractions of nodes [10]. Tariq et al. have proposed a flooding based Diagonal and Vertical Routing Protocol (DVRP) to minimize end to end delay. On energy basis flooding is made and next node forwards the nodes to sink node. This protocol also improved end to end delay and reduced energy consumption. In [11], a protocol has been proposed to minimize packets collision due to which it is called Geographical and Partial Network Coding (GPNC) based routing protocol. Simulation and results evaluation proved that GPNC achieved the desired improvement in the form of efficient delivery.

Four suitable functions are similar to underwater sensor networks data aggregation technique. After examination and associating of functions, the results gave the outstanding features of underwater sensor networks. In UWSN cosine distance and Euclidean distance are more appropriate on collected data. All former works focused on data aggregation; they only discussed existing techniques of data aggregation in UWSN & WSN.

10.3 Proposed Scheme

Logic behind our proposed node skipping technique is to conserve energy and prolong network lifetime. We will briefly explain it with respect to energy efficiency. We have set a threshold time that normally takes place to form a cluster, collect data, and pass it to cluster head. We have also gathered the speed of moving nodes and compared them with the threshold speed.

10.3.1 Network Architecture

As mentioned earlier, we are following RBC protocol and assuming the nodes are deployed in 3D. All sink node sensors deployed at water surface are equipped with both radio antenna to receive data from underwater sensors and acoustic modems to pass it to on-shore stations. Underwater sensor nodes deployed in 3D are also equipped with two types of modems for optical and acoustic communication. Each of them could be a data source or relay node to pass data to sink node. Using radio antennas all the sink nodes efficiently may communicate with other nodes and with the base stations on land. Furthermore, in our case it is supposed that every deployed sensor under the water is equipped with depth sensor and knows its exact depth from the sink node on surface. We have divided network into two layers, bottom layer is to form clusters and source data and upper layer is to find best suitable forwarder to forward that data to sink node. Residual energy level of each sensor is being changed continuously.

10.3.2 Depth Threshold

We will follow the depth threshold from Depth Based Routing (DBR). In [12] they have introduced a new parameter named as depth threshold dth, and it is a value we initially set to find the relay or forwarding nodes. Any node will only select the next node as forwarder if the difference between two hops is greater than the threshold value. If resultant value is equivalent to threshold, then residual energy will be the parameter to get the result.

10.3.3 Cluster Formation

Clustering is a concept in which we divide our network into small portions of nodes called clusters. Each cluster may consist of only one cluster head (CH) and a number of cluster nodes. Clustering technique is used to stabilize and precise a big network. Each node is equipped with two types of modems to communicate within clusters and between clusters. Nodes in a cluster may be the source nodes, which actually collect data and pass it to their respective CHs, which are later transmitted to sink nodes on surface. In our proposed scheme we are following RBC model to form clusters again upon any change in the network. To form a cluster formation, we have designed an algorithm to select cluster head. Algorithm is designed to form the entire cluster. High data rate transmission from source nodes consumes more energy, create delay, and increase data redundancy. In order to counter these issues, we are using data aggregation techniques on CH level to reduce size of data that is to be transmitted to sink node and then to base stations on land. These techniques are commonly used in support of routing protocols to conserve energy, reduce data redundancy and delay rates, etc. CH should be the strongest node and closest to sink node, because aggregation techniques are performed on CH level. It is the responsibility of a CH to collect data and perform aggregation and further pass the data to sink node. All these processes consume a lot of energy, and if a node once acts as CH in first round, its chances of being selected for CH again in second round are very low.

We are considering the following two basic parameters to select CH:

- Within a cluster, select a node that has less depth among all nodes.
- Compare residual energy of all nodes and if more than one node have same energy level, we will compare the depth of those nodes.

The cluster formation process is shown in below provided Algorithms.

Algorithm 1 Cluster formation

Sn : Sensor Node
for each Sn **do** create an empty cluster Ck
    for m=1; m<CH; m++ **do** CH broadcast cluster join message to Sn
    end for
    for m=1; m<=num; m++ **do** Distance cost = sensor node to CH distance + distance from CH to sink
        if distance cost < previous distance cost **then** Si join with CH by sending reply message

Algorithm 2 Cluster head selection

Node = $N_i; Cluster = C_k; Depth = D$
for each S **do** $N_i is the first node member of Ck$
    for each N_i **do**
find depth and residual energy of N_i
        if $D(ij) > within threshold area$ **then** $N_i is a member of C_k$

10.3.4 Skipping Nodes

We have designed Algorithm to skip unsuitable nodes at the time of cluster formation. At first step we set a threshold distance Dth between any two nodes of a cluster, which means if any two nodes came closer enough and distance reduced to less than threshold, we have to skip one node. We have considered following parameters to evaluate for a node to add in a cluster or skip:

- No node should exist within threshold distance Dth.
- If any second node is there with that specific range, its residual energy ER should be higher than that node.
- Moving speed of the node should not be greater than the threshold speed Sth.

Algorithm 3 Skipping fast moving and close neighbor nodes

Node=N_i; $Cluster = C_k; Distance = D; ResidualEnergy = ER$;
Threshold Distance=D_{th}
for each Sn **do** create an empty cluster C_k
$S_i$is the first node member of C_k
  for each node Ni **do** Measure distance D from Ni to $N_{i+1}$and$NitoN_{i-1}$
    if Di,i+1>Dth Di,i+1<Dth **then**
Compare residual energy ER of $N_i toN_{i+1} andN_i toN_{i-1}$
      if Ei>Ei+1 Ei>Ei-1 **then**
add $N_i toC_k$
else skip N_{i+1}

10.3.5 Inter-Cluster and Intra-Cluster Communication

Acoustic method is used for short distance underwater communication that consumes less energy, so we will use acoustic method to communicate within cluster to send data from nodes to cluster head. After applying similarity functions of aggregation, we will use optical method to forward data from CH to sink node, as optical method is suitable for long distance communication, but it consumes more energy. Energy of CH will drain fast because of sending data to sink node on a long distance using optical method. Table 10.1 shows different properties for Inter-Cluster and Intra-Cluster Communication.

Table 10.1 Comparison of communication medium in UWSNs

Parameter	Acoustic	Optical
Mode	Omni-directional	Directional
Propagation speed	1500 m/s	2.55 × 108 m/s
Data rate	Low data rate for long distance	Large data for small distance only
Energy consumption	Not efficient	Energy efficient

10.3.6 Aggregation Techniques

Aggregation is the method to remove redundant data while maintaining data quality, and there are many aggregation techniques being used. IDACB is used to avoid collision and to reduce data duplication.

10.3.6.1 Flow Chart

Figure 10.3 is the flow chart of our proposed scheme. Initially sink node will broadcast a control message to get the information of all nodes. In reply to that control message all nodes will reply with their depth information, residual energy level, and their movement speed. The network is divided logically into two layers; upper layer contains the relay nodes and lower layer is the layer of source node. In lower layer a number of cluster heads are selected initially to form a cluster around. Source nodes collect data and pass to respective CH, which perform aggregation function and find a relay/ forwarding node based on high energy level and closer to sink. In the next round, process of cluster formation and CH selection is repeated. After each successful data transmission, energy level of CH and relay node is decreased, and the same nodes cannot be selected in the next round consecutively.

In Fig. 10.4 we have shown the flow of node skipping. Initially we broadcast a message to form a cluster and get the information of nearby nodes. Based on the information we will calculate distance of a node from all its neighbor nodes. If it is very close to any other node, then we will compare their residual energy level to decide which of them to skip from this cluster. Secondly, we measure movement speed of each node at the time of adding them to cluster. If a node is moving fast enough that it cannot stay longer in a cluster to collect data and pass it to CH, it will be skipped to conserve energy. The same steps will be followed in each round of cluster formation.

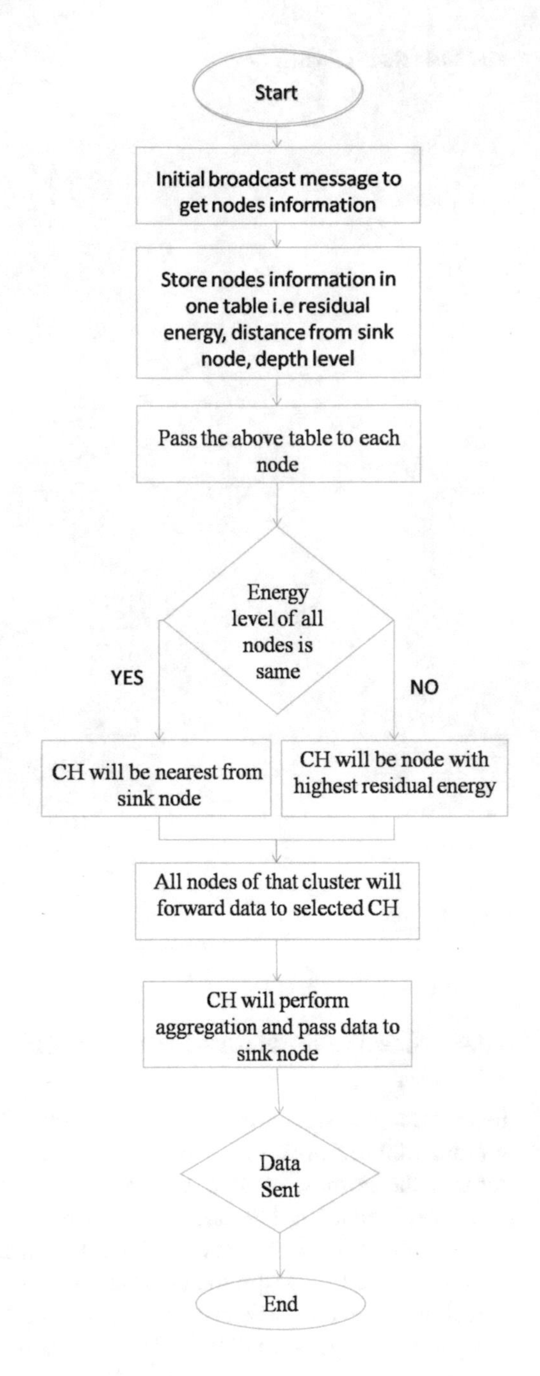

Fig. 10.3 Cluster formation

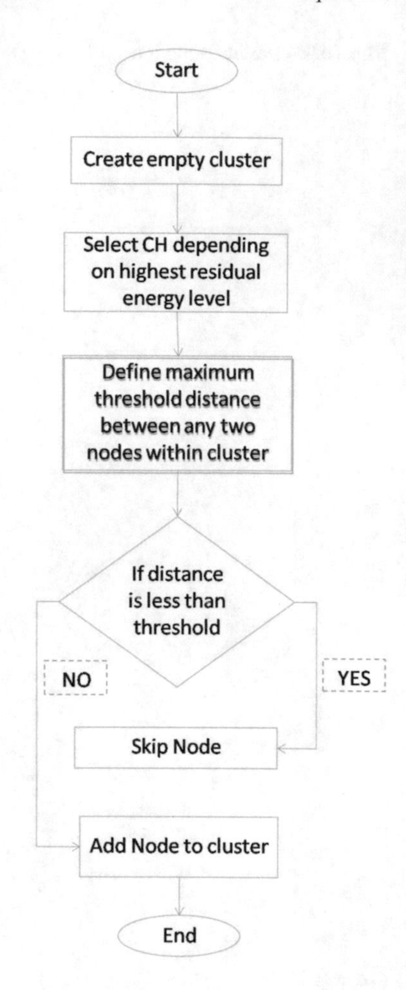

Fig. 10.4 Node skipping

10.4 Conclusions and Future Work

In this chapter we have presented a protocol which is energy efficient, and it prolongs the network life. We have measured movement speed of nodes before forming the clusters. We took into account two major parameters of nodes at the time of cluster formation, i.e., node residual energy, currently moving speed, distance from neighbor nodes. We have defined threshold distance from neighbor nodes and threshold speed for fast moving nodes. If any node comes closer enough to other neighbor nodes and their distance minimized to threshold, one of them would be skipped from cluster. In case of fast moving node, if a node is moving fast enough and crossed speed threshold, that node would not be considered suitable for cluster formation.

References

1. Ghoreyshi, S. M., Shahrabi, A., & Boutaleb, T. (2017) Void-handling techniques for routing protocols in underwater sensor networks: Survey and challenges. *IEEE Communications Surveys & Tutorials, 19*, 800–827.
2. Lurton, X. (2010). *An introduction to underwater acoustic principles and applications*. Springer
3. Oh, S. H., & Tran, K. T. M. (2013). A comparative analysis of similarity functions of data aggregation for underwater wireless sensor networks. *International Journal of Digital Content Technology and Its Applications, 7*(2), 830.
4. Liu, G., & Wei, C. (2011). A new multi-path routing protocol based on cluster for underwater acoustic sensor networks. In *IEEE International Conference on Multimedia Technology (ICMT)* (pp.91–94), China.
5. Tran, K. T.-M., & Oh, S.-H. (2014). UWSNs: A round-based clustering scheme for data redundancy resolve. *International Journal of Distributed Sensor Networks, 10*, Art. ID 383912, 6.
6. Tran, K. T.-M., Oh, S.-H., & Byun, J.-Y. (2013). Well-suited similarity functions for data aggregation in cluster-based underwater wireless sensor networks. *International Journal of Distributed Sensor Networks, 9*, Art. ID 645243, 7.
7. Zhang, S., Li, D., & Chen, J. (2013). A link-state based adaptive feedback routing for underwater acoustic sensor networks. *IEEE Sensors Journal, 13*(11), 4402–4412.
8. Han, L. I. (2010). An energy efficient routing algorithm for heterogeneous wireless sensor networks. *Proceedings of the International Conference on Computer Application and System Modeling (ICCASM 2010)* (pp. V3-612–V3-616). IEEE.
9. Nowsheen, N., Karmakar, G., Kamruzzaman, J. (2016). PRADD: a path reliability aware data delivery protocol for under water acoustic sensor networks. *Journal of Network and Computer Applications, 75*, 385–397.
10. Tran, K. T. M., Oh, S. H., & Byun, J. Y. (2013). An efficient data aggregation approach for underwater wireless sensor networks. In *ICCA 2013, ASTL 2014* (Vol. 24, pp. 46–48).
11. Gholami, E., Rahmani, A. M., & Fooladi, M. D. T. (2015). Adaptive and distributed TDMA scheduling protocol for wireless sensor networks. *Wireless Personal Communications, 80*(3), 947–969.
12. Hong, L., Hong, F., Yang, B., & Guo, Z. (2013). ROSS: receiver oriented sleep scheduling for underwater sensor networks. In *Proceedings of the 8th ACM International Conference on Underwater Networks and Systems*.

Chapter 11
Advanced Signal Processing for Autonomous Transportation Big Data

Haibin Lv, Dongliang Chen, Jinkang Guo, and Zhihan Lv

11.1 Introduction

With the rapid development of information technology, the Internet, big data, and cloud computing are gradually penetrating various areas of life and are gradually integrating with or replacing traditional lifestyles. Since its introduction, "Industry 4.0" has attracted widespread attention from various countries. For example, France's "New Industry" strategy and the USA's "Advanced Manufacturing" plan are all positive responses to "Industry 4.0" [1]. In response to "Industry 4.0," China also proposed the "China Manufacturing 2025" strategy in 2015, which has enabled China's industrial reality to continuously change from "Made in China" to "Manufactured in China" [2]. Therefore, to achieve the rapid development and rapid transformation of industrialization, the efficient collection, transmission, and processing of a large amount of industrial data have become a key part of the industry, which has also attracted the attention of scientific researchers in relevant fields.

In the Internet era, mutual assistance between traditional manufacturing and Internet companies has become the new normality. For example, in the field of home

This work was supported in part by the National Natural Science Foundation of China (No. 61902203) and Key Research and Development Plan—Major Scientific and Technological Innovation Projects of ShanDong Province (2019JZZY020101).

H. Lv
North China Sea Offshore Engineering Survey Institute, Ministry of Natural Resources North Sea Bureau, Qingdao, China

D. Chen · J. Guo · Z. Lv (✉)
College of Computer Science and Technology, Qingdao University, Qingdao, China
e-mail: cdlord@qq.com; 2020025951@qdu.edu.cn

S. Garg et al. (eds.), *Intelligent Cyber-Physical Systems for Autonomous Transportation*, Internet of Things, https://doi.org/10.1007/978-3-030-92054-8_11

appliances, enterprises such as Midea have collaborated with Xiaomi, JD.com, and Dangdang.com to realize the smart home and channel sharing strategy. In the automotive field, Dongfeng, Changan, and Huawei cooperate to promote the continuous transformation of "functional cars" to "smart cars." All these examples show the progress of scientific information technology [3]. The field of industrial production is always in contact with data, such as temperature data and humidity data received by sensors. Therefore, factories can be regarded as the typical big data systems, and they are the specific applications of big data in the industry [4, 5]. Industrial big data usually comes from three sources: first, internal business data related to the enterprise; second, various types of data collected by the equipment in the enterprise under the state of information interconnection; and external data, i.e., the external Internet data related to the production and operation of the enterprise [6]. Industrial big data technology, as one of the key technologies of intelligent manufacturing, is mainly to promote the two-way communication between the virtual information world and the real physical world, thereby promoting the traditional manufacturing industry to enter the modern manufacturing service industry [7].

The core and symbol of the big data signal processing technology is the digital signal processor. Generally, traditional signal processing mainly includes three methods: time-domain analysis, frequency-domain analysis, and time-frequency analysis. The platform often cannot meet the real-time processing of large amounts of data due to the low computing capacity of the main processor and the data throughput rate [8, 9]. With the upgrade of the signal processor, it can perform a series of tasks such as high-speed data distribution, access, real-time signal processing, and large-scale data exchange [10].

In summary, with the rapid development of scientific information technology, the emergence of industrial big data has become inevitable. To make industrial big data play a greater role, this study provides an experimental basis for later industrial production by constructing an advanced signal processing system for industrial big data and analyzing the performance of Kafka clusters appearing in this study.

11.2 Related Work

At present, the rapid development of information science and technology has greatly affected the changes in the industrial field. The emergence of industrial big data is the most obvious change, and many researchers have studied it. Yan et al. [11] proposed a new multi-source heterogeneous information structure framework, which considered the spatiotemporal characteristics to describe structured data and modeled invisible factors. Finally, it was found that this model can make the production process transparent and realize the predictive maintenance and energy-saving of equipment in the era of Industry 4.0, which verified the effectiveness of the model. Li et al. [12] proposed an incremental deep convolutional computation model (DCCM) algorithm. They used this algorithm to introduce tensor convolution, pool, and update rules of the fully connected layer to transfer the previous knowledge and extended the packet loss strategy to the tensor fully connected layer. In the end, it was found that the robustness of this model was improved significantly, and its adaptability, conservativeness, and convergence were verified. Wan et al. [13] established producible intelligence to cope with increasing production complexity. The production process of products was transformed to a deeper level. Finally, it was found that this intelligent production could cope with the booming global economy and the demand for customized products. McMahon et al. [14] used big data to manage relevant information in enterprises, and the challenges encountered during its application were analyzed.

With the rapid development of signal processing technology, it has become popular in all walks of life. Patole et al. [15] studied various aspects of automotive radar signal processing technology, including waveform design, possible radar structures, estimation algorithms, implementation of complexity-resolution trade-offs, adaptive processing in complex environments, and unique problems associated with automotive radar; the development of signal processing technology played a key role in automotive radar systems. Allwood et al. [16] expounded the progress of intestinal sound analysis by understanding the relevant content of intestinal sound signal processing and finally found that the combination of advanced sound signal processing technology and artificial intelligence could better the intestinal sound information monitoring understanding. Mishra et al. [17] used the advantages of joint radar communications (JRC) model of low cost, small size, low power consumption, spectrum sharing, improved performance, and enhanced information sharing; they found that the signal processing technology was to realize the millimeter-wave JRC system. Later, the latest advances in cognition, compressed sensing, and machine learning were used to reduce the required resources, and these resources were dynamically allocated with lower overhead, indicating that the signal processing technology had good application prospects.

In summary, through the results of the aforementioned researches, it is found that there are many studies on industrial big data technology and signal processing technology, but there are few studies on applying signal processing technology to industrial big data analysis. Therefore, this study combines advanced signal processing technology with industrial big data to analyze the industrial big data and promote the transformation and upgrading of China's traditional manufacturing industry, as well as its healthy and stable development.

11.3 Method

11.3.1 Industrial Big Data

With the rapid development of science and technology, the concept of big data has appeared in all aspects of human lives. In the industrial field, big data is divided into the broad and narrow sense. Broad industrial big data refers to production data collected during the co-malicious manufacturing process [18]. It has the characteristics of hierarchy, scenario, and time domain. The narrow sense of industrial big data refers to data that has a clear purpose in the industrial production process, mainly from manual collection and sensor collection [19, 20]. To improve the flexibility, versatility, and expandability of industrial big data collection systems, this study proposes data processing modules for data collection systems. Also, database operations are associated with specific data processing modules; therefore, they also have a pluggable characteristic. Based on the development of Kafka, data processing nodes have distributed characteristics. After a data processing node crashes, relevant data will be processed by machine data processing nodes [21]. The topology of the data collection system is shown in Fig. 11.1.

It shows that the data collection node receives the datagram sent by the sensor terminal and sends commands such as control and debugging to the sensor terminal. The data processing node processes the received data, generates response messages, control, and debugging commands, and the Kafka cluster forwards messages between these two nodes [22, 23].

11.3.2 Signal Processing Technology

The development of practical algorithms such as Fourier transform has also promoted the development of digital signal processing technology. The related signal processing algorithm is shown in Fig. 11.2 [24]. Singular value decomposition (SVD), as a very effective method of matrix decomposition, initially only decomposed the real square matrix [25–27]. Then, some scholars have extended it to the compound matrix and arbitrary dimension matrix [28]. When processing signals, because the signals are unique, and SVD is for matrices, a one-dimensional signal vector must be transformed into a two-dimensional matrix before it can be decomposed. The signal is decomposed into different vector spaces, and appropriate components are selected as required. After reconstruction, the signal can be denoised and its feature can be extracted. In addition, SVD has unique advantages over other noise processing methods. There is no phase shift in the processing results, and the waveform distortion is small and the signal-to-noise ratio is high [29].

In the signal processing algorithm, the relationship between the data volume and running time of different algorithms is shown in Fig. 11.3.

Generally, SVD has a bilinear function with the following equation:

$$f(x, y) = x^T A y \tag{11.1}$$

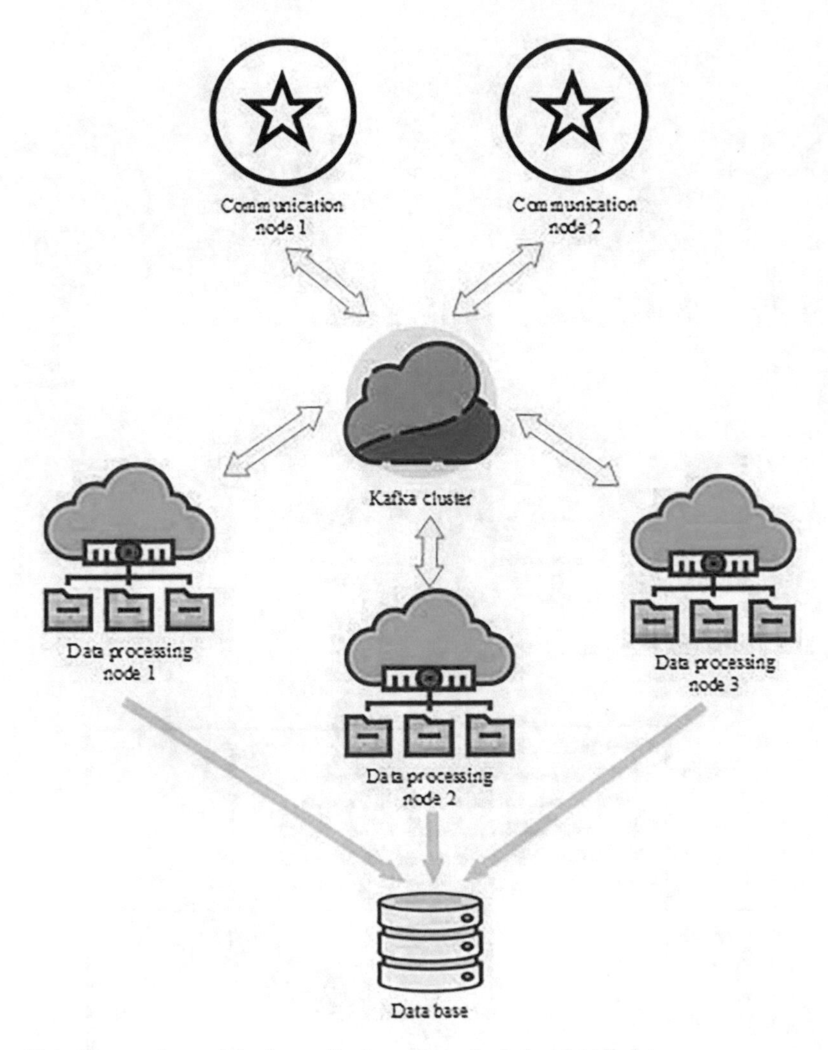

Fig. 11.1 The topology of the data collection system for industrial big data

Where, x, $y \epsilon R^{n \times 1}$, $A \epsilon R^{n \times n}$, and T represents the transpose of the matrix. By introducing a linear transformation $x = U\xi$, $y = V\eta$, $U, V \epsilon R^{n \times n}$, the above equation can be written as

$$f(x, y) = (U\xi)^T A(V\eta) = \xi^T U^T AV\eta \quad (11.2)$$

If $S = U^T AV$, then it can be written as

$$f(x, y) = \xi^T S\eta \quad (11.3)$$

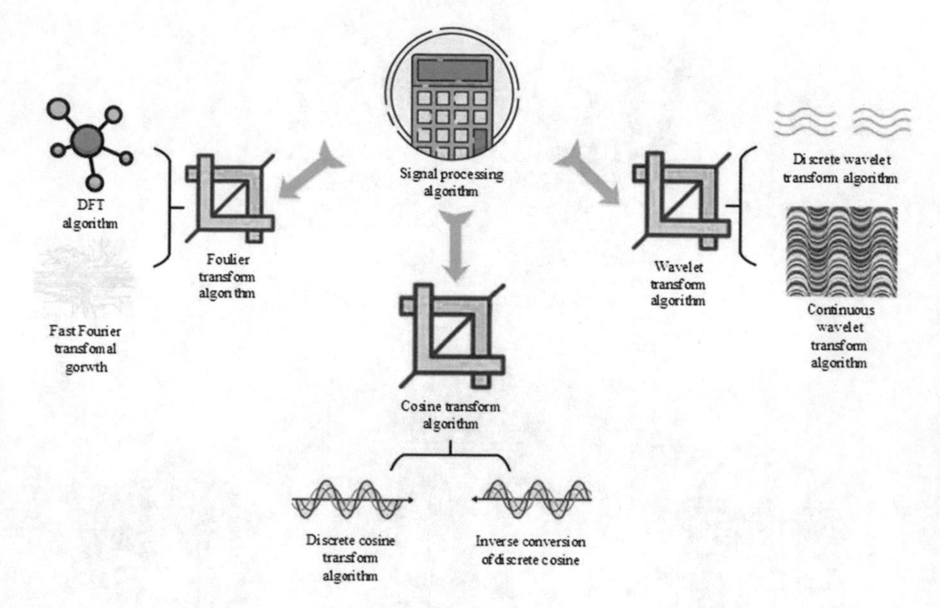

Fig. 11.2 Signal processing algorithm

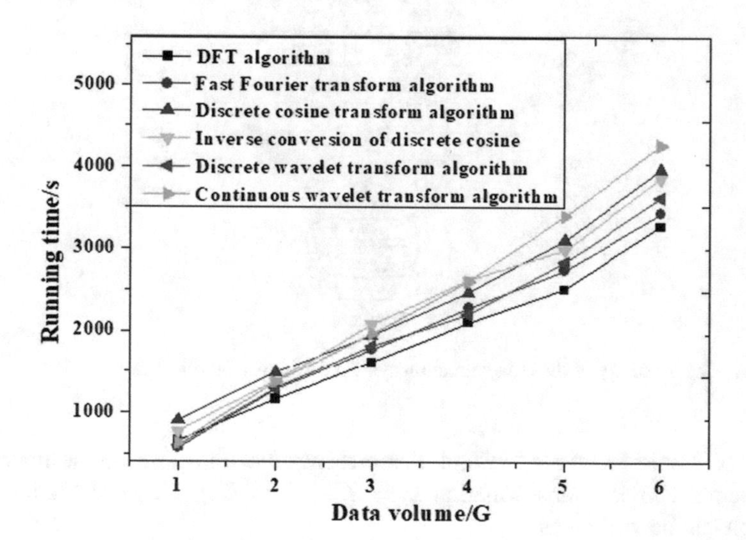

Fig. 11.3 Relationship between data volume and running time of different algorithms

Specifically, if U and V are orthogonal matrices, S will be a diagonal matrix by choosing the appropriate U and V, i.e.,

$$S = diag(\sigma_1, \sigma_2, \ldots, \sigma_n) \tag{11.4}$$

In addition, if $A, B \epsilon R^{m \times n}$, the singular values of A and B are respectively $\sigma_1 \leq \sigma_2 \leq \cdots \leq \sigma_n$ and $\tau_1 \leq \tau_2 \leq \cdots \leq \tau_n$, where, $q = min(m, n)$. Then:

$$|\sigma_i - \tau_i| \leq ||A - B||_2 \tag{11.5}$$

When matrix A has a small disturbance, its singular value change will not be greater than the L2 norm of the disturbance matrix. As inferred from the nature of the norm, the L2 norm of the disturbance matrix is small, i.e., when the matrix has a small change, the singular value is affected minimally [30]. This shows that for the matrix of signal construction, small changes in the signal will not cause large changes in singular values, which is crucial for signal denoising.

While analyzing the properties of the SVD, two types of invariance and rotation invariance are also included. In proportional invariance, if the singular value of the matrix $A \epsilon R^{m \times n}$ is $\sigma_1 \leq \sigma_2 \leq \cdots \leq \sigma_q$, and the matrix A is multiplied by a constant to get aA, the singular value will be $\sigma_1^* \leq \sigma_2^* \leq \cdots \leq \sigma_q^*$, where $q = min(m, n)$. Then, the relationship between the two satisfies the following equation:

$$|a|\sigma_i = \sigma_i^*, \quad i = 1, 2, \ldots, q \tag{11.6}$$

This equation shows that after the matrix is scaled, its singular values are also scaled by the same scale. In other words, when the matrix is scaled in the same proportion, its singular value only changes in size, and the characteristic proportion of the singular value and its changing trend remain unchanged [31].

In the analysis of rotation invariance, if there is a matrix $A \epsilon R^{m \times n}$, the singular value decomposition of its conjugate transposed matrix A^T is

$$A^T = (USV^T)^T = VS^TU^T \tag{11.7}$$

The matrices S and ST are analyzed. The elements on the first $q = min(m, n)$ diagonals are the same, i.e., A and A^T have the same singular value, which indicates that when the matrix is rotated, its singular value characteristics is constant. In the process of signal processing, SVD often includes three steps: matrix modeling and decomposition, component selection, and signal recovery [32, 33]. Since the object of singular value decomposition is a matrix, how to construct a one-dimensional time-series signal into a two-dimensional space matrix form is critical when applying SVD to signal processing. Because the component signals obtained by the Hankel matrix method do not have orthogonality, this study is introduced in this way. If there is an original ideal signal x of length $L = [x(1), x(2), \ldots, x(L)]$, and the signal is superimposed with noise $w = [w(1), w(2), \ldots, w(L)]$; then, the noisy signal $y = [y(1), y(2), \ldots, y(L)]$ can be obtained, and their relationship is

$$y = x + w \tag{11.8}$$

The Hankel matrix Y constructed from the noisy signal y is as follows:

$$Y = \begin{bmatrix} y(1) & y(2) & \ldots & y(n) \\ y(2) & y(3) & \ldots & y(n+1) \\ \cdot & \cdot & \cdot & \cdot \\ \cdot & \cdot & \cdot & \cdot \\ \cdot & \cdot & \cdot & \cdot \\ y(m) & y(m+1) & \ldots & y(L) \end{bmatrix} = X + W \tag{11.9}$$

Where: X and W are Hankel matrices composed of ideal signal x and noise w, $L = m + n - 1$, $m \leq n$. In addition, the closer the m and n are, the better the signal processing effect is. Therefore, when L is even, $m = L/2$, $n = L/2 + 1$; when L is odd, $m = (L+1)/2$, $n = (L+1)/2+1$. While selecting the component matrix, SVD decomposition is performed on the matrix Y to obtain each component matrix Yi. The Hankel matrix X constructed by the ideal signal x has only one data difference between two adjacent rows. The entire matrix is an ill-conditioned matrix. The ill-conditioned matrix is characterized by the large first singular values, and the energy is mainly distributed in the first few component matrices. The singular value of N2 tends to 0, and l is the rank of matrix X. Although the Hankel matrix W constructed by the noise w has two adjacent rows with different data, they are not related to each other. Thus, the matrix is a full-rank matrix. The autocorrelation function of noise w satisfies the following equation:

$$R_{ww}(\tau) = \begin{cases} d^2 & \tau = 0 \\ 0 & \tau \neq 0 \end{cases} \tag{11.10}$$

Where: d is the standard deviation of the noise sequence w. When the ideal noise strictly satisfies the above equation, the singular values of the constructed Hankel matrix are equal and generally smaller than the singular values of the ideal signal matrix. The singular value of the Hankel matrix Y constructed from the noisy signal sound y, the singular value $\sigma_i(X)$ of the matrix X, and the singular value $\sigma_i(W)$ of the matrix W satisfy the following relationship:

$$\sigma_i(X) \leq \sigma_i(Y) \leq \sigma_i(X) \leq \sigma_i(W) \tag{11.11}$$

The above equation shows that the singular value distribution of the Hankel matrix Y constructed with the noisy signal y is similar to the singular value distribution of the matrix X. The first r singular values are much larger than the following m-r singular values. Therefore, the first r singular values are reconstructed. If the sequence of singular values of matrix Y is $S = (\sigma_1, \sigma_2, \ldots, \sigma_m)$, the following equation is defined:

$$b_i = \sigma_i - \sigma_{i+1}, \quad i = 1, 2, \ldots, m - 1 \tag{11.12}$$

Then, the sequence $b = (b_1, b_2, b_{m-1})$ composed of b is a singular value difference spectrum sequence. According to the sequence number of the largest point of the differential spectrum sequence, the value of r can be determined. Therefore, the first r component matrices can be selected for reconstruction, and the matrix form is written as

$$\hat{Y} = U_r S_r U_r^T \tag{11.13}$$

Where: $U_r = [u_1 \ u_2 \ \ldots \ u_r] \in R^{m \times r}$ is the first r columns of the left singular matrix U. In signal recovery, a matrix of $\hat{Y}$ obtained by selecting the first r component matrices for reconstruction is actually a low-rank matrix obtained by performing rank reduction processing on the matrix Y formed by the noisy signal. Then, the simple method and the average method are used to get the component signal from the component matrix. The appropriate component signal is selected and accumulated to achieve the approximate estimation of the ideal signal [34].

11.3.3 Realization and Application of Autonomous Driving Based on Signal Processing

One of the most common applications of high-performance image processors is autonomous driver. With the development of the autonomous driving vehicle and the next generation security system, image signal processors are becoming more and more important in these applications. Figure 11.4 presents the key modules of autonomous driving technology. Assisted by autonomous driving technology, the logistics operation processes such as loading and unloading, transportation, receiving, warehousing, and transportation will gradually be unmanned and mechanized, which will promote the cost reduction and efficiency improvement, innovation, and upgrading of the entire industrial chain in the field of logistics distribution. Since the beginning of the autonomous driving industry, unmanned logistics has been a must for major enterprises, especially e-commerce express enterprises. Autonomous driving terminal distribution car can solve the above problems. It integrates hardware, software, algorithm, communication, and other technologies, which can more accurately determine the distribution location, and is not affected by the weather. It improves the timeliness of goods and protects goods from damage. Meanwhile, it also solves the problem of insufficient labor supply in logistics industry. The uniqueness of the autonomous driving terminal distribution car is that the distribution scene is characterized by "small, light, slow, and heavy," so

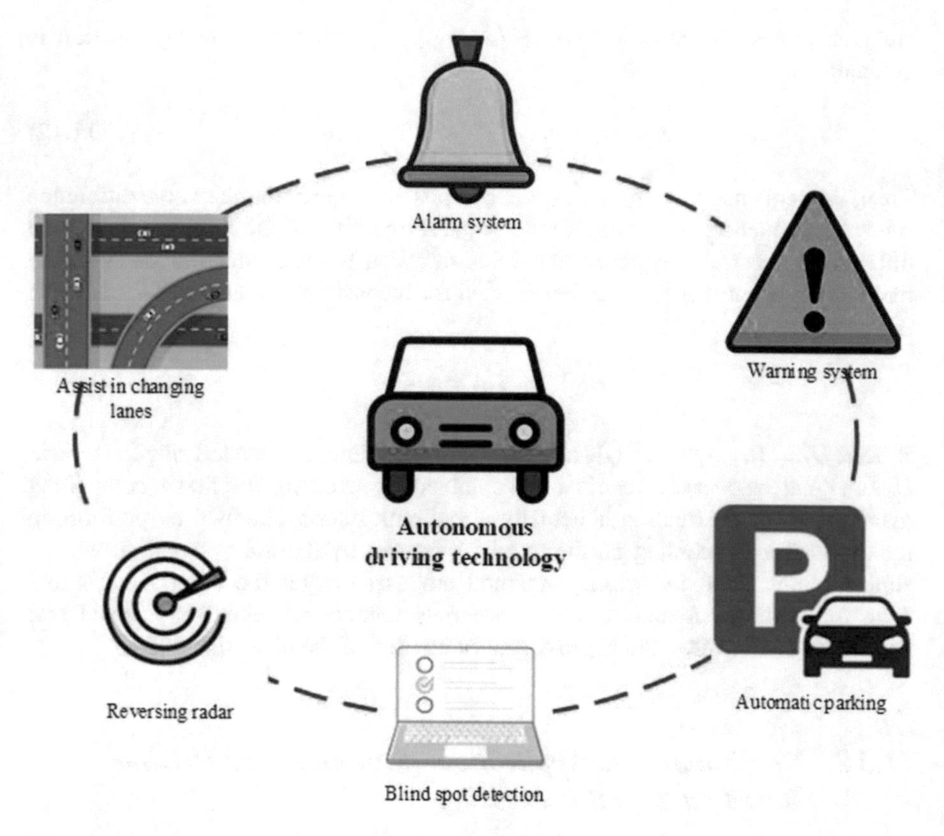

Fig. 11.4 Key modules of autonomous driving technology

the reliability requirement of the unmanned driving technology is relatively low. At this stage, in the case of unmanned driving technology not fully mature, the terminal distribution scene is relatively closed, the technical implementation is less difficult, and its market scale is considerable, which provides a solid foundation for the application of autonomous driving.

Automatic driving is also an innovation for the taxi mode. The top of the self-driving taxi is equipped with an unmanned sensor kit, which is composed of lidar, camera, millimeter-wave radar, high-precision positioning, and other sensors, with a sensing range of 360°. In the process of driving, these sensing devices scan the surrounding environment of the car body, driving, and pedestrians in real time, and display the driving track, vehicle information, pedestrian information, and traffic signals in the form of three-digit digital model on the high-definition display screen in the car, and autonomously select the optimal route and operation scheme. After entering the car, passengers need to scan the code on the mobile phone to confirm that the order is correct, then they can click the "start journey" button on the touch screen to start the experience journey. The passengers at rear row can know the route, road condition, speed, and other conditions in real time through the on-board screen in front of them. When there are vehicles and pedestrians around, the screen will be presented in the form of icons and color blocks for the passengers. When the body shakes slightly due to braking, the voice assistant on the vehicle will sound "sorry" and others prompt. According to the relevant requirements, at present, every Baidu self-driving taxi has a professional safety officer who is always concerned about the driving situation and can take over the "driving right" of the vehicle at any time.

Moreover, the application of autonomous driving technology in sweeper is also a new trend. The working environment of environmental protection is bad, the workers are insufficient, and the aging is serious, and the continuous urbanization process also puts forward more and more demands for environmental sanitation work. For the electronic and electrical architecture and sensor design, it is essential to closely around the two requirements of driverless + cleaning to collaborative design, thus improving product function and quality. According to the design idea of electronic and electrical architecture of passenger cars, the autonomous driving sweeper is divided into the following domains: autonomous driving, power system, body system, steering system, cleaning system, and information system. Among them, power, body, steering, and information belong to the support domain, and autonomous driving and cleaning belong to the functional domain. In the positioning module, the autonomous driving sweeper integrates the speed and attitude information, magnetometer, and inertial navigation technology. At the perception level, the multi-sensor fusion schemes such as 16-line lidar, visible light camera, millimeter-wave radar, ultrasonic radar, and reversing radar are selected. Compared with the traditional cleaning mode, the utilization rate of the intelligent environmental sanitation sweeper is 2–3 times higher than that of the original one. The cleaning effect can achieve accurate edge cleaning within 5 cm and achieve everywhere sweeping.

11.3.4 The Advanced Signal Processing System for Industrial Big Data

In the enterprise, to ensure the normal operation of the enterprise, a multi-processor and multi-core computing platform is required to realize the cooperative work of different processors. Figure 11.5 suggests the main architecture of industrial big data processing. It is extremely difficult to achieve synchronization between different processing cores by using a single-ended synchronization signal. The data are transmitted in a data-driven model by packaging the echo from the signal processor. In the data flow driving mode, the entire advanced signal processing system does not have a control bus, and all signal collection and parameters are packaged together with the echo data. Therefore, software command analysis and data shunting need to be implemented in software design. Since the data packet is packed according to the protocol requirements, the command and parameters come first, and the echo data follows immediately.

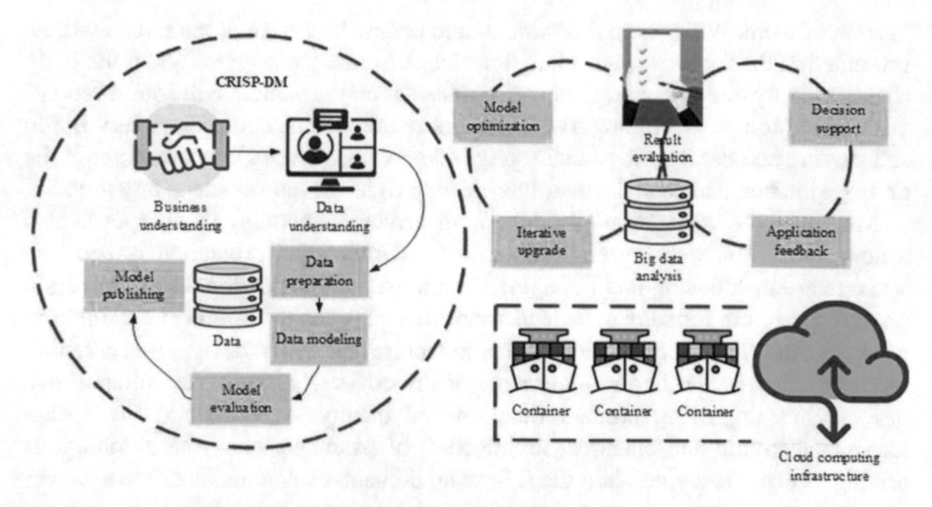

Fig. 11.5 The main architecture of industrial big data processing

In the analysis of industrial big data, to detect the dual requirements of distance and distance resolution, the advanced signal processing system needs to use a large time and wide bandwidth product signal. Besides, the pulse compression processing compresses the transmitted wide pulse signal into a narrow pulse signal. The large time-width and wide-band product of a pulse compression signal are mostly obtained by the non-linear phase modulation of the signal, such as linear frequency modulation, non-linear frequency modulation, frequency coding, and phase coding within the pulse width. Pulse compression can be performed in either the time domain or the frequency domain. The time-domain matching filtering method is equivalent to the complex correlation operation (convolution operation) between the discrete received signal and the discrete samples of the transmitted waveform. This method is simple in the circuit and easy to implement and is suitable for small compression ratios and big distance units. While in data transmission, if the collection point and the receiving point follow an exponential distribution with parameter λ in the signal transmission, the probability that information can be successfully propagated is p, and the probability of transition from state M_j to M_{j+1} is $(n-j)j\lambda\lambda$; then, the time that it stays in the state M_i is

$$t_i = (n-1)i\lambda\lambda p, \quad 1 \le i \le n-1 \tag{11.14}$$

Then, the probability density function is

$$f(t_i) = (n-i)i\lambda\lambda p^{-(n-i)i\lambda p t_i} \tag{11.15}$$

Table 11.1 Tools for model construction

Tool name	Version
Simulation platform	MATLAB
Matrix transportation	Numpy 1.12.6; Pandas 0.23.0
Programming language	Python 3.2
Development platform	PyCharm
Operating system	Linux

Thus, the time T required for the information to be transmitted to all n vehicles is

$$T = \sum_{i=1}^{n-1} t_i = \sum_{i=1}^{n-1} (n-1)i\lambda p \tag{11.16}$$

11.3.5 Simulation Analysis

This study predicts and analyzes the system on the Matlab network simulation platform. The industrial big data uses Kafka clusters, which are composed of three Kafka agent nodes and run on three Linux servers. In the test environment, even if the multi-processor is within the wireless signal transmission range, the signal cannot be successfully transmitted due to obstacles, signal interference, channel conflicts, and other factors. In the simulation, the CPU model of the computer is CORE-i7-4720HQ-2.6GHz. Matrix operations use Numpy and Pandas open-source toolkit; the Numpy library is an open-source matrix processing library, and the Pandas library provides good help for data cleaning and data preprocessing in data analysis. The specific modeling tools are shown in Table 11.1.

11.4 Results and Discussions

11.4.1 Analysis of the Transmission Accuracy of Signal Processing

In the analysis of the transmission performance of the signal processing in this study, under different successful transmission probability p, when λ is 0.05 and 0.001, the data propagation delay is shown in Figs. 11.6 and 11.7. Furthermore, the correlation between different λ values and actual propagation delay when the successful transmission probability p is 100% and 80% is analyzed, as shown in Figs. 11.8 and 11.9.

As shown in Figs. 11.6 and 11.7, when $\lambda = 0.05$ and $\lambda = 0.001$, the data propagation delay decreases as the probability of successful propagation increases, and the delay time is the shortest when the probability of successful propagation is

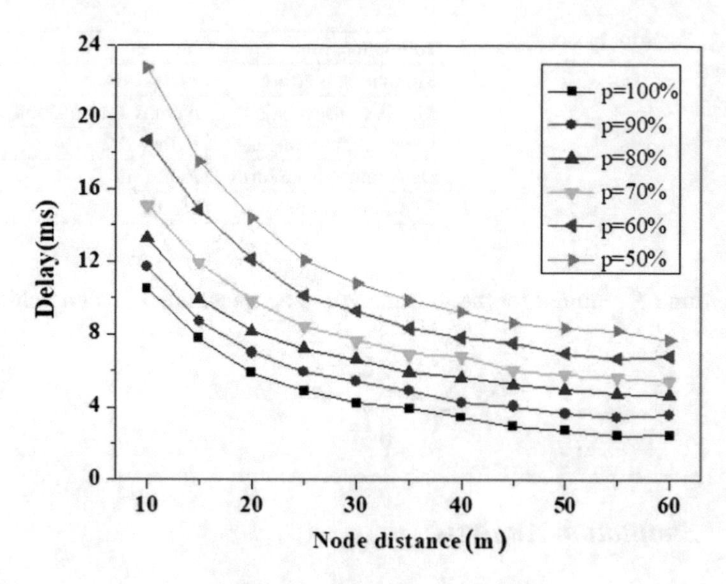

Fig. 11.6 Propagation delay with different successful transmission probability p when $\lambda = 0.05$

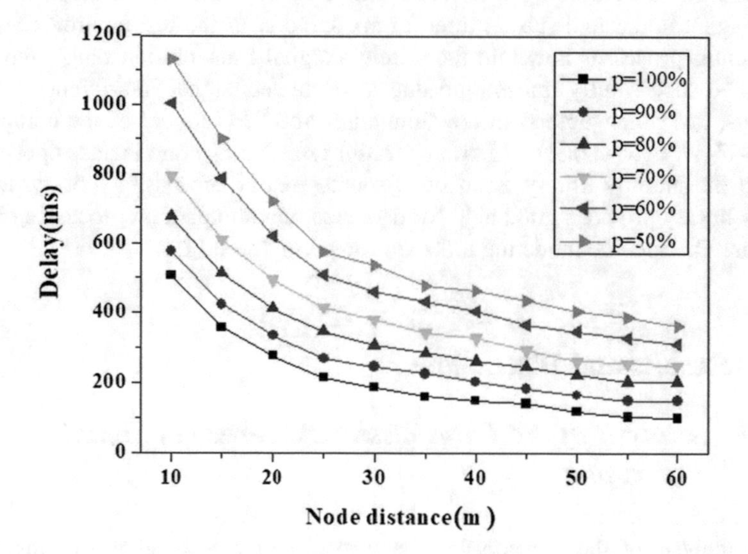

Fig. 11.7 Propagation delay with different successful transmission probability p when $\lambda = 0.001$

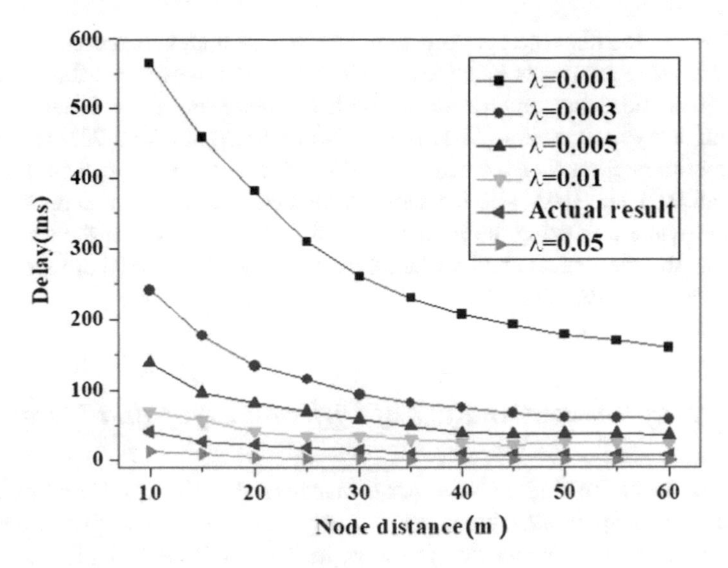

Fig. 11.8 Experimentally measured propagation delay at $p = 100\%$

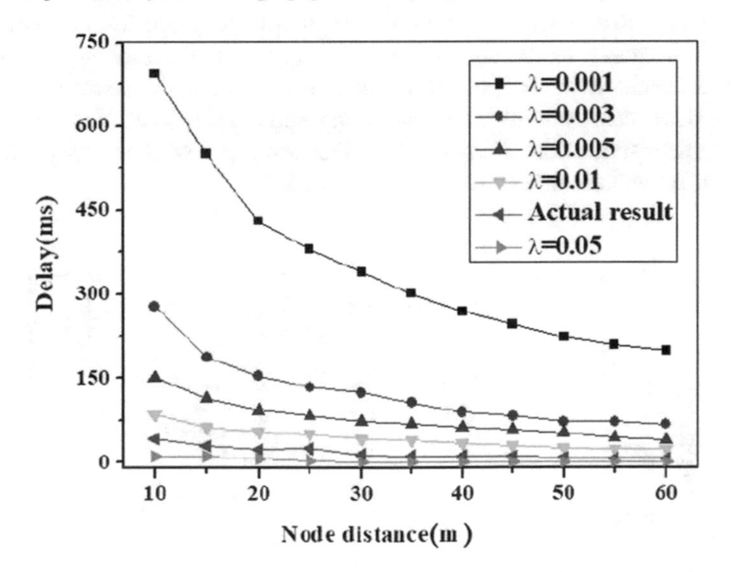

Fig. 11.9 Experimentally measured propagation delay at $p = 80\%$

100%. Therefore, in analyzing the comparison of different successful propagation probability p versus propagation delay, the higher the successful propagation probability is, the shorter the propagation delay is, and the magnitude of the λ value will not affect this correlation.

As shown in Figs. 11.8 and 11.9, when the successful propagation probability is 100% and 80%, different values of λ have different effects on the transmission delay time. As the distance between signal collection nodes increases, the delay of the same λ value continues to decrease; when the delay between different λ values is compared, the delay time decreases with the increase of the λ value, and the maximum delay time is about 570 ms and about 670 ms at $\lambda = 0.001$, respectively. The delay time approaches 0 when $\lambda = 0.05$, and it is found that when the λ value is between 0.01 and 0.05, it is the same as the actual delayed transmission result. Therefore, while analyzing the influence of different λ values on the transmission delay time, the theoretical results when the λ value is 0.01 to 0.05 are most similar to the actual transmission result.

11.4.2 Performance Analysis of Different Algorithm Libraries

In the analysis of the Kafka cluster performance of this study, it is compared with MapReduce and Spark algorithms, respectively, from the perspective of data scale and number of nodes. The results are shown in Figs. 11.10 and 11.11.

As shown in Figs. 11.10 and 11.11, from the perspective of data scale, as the data scale increases, the running time shows an upward trend; from the perspective of computing nodes, as the number of nodes increases, the running time shows a downward trend. However, the Kafka cluster requires the shortest running time in the same data scale and the same computing node. Therefore, the analysis of the system performance of this study indicates that compared with other algorithms, the Kafka cluster performance of this study is optimal.

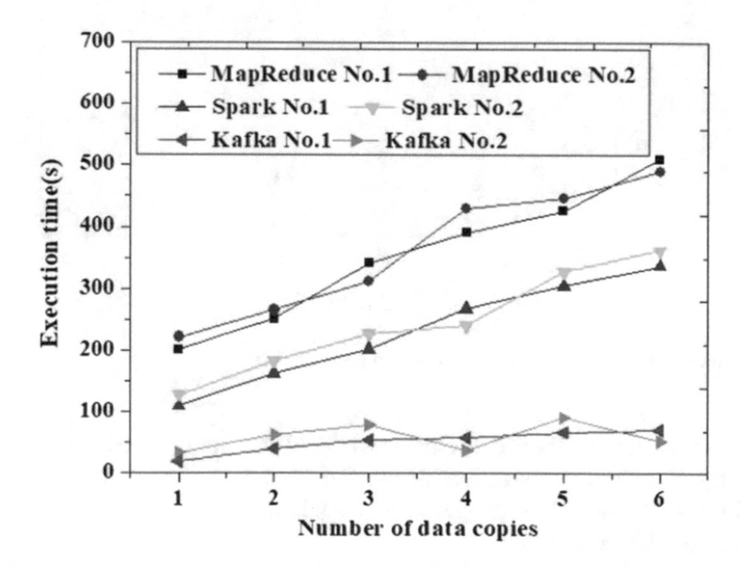

Fig. 11.10 Comparative analysis of data scale in different algorithms

11.4.3 Performance Analysis of Packet Loss Rate in the Data Signal Processing System

By comparing and analyzing the performance of the data signal processing system in this study, Figs. 11.11, 11.12, 11.13, and 11.14 are obtained, where Fig. 11.13 includes the throughput of Kafka and Fig. 11.14 shows the packet loss statistics.

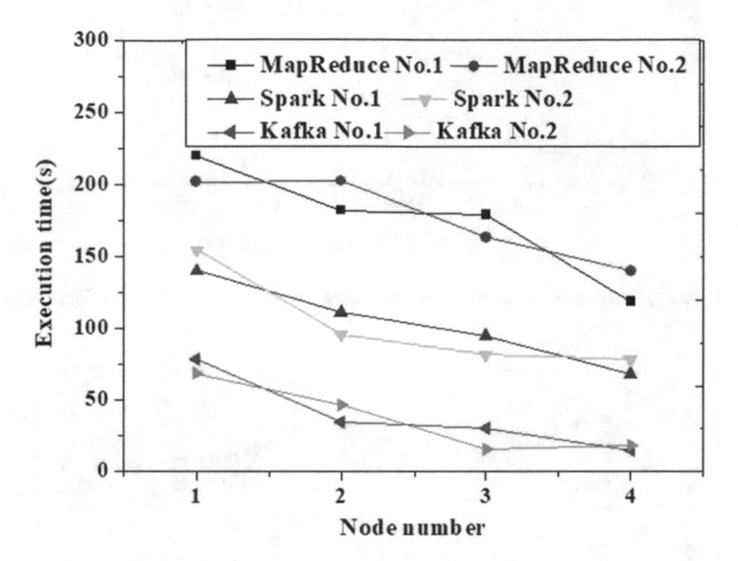

Fig. 11.11 Comparative analysis of the number of computing nodes in different algorithms

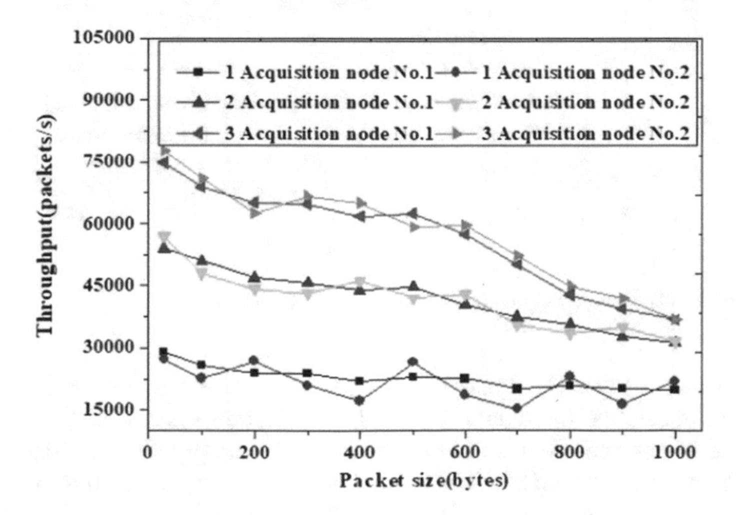

Fig. 11.12 Analysis of data collection system throughput of different nodes (number of packets per unit time)

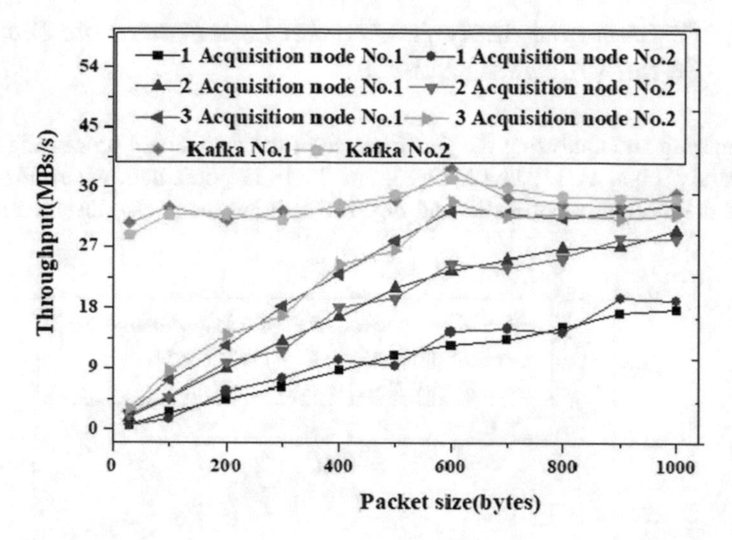

Fig. 11.13 Analysis of data collection system throughput of different nodes (data size per unit time)

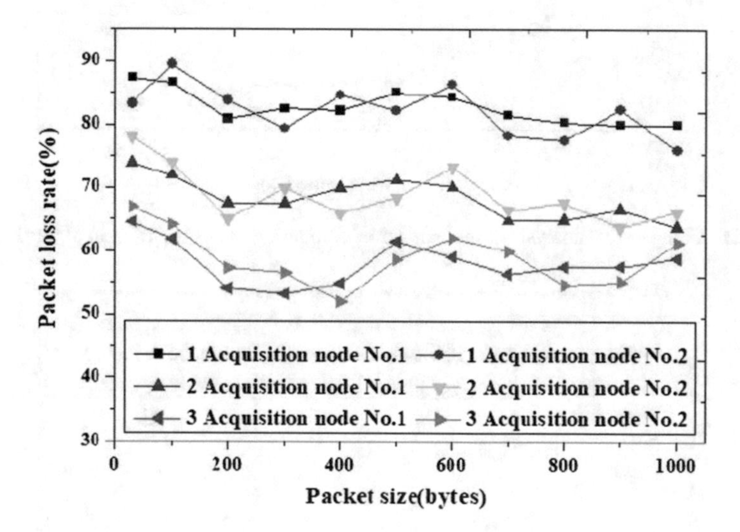

Fig. 11.14 Analysis of packet loss rate at different collection points

As shown in Figs. 11.12 and 11.13, when the data collection system has two collection nodes, the throughput of the data collection system is only about 1.7 times that of one node. When three collection nodes are provided, the data packets are different as the throughput changes. However, the amount of transmitted data per unit time has only slightly increased. The reason is that the throughput of the data collection system at this moment is close to the maximum throughput of Kafka. During the data transmission process, there will be packet losses. As shown in Fig. 11.14, when the data collection system works at the maximum receiving amount, the packet loss rate will be very high. When data collection nodes are added, the packet loss rate will decrease, but it is still at a high level. The possible reason is that the packet loss rate increases due to reasons such as exceeding the signal processing capability.

11.5 Conclusions

With the gradual popularization of the Internet of Things, this technology has also been increasingly used in industrial production. In the meantime, the rise of the Internet of Things has made industrial production gradually enter a new revolutionary era. In this study, an advanced signal processing system for industrial big data is constructed, and its performance is analyzed. The results show that for data transmission, when the successful propagation probability is 100% and the λ value is 0.01–0.05, it is closest to the actual result, and the data propagation delay is gradually smallest. Through a comparative analysis of system performance, it is found that compared to MapReduce and Spark algorithms, Kafka clusters require the shortest running time at the same data scale and the same computing nodes. Further analysis of their packet loss rate indicates that as the number of collection points increases, the amount of transmitted data has only increased slightly, but the packet loss rate has not changed significantly.

In summary, this study suggests that the system constructed in this study can reduce the delay of data transmission and the running time of this system significantly, which provides experimental references for later industrial production and development. However, there are some shortcomings in the research process. The data collection nodes in this study do not have distributed characteristics. Therefore, in the subsequent study, a good solution will be designed to make the data collection nodes have distributed characteristics, which will make the availability and fault tolerance of this system to greatly increase.

References

1. Xu, Y., Sun, Y., Wan, J., Liu, X., & Song, Z. (2017). Industrial big data for fault diagnosis: Taxonomy, review, and applications. *IEEE Access, 5*, 17368–17380.
2. Psannis, K. E., Stergiou, C., & Gupta, B. B. (2018). Advanced media-based smart big data on intelligent cloud systems. *IEEE Transactions on Sustainable Computing, 4*(1), 77–87.
3. Wan, J., Tang, S., Li, D., Wang, S., Liu, C., Abbas, H., & Vasilakos, A. V. (2017). A manufacturing big data solution for active preventive maintenance. *IEEE Transactions on Industrial Informatics, 13*(4), 2039–2047.
4. Khan, M., Wu, X., Xu, X., & Dou, W. (2017, May). Big data challenges and opportunities in the hype of Industry 4.0. In: *2017 IEEE International Conference on Communications (ICC)* (pp. 1–6). IEEE.

5. Lv, Z., Song, H., Basanta-Val, P., Steed, A., & Jo, M. (2017). Next-generation big data analytics: State of the art, challenges, and future research topics. *IEEE Transactions on Industrial Informatics, 13*(4), 1891–1899.
6. He, Y., Guo, J., & Zheng, X. (2018). From surveillance to digital twin: Challenges and recent advances of signal processing for industrial internet of things. *IEEE Signal Processing Magazine, 35*(5), 120–129.
7. Li, P., Chen, Z., Yang, L. T., Zhang, Q., & Deen, M. J. (2017). Deep convolutional computation model for feature learning on big data in internet of things. *IEEE Transactions on Industrial Informatics, 14*(2), 790–798.
8. Yan, H., Wan, J., Zhang, C., Tang, S., Hua, Q., & Wang, Z. (2018). Industrial big data analytics for prediction of remaining useful life based on deep learning. *IEEE Access, 6*, 17190–17197.
9. Wollschlaeger, M., Sauter, T., & Jasperneite, J. (2017).The future of industrial communication: Automation networks in the era of the internet of things and industry 4.0[J]. *IEEE Industrial Electronics Magazine, 11*(1), 17–27.
10. He, M., & He, D. (2017). Deep learning based approach for bearing fault diagnosis. *IEEE Transactions on Industry Applications, 53*(3), 3057–3065.
11. Yan, J., Meng, Y., Lu, L., & Li, L. (2017). Industrial big data in an industry 4.0 environment: Challenges, schemes, and applications for predictive maintenance. *IEEE Access, 5*, 23484–23491.
12. Li, P., Chen, Z., Yang, L. T., Gao, J., Zhang, Q., & Deen, M. J. (2018). An incremental deep convolutional computation model for feature learning on industrial big data. *IEEE Transactions on Industrial Informatics, 15*(3), 1341–1349.
13. Wan, J., Hong, J., Pang, Z., Jayaraman, B., & Shen, F. (2019). IEEE ACCESS Special section editorial: Key technologies for smart factory of Industry 4.0. *IEEE Access, 7*, 17969–17974.
14. Mcmahon, P., Zhang, T., & Dwight, R. (2020). Requirements for big data adoption for railway asset management. *IEEE Access, 8*, 15543–15564.
15. Patole, S. M., Torlak, M., Wang, D., & Ali, M. (2017). Automotive radars: A review of signal processing techniques. *IEEE Signal Processing Magazine, 34*(2), 22–35.
16. Allwood, G., Du, X., Webberley, K. M., Osseiran, A., & Marshall, B. J. (2018). Advances in acoustic signal processing techniques for enhanced bowel sound analysis. *IEEE Reviews in Biomedical Engineering, 12*, 240–253.
17. Mishra, K. V., Shankar, M. B., Koivunen, V., Ottersten, B., & Vorobyov, S. A. (2019). Toward millimeter-wave joint radar communications: A signal processing perspective. *IEEE Signal Processing Magazine, 36*(5), 100–114.
18. Aujla, G. S., Prodan, R., & Rawat, D. B. (2021). Big data analytics in Industry 4.0 ecosystems. https://doi.org/10.1002/spe.3008
19. Ghorbanian, M., Dolatabadi, S. H., & Siano, P. (2019). Big data issues in smart grids: A survey. *IEEE Systems Journal, 13*(4), 4158–4168.
20. Deutsch, J., & He, D. (2017). Using deep learning-based approach to predict remaining useful life of rotating components. *IEEE Transactions on Systems, Man, and Cybernetics: Systems, 48*(1), 11–20.
21. Zhu, L., Yu, F. R., Wang, Y., Ning, B., & Tang, T. (2018). Big data analytics in intelligent transportation systems: A survey. *IEEE Transactions on Intelligent Transportation Systems, 20*(1), 383–398.
22. Kibria, M. G., Nguyen, K., Villardi, G. P., Zhao, O., Ishizu, K., & Kojima, F. (2018). Big data analytics, machine learning, and artificial intelligence in next-generation wireless networks. *IEEE Access, 6*, 32328–32338.
23. Rani, S., Ahmed, S. H., Talwar, R., & Malhotra, J. (2017). Can sensors collect big data? An energy-efficient big data gathering algorithm for a WSN. *IEEE Transactions on Industrial Informatics, 13*(4), 1961–1968.
24. Marjani, M., Nasaruddin, F., Gani, A., Karim, A., Hashem, I. A. T., Siddiqa, A., & Yaqoob, I. (2017). Big IoT data analytics: Architecture, opportunities, and open research challenges. *IEEE Access, 5*, 5247–5261.

25. Kaur, D., Aujla, G. S., Kumar, N., Zomaya, A. Y., Perera, C., & Ranjan, R. (2018). Tensor-based big data management scheme for dimensionality reduction problem in smart grid systems: SDN perspective. *IEEE Transactions on Knowledge and Data Engineering, 30*(10), 1985–1998.
26. Singh, A., Aujla, G. S., Garg, S., Kaddoum, G., & Singh, G. (2019). Deep-learning-based SDN model for Internet of Things: An incremental tensor train approach. *IEEE Internet of Things Journal, 7*(7), 6302–6311.
27. Aujla, G. S., & Jindal, A. (2020). A decoupled blockchain approach for edge-envisioned IoT-based healthcare monitoring. *IEEE Journal on Selected Areas in Communications, 39*(2), 491–499.
28. Liu, H., Ong, Y. S., Shen, X., & Cai, J. (2020). When Gaussian process meets big data: A review of scalable GPs. *IEEE Transactions on Neural Networks and Learning Systems, 31*(11), 4405–4423.
29. Xu, W., Zhou, H., Cheng, N., Lyu, F., Shi, W., Chen, J., & Shen, X. (2017). Internet of vehicles in big data era. *IEEE/CAA Journal of Automatica Sinica, 5*(1), 19–35.
30. Tang, B., Chen, Z., Hefferman, G., Pei, S., Wei, T., He, H., & Yang, Q. (2017). Incorporating intelligence in fog computing for big data analysis in smart cities. *IEEE Transactions on Industrial Informatics, 13*(5), 2140–2150.
31. Zhang, Q., Yang, L. T., Chen, Z., Li, P., & Bu, F. (2018). An adaptive dropout deep computation model for industrial IoT big data learning with crowdsourcing to cloud computing. *IEEE Transactions on Industrial Informatics, 15*(4), 2330–2337.
32. Chen, B., Wan, J., Shu, L., Li, P., Mukherjee, M., & Yin, B. (2017). Smart factory of industry 4.0: Key technologies, application case, and challenges. *IEEE Access, 6*, 6505–6519.
33. Sharma, S. K., & Wang, X. (2017). Live data analytics with collaborative edge and cloud processing in wireless IoT networks. *IEEE Access, 5*, 4621–4635.
34. Aazam, M., Zeadally, S., & Harras, K. A. (2018). Deploying fog computing in industrial internet of things and industry 4.0. *IEEE Transactions on Industrial Informatics, 14*(10), 4674–4682.

Chapter 12
Deep Neural Network-Based Prediction of High-Speed Train-Induced Subway Track Vibration

Michał Wieczorek, Jakub Siłka, and Marcin Woźniak

12.1 Introduction

Deep learning systems are widely applied in many fields of modern technology due to high precision and many possible applications [1, 2]. Intelligent transportation is related to the variety of developments in technology.

We can read about many interesting models used to predict or simulate vibrations. In [3] was proposed a model of vibration analysis for elements joined by friction joints. Neural networks were used to predict ground vibration in [4]. As a result, warning system for geological activities was developed. Similarly in [5] was presented how to use heuristic model to help in the prediction of blast-produced ground vibration. A model composed for open-pit mine vibrations prediction presented in [6] was also based on neural network.

There are also very interesting models of machine learning devoted to the topic of vibration analysis in means of transport. In [7] was presented an idea to evaluate vibration of high-speed railway by using deep learning. The model proposed in [8] was oriented on the effect of vibrations that come from trains to urban areas. This topic was explored in [9] to show various aspects of such vibrations. Results presented in [10] described how energy consumption analysis may help in the prediction of interior noise in a high-speed train. A wide spectrum of numerical experiments for various urban trains was presented in [11]. In [12] was presented an analytical approach to use neural networks for simulation of dynamical elements in moving vehicles.

In this chapter we want to discuss the use of deep learning model to predict potential vibrations of high-speed trains. Our proposed model is developed using

M. Wieczorek · J. Siłka · M. Woźniak (✉)
Faculty of Applied Mathematics, Silesian University of Technology, Gliwice, Poland
e-mail: marcin.wozniak@polsl.pl

S. Garg et al. (eds.), *Intelligent Cyber-Physical Systems for Autonomous Transportation*, Internet of Things, https://doi.org/10.1007/978-3-030-92054-8_12

Recurrent Neural Network type. We propose to use NAdam training algorithm. Tests were done for different configurations of developed classifier. We have tested how many readings in a time proposed classifier are needed to work with the highest potential efficiency. We have also tested how much time would be necessary to train the classifier to work with the assumed efficiency. The tests have shown that the proposed model is able to predict a long series of readings.

12.2 Deep Neural Network Model

Because in our work we are working with time-step data as we have chosen empirically to use Recurrent Neural Network with Long Short-Term Memory (LSTM) neurons. However to ensure that our choice was correct we have done a preliminary comparison between RNN and ANN. Results are 99.12% for RNN and 88.11% for ANN. As we can see, applied Recurrent Neural Network scored much better making it in this case better than the classical approach, and therefore we oriented our research on this architecture.

12.2.1 Architecture

In our final model we have used 8 LSTM layers: 1 input layer and 7 hidden layers. The final decision, however, is made on 2 standard, fully connected layers. For all hidden layers, we have used hyperbolic tangent activation function and for the output layer we have decided to use ReLU as we are dealing with value prediction so we cannot be restrained by -1 to 1 range. That is very important, because if we would not use ReLU or some other activation function that can reach values over 1, we would not be able to predict peaks over the maximum value found in the data set. In some cases better results could be performed using leaky ReLU as that would allow us to also predict values under 0; however, in this data set it is not needed because vibrations could not be negative and, what is more, leaky ReLU is more computationally heavy so these two factors made us use the standard ReLU function for the output layer. The final architecture applied in our research is shown in Fig. 12.1.

12.2.2 Model Training

We have used a variation of Adaptive Moment Estimation Algorithm—NAdam. This algorithm is widely used in AI research because of its high accuracy and short training times. To improve model performance even more, we have used learning rate decay. It allowed our model to quickly roughly adapt to the problem. NAdam

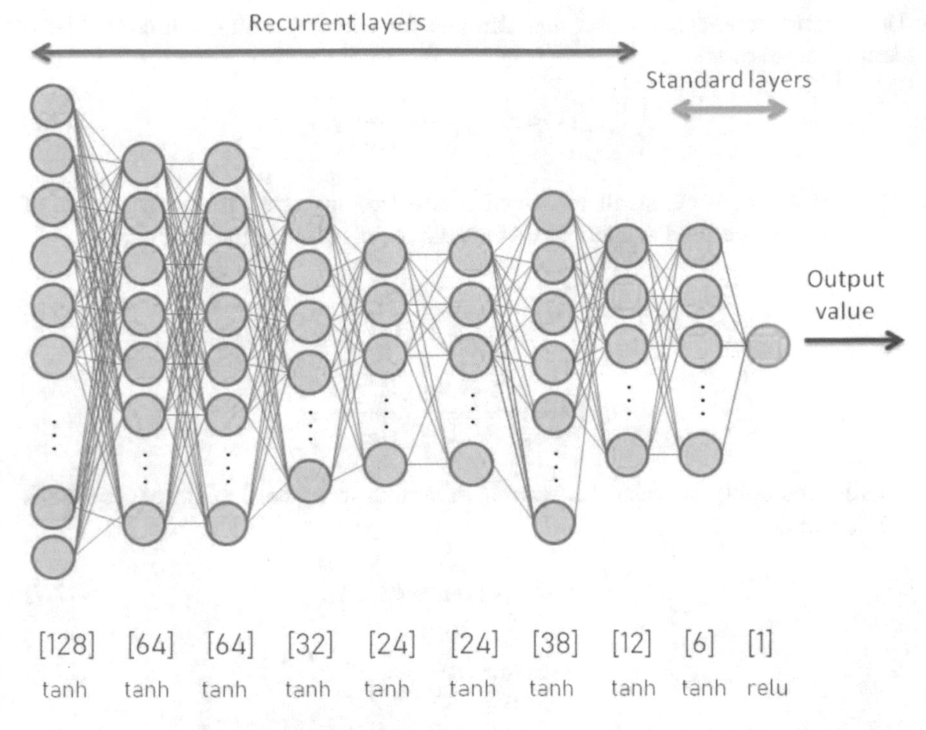

Fig. 12.1 Applied RNN model that we used to evaluate input vibration readings from underground train data set

formula can be described as follows:

$$c_a = \theta_1 c_{a-1} + (1 - \theta_1) g_a, \tag{12.1}$$

$$v_a = \theta_2 v_{a-1} + (1 - \theta_2) g_a^2, \tag{12.2}$$

where θ are constant hyper-parameters and g is a current gradient value of the error function. Values a_m and c_a are used for computing correlations marked as $\hat{m}_a$ and $\hat{v}_a$ and calculated as follows:

$$\hat{c}_a = \frac{c_a}{1 - \theta_1^a}, \tag{12.3}$$

$$\hat{v}_a = \frac{v_a}{1 - \theta_2^a}. \tag{12.4}$$

Using results computed above, final formula for updating weights in our model is defined as follows:

$$z_{a+1} = z_a - \frac{\eta}{\sqrt{\hat{v}_a} + \gamma}\hat{c}_a, \tag{12.5}$$

where γ is a constant small value and η is a learning rate (in our case value of 0.0008). Next we need to apply NAG formula to Adam using these equations:

$$z_a = z_{a-1} - \eta \frac{\theta_1 c_{a-1}}{\theta_2 v_{a-1} + (1-\theta_2)g_a^2 + \gamma} \tag{12.6}$$

$$-\eta \frac{(1-\theta_1)g_a}{\sqrt{\theta_2 n_{a-1} + (1-\theta_2)g_a^2} + \gamma}.$$

Finally we modify the standard Adam update rule, so we need to change equations for $\hat{c}_a$ and z_a:

$$\hat{c}_a = (1-\theta_1)g_a + \theta_{a+1}c_a, \tag{12.7}$$

$$z_a = z_{a-1} - \eta \frac{\hat{c}_a}{\sqrt{\theta_{2a}} + \gamma}. \tag{12.8}$$

12.3 Numerical Experiments

Let us now present results of our experiments on data set provided by Cui [13]. It contains 200 h of metro train vibration energy harvesting data recorded at intervals of 2 min. All data were normalized by using the standard Min–Max approach to fit them into the 0 to 1 range, and final results are denormalized by reversed Min–Max algorithm. In order to test our model's performance we have divided our data set into test and train data with 30%:70% proportion in a such way that we train on older data and test on the most current ones. Because of complexity, our deep learning model needs to be trained on a decent machine if we want to get reasonable training times. Specification of our training server is CPU: AMD Ryzen Threadripper 2950X, 16 core/32 threads, RAM: 64GB, GPU: 2x RTX 2080 8GB.

12.3.1 Searching for the Best Time Step

We have done some experiments to find the best time step for our RNN predictor. Results can be seen in Table 12.1, while charts of these characteristics are presented in Fig. 12.2. As shown, our RNN architecture is very sensitive to changing the time-

Table 12.1 Comparison of using different time steps during RNN training

Time step	Accuracy
10	96.41%
11	89.27%
12	97.26%
13	90.48%
14	49.41%
15	91.22%
16	97.59%
17	89.94%
18	97.56%
19	49.44%
20	98.29%
21	93.07%
22	98.20%
23	50.50%
24	50.44%
25	91.71%
26	98.17%
27	50.49%
28	50.21%
29	50.28%
30	49.96%

step value. For example, changing it from 20 to 19 caused almost 50% reduction in accuracy, but further reduction to 18 increased accuracy again. Because of that, we have tested some different configurations and finally we have chosen time-step value of 20 as the one giving the best results for our deep learning model.

12.3.2 How Far in the Future Can We Correctly Predict?

During our experiments, we were trying to answer the question "How many values in the future we can correctly predict?". We were trying to solve it by analyzing changes to the accuracy metric of our model. Sample results are shown in Table 12.2. Results of various tests are presented in Fig. 12.3. As we can see, in our case changing the main parameter did not have a huge impact on accuracy, actually making it oscillate around 99%. Because of that, we have selected the value of 5000. We have decided empirically that this value is far enough without losing performance.

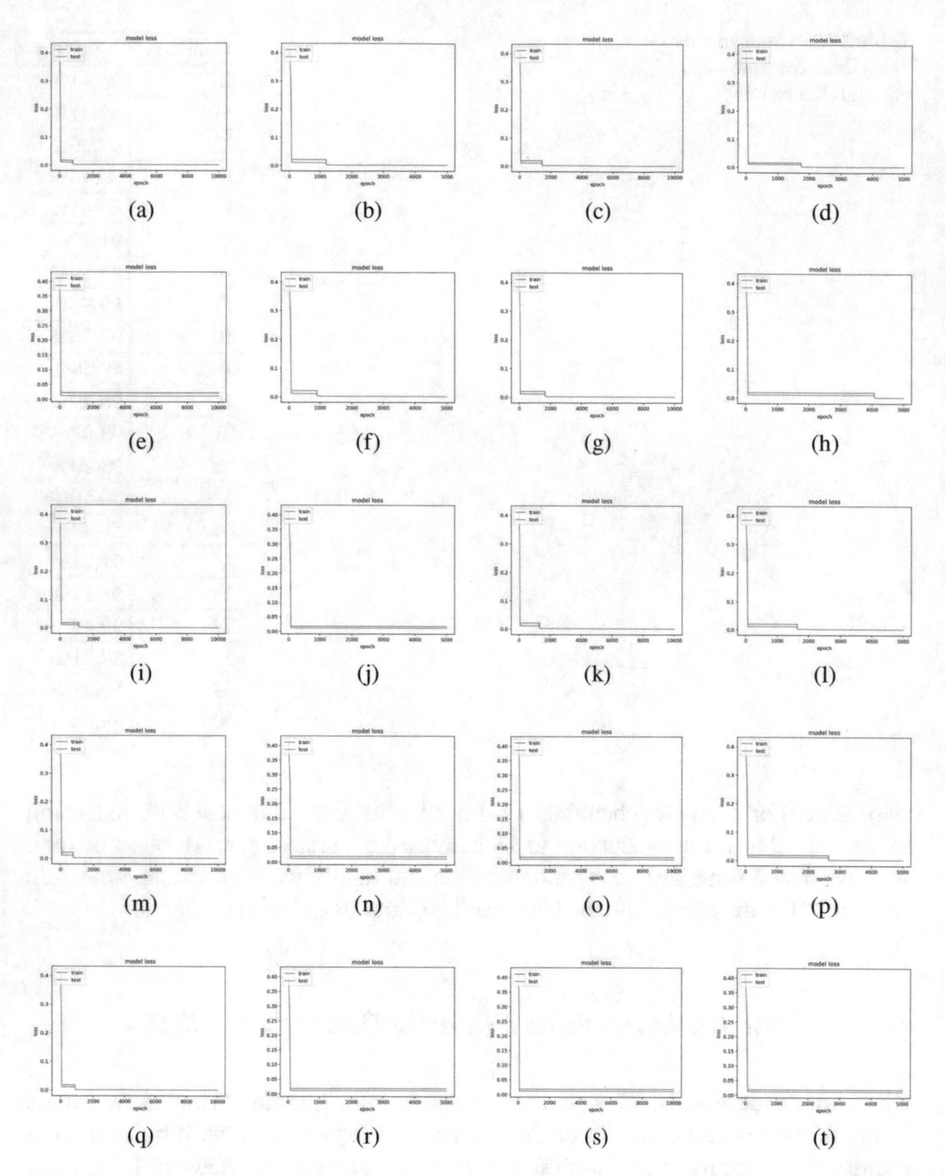

Fig. 12.2 Results of loss using different time steps for the proposed deep neural network architecture. (**a**) Loss for 10. (**b**) Loss for 11. (**c**) Loss for 12. (**d**) Loss for 13. (**e**) Loss for 14. (**f**) Loss for 15. (**g**) Loss for 16. (**h**) Loss for 17. (**i**) Loss for 18. (**j**) Loss for 19. (**k**) Loss for 20. (**l**) Loss for 21. (**m**) Loss for 22. (**n**) Loss for 23. (**o**) Loss for 24. (**p**) Loss for 25. (**q**) Loss for 26. (**r**) Loss for 27. (**s**) Loss for 28. (**t**) Loss for 29

Table 12.2 Comparison of how many reading values forward the network predicts correctly

Values forward	Accuracy
5	97.77%
10	98.81%
15	97.81%
20	98.01%
25	97.36%
50	98.20%
100	98.21%
250	97.35%
500	97.23%
1000	98.22%
1500	98.31%
2500	98.26%
5000	99.12%
10,000	94.60%

12.3.3 Computing the Overall Accuracy Based on Error Margin

Our network predicts the future values with a very high accuracy. As we can see in Table 12.3, if we allow the network to have an error margin of 5% of the real value, the accuracy reaches about 75.66%. However if we step up the margin by another 5%, the accuracy is around 99.12%. With that, we can conclude that about 3/4 of the values are almost ideal and most of them are only up to 10% off. Stepping up to the error margin of 30% gives us a perfect accuracy of 100% so, as we can see, no value was predicted with a higher error making our model a good fit to the given problem. Results from our experiments are shown in Fig. 12.4. Despite the fact that our model was trained using modern, high-powered GPU, the trained model forward pass is not computationally demanding and can be run on a single machine with medium specification and using just a CPU.

12.3.4 Training Times

During our experiments, we have tested time performance. Average time is 48 min 35.86 s. Comparisons for different future predictions are presented in Table 12.4. However, as we can see in any case it stays around 1 h of training (Fig. 12.5).

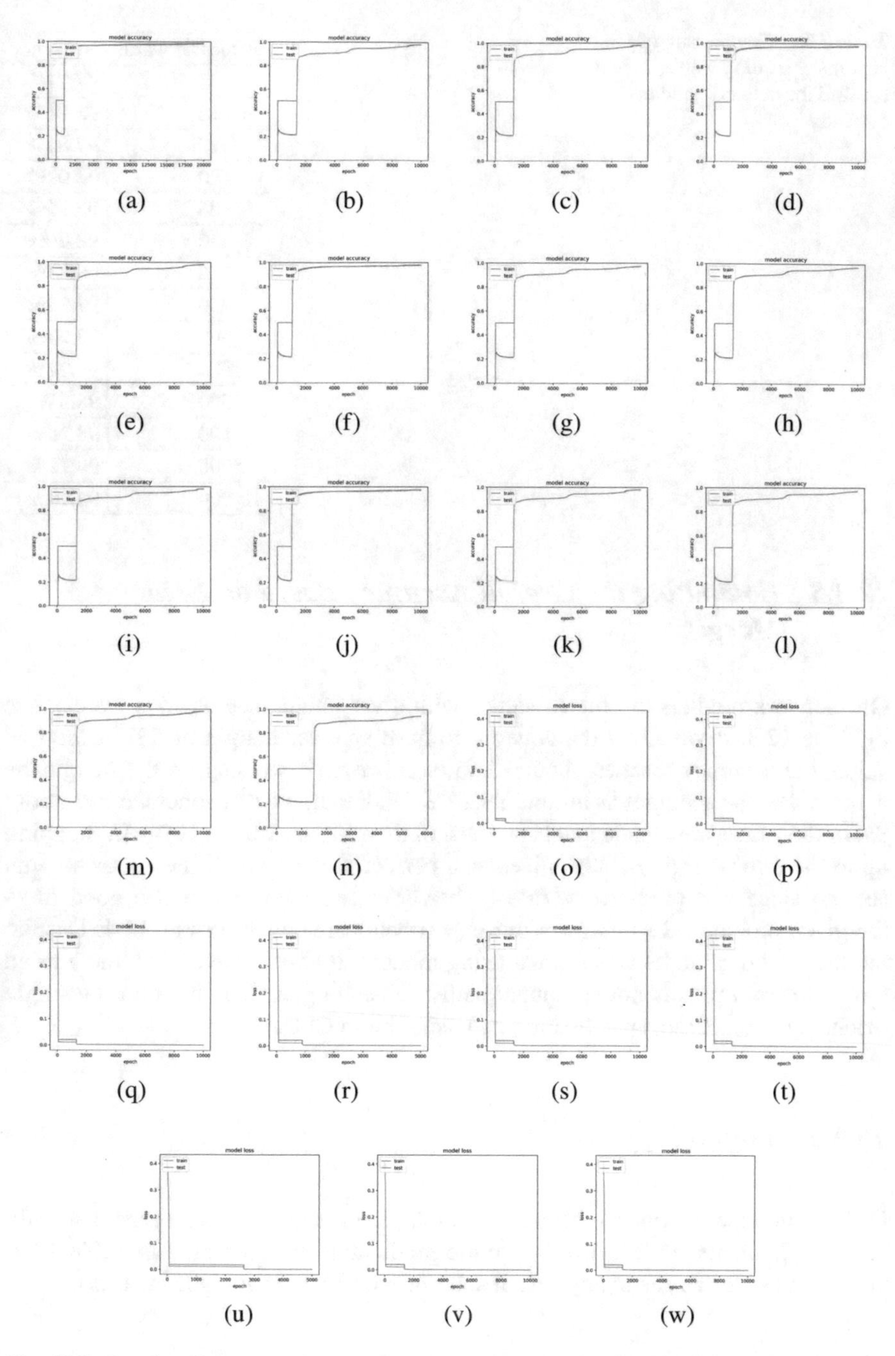

Fig. 12.3 Results of how many future values should the network predict with the highest potential efficiency. (**a**) Loss for 10. (**b**) Loss for 100. (**c**) Loss for 1000. (**d**) Loss for 15. (**e**) Loss for 1500. (**f**) Loss for 20. (**g**) Loss for 25. (**h**) Loss for 250. (**i**) Loss for 2500. (**j**) Loss for 5. (**k**) Loss for 50. (**l**) Loss for 500. (**m**) Loss for 5000. (**n**) Loss for 10,000. (**o**) Loss for 10. (**p**) Loss for 100. (**q**) Loss for 1000. (**r**) Loss for 15. (**s**) Loss for 1500. (**t**) Loss for 20. (**u**) Loss for 25. (**v**) Loss for 250. (**w**) Loss for 2500

Table 12.3 Comparison of using different error margins

Error margin	Accuracy
0.05	75.66%
0.1	99.12%
0.15	99.91%
0.2	99.99%
0.3	100.0%
0.4	100.0%
0.5	100.0%
0.6	100.0%
0.7	100.0%
0.8	100.0%

12.3.5 Possible Applications of Proposed Deep Learning Vibration Estimation

Proposed model of vibration prediction can be applied in various transportation events, both for passenger and cargo types. Let us now discuss some examples:

- **Chemicals containing dangerous substances.** Although most of the hazardous chemicals can be roughly preserved while undergoing through a plenty of disrupting vibrations, some of them, such as 1-diazidocarbamoyl-5-azidotetrazole, can bring appreciable risk of explosion. The mentioned compound (also known as an azidoazide azide) is extremely reactive—even microscopical vibrations can result in an explosion, as well as tiny rise or drop in the temperature. This compound is also susceptible to any radiation. Our system might be utterly helpful in such case, allowing to safely transport this sort of substances by trains. As a result self-evidently it refers also to other, less dangerous compounds, where occurs a necessity of assuring safe transport of either—substance, workers, and area within which the transport would be conducted.
- **Fragile electronics and Others.** Such predictor can also be used for materials exposed to fatigue, which are easy to find among electronic systems, caused by continuous vibrations of a given device. The components such as condensers and transformer are heavy (comparatively to the fastening elements), what has a relevance to fragile prototypes—often not well-optimized for long drives.
- **Increasing passenger comfort.** There exists a possibility to apply such prediction model to transportation of passengers to increase not only traveling comfort but also safety.

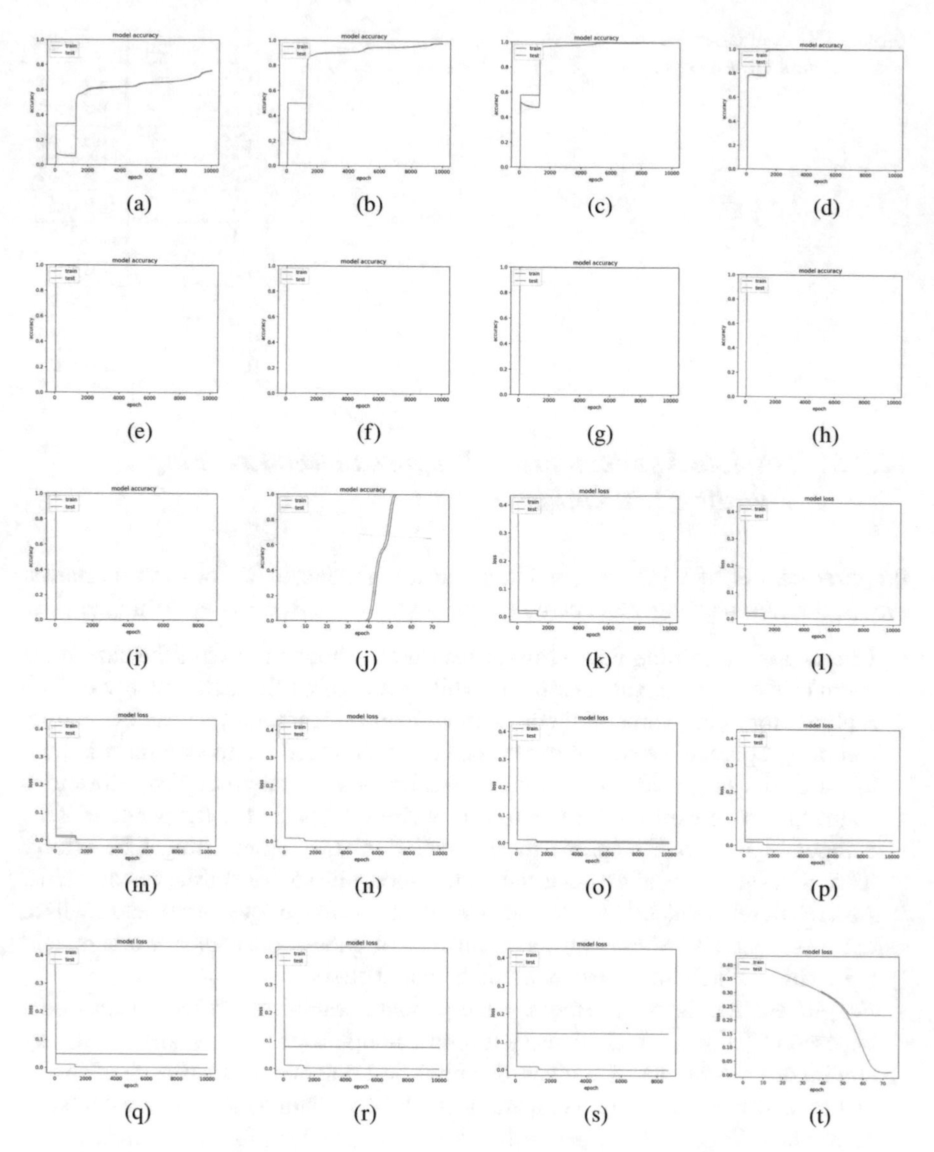

Fig. 12.4 Results of using different error margins for our developed predictor. (**a**) Accuracy for 0.05. (**b**) Accuracy for 0.1. (**c**) Accuracy for 0.15. (**d**) Accuracy for 0.2. (**e**) Accuracy for 0.3. (**f**) Accuracy for 0.4. (**g**) Accuracy for 0.5. (**h**) Accuracy for 0.6. (**i**) Accuracy for 0.7. (**j**) Accuracy for 0.8. (**k**) Loss for 0.05. (**l**) Loss for 0.1. (**m**) Loss for 0.15. (**n**) Loss for 0.2. (**o**) Loss for 0.3. (**p**) Loss for 0.4. (**q**) Loss for 0.5. (**r**) Loss for 0.6. (**s**) Loss for 0.7. (**t**) Loss for 0.8

Table 12.4 Comparison of training times for reading values used in forward predictions

Values forward	Training time
5	66 min 17.73 s
10	66 min 4.61 s
15	66 min 10.35 s
20	62 min 52.59 s
25	64 min 6.17 s
50	62 min 40.76 s
100	60 min 45.91 s
250	64 min 55.96 s
500	62 min 39.45 s
1000	60 min 52.66 s
1500	62 min 9.79 s
2500	60 min 33.61 s
5000	56 min 26.67 s
10,000	46 min 20.24 s

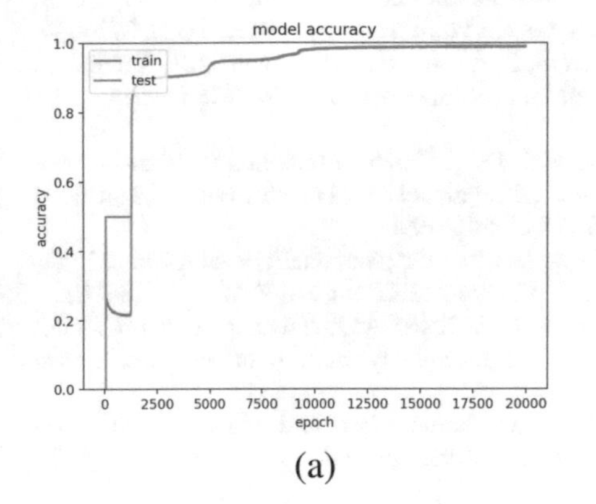

(a)

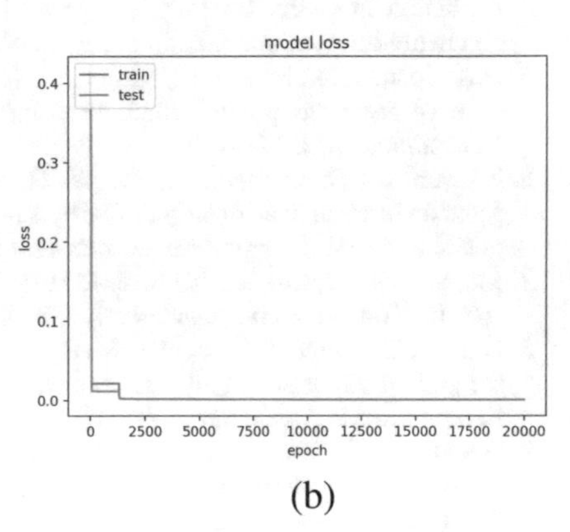

(b)

Fig. 12.5 Results of training metrics for applied model. (**a**) Accuracy for NAdam. (**b**) Loss for NAdam

12.4 Final Remarks

In this chapter we proposed vibration prediction model by the use of LSTM-RNN deep learning architecture. We collaterally presented optional applications of those predictions to create a system allowing to decrease sensible vibrations to satisfying extent. As the entire architecture was developed based on the data from telegraphic sensors during a drive, the model was devised in regard to that sort of transport. However, being obtained with accurate measures and equipment, we will be able to recreate the systems on the rules of a different transport, including air transport, road transport, and water transport. In the future, we are willing to focus on adaptive ability, which will yield in further development in deep learning models for smart transport.

Acknowledgments The authors would like to acknowledge the contribution to this research from the Rector of the Silesian University of Technology, Gliwice, Poland, under program Excellence Initiative—Research University no. 08/IDUB/2019/84.

References

1. Aujla, G. S., & Jindal, A. (2020). A decoupled blockchain approach for edge-envisioned IoT-based healthcare monitoring. *IEEE Journal on Selected Areas in Communications, 39*(2), 491–499.
2. Singh, A., Aujla, G. S., Garg, S., Kaddoum, G., & Singh, G. (2019). Deep-learning-based SDN model for Internet of Things: An incremental tensor train approach. *IEEE Internet of Things Journal 7*(7), 6302–6311.
3. Krack, M., Salles, L., & Thouverez, F. (2017). Vibration prediction of bladed disks coupled by friction joints. *Archives of Computational Methods in Engineering, 24*(3), 589–636.
4. Hajihassani, M., Armaghani, D. J., Marto, A., & Mohamad, E. T. (2015). Ground vibration prediction in quarry blasting through an artificial neural network optimized by imperialist competitive algorithm. *Bulletin of Engineering Geology and the Environment, 74*(3), 873–886.
5. Hasanipanah, M., Naderi, R., Kashir, J., Noorani, S. A., & Qaleh, A. Z. A. (2017). Prediction of blast-produced ground vibration using particle swarm optimization. *Engineering with Computers, 33*(2), 173–179.
6. Nguyen, H., Drebenstedt, C., Bui, X. N., & Bui, D. T. (2020). Prediction of blast-induced ground vibration in an open-pit mine by a novel hybrid model based on clustering and artificial neural network. *Natural Resources Research, 29*(2), 691–709.
7. Ma, S., Gao, L., Liu, X., & Lin, J. (2019). Deep learning for track quality evaluation of high-speed railway based on vehicle-body vibration prediction. *IEEE Access, 7*, 185099–185107.
8. Paneiro, G., Durão, F. O., e Silva, M. C., & Neves, P. F. (2018). Artificial neural network model for ground vibration amplitudes prediction due to light railway traffic in urban areas. *Neural Computing and Applications, 29*(11), 1045–1057.
9. Weber, C., & Karantonis, P. (2018). Rail ground-borne noise and vibration prediction uncertainties. In: *Noise and vibration mitigation for rail transportation systems* (pp. 307–318). Springer.
10. Zheng, X., Dai, W., Qiu, Y., & Hao, Z. (2019). Prediction and energy contribution analysis of interior noise in a high-speed train based on modified energy finite element analysis. *Mechanical Systems and Signal Processing, 126*, 439–457.
11. Kouroussis, G., Vogiatzis, K. E., & Connolly, D. P. (2017). A combined numerical/experimental prediction method for urban railway vibration. *Soil Dynamics and Earthquake Engineering, 97*, 377–386.
12. Woźniak, M., & Połap, D. (2017). Hybrid neuro-heuristic methodology for simulation and control of dynamic systems over time interval. *Neural Networks, 93*, 45–56.
13. Cui, E. (2020). Metro vehicle vibration energy harvesting dataset. *IEEE Dataport*. https://doi.org/10.21227/0jcz-t470

Chapter 13
Advanced Complex Data Analysis of Autonomous Transportation for Smart City Industrial Environment

Zhihan Lv, Liang Qiao, Jingyi Wu, and Haibin Lv

13.1 Introduction

Currently, with the development of the Internet of Things (IoT) technology and information technology, as well as the continuous advancement of national governance systems and governance capabilities in the modernization process, smart cities have entered a new period of construction [1, 2]. As a deep integration of urbanization and informationization, smart cities integrate big data, cloud computing, IoT technology, and spatial geographic information integration technologies to realize the intelligent general coordination in urban planning, construction, management, and services, which is the direction of future urban development [3]. Globally, most countries have now entered the "new normal" of economic development, and the new momentum for economic growth has been shaping gradually. The planning and construction levels of a smart city directly affect the level of urban core competitiveness; therefore, it is also an effective way to achieve sustainable urban development in this circumstance [4, 5].

Under the smart city environment, information technology and industrialization are accelerating. The industrial model innovation of "Industry 4.0" [6, 7] proposed by Germany has triggered a new wave of industrial transformation around the world; its core is to transform industrial production to intelligent manufacturing [8]. By applying advanced modern information technology and taking the optimization of intelligent production processes as the core, promoting intelligent production is significant for improving the quality and efficiency of China's process industry and

Z. Lv (✉) · L. Qiao · J. Wu
College of Computer Science and Technology, Qingdao University, Qingdao, China

H. Lv
North China Sea Offshore Engineering Survey Institute, Ministry of Natural Resources North Sea Bureau, Qingdao, China

S. Garg et al. (eds.), *Intelligent Cyber-Physical Systems for Autonomous Transportation*, Internet of Things, https://doi.org/10.1007/978-3-030-92054-8_13

promoting the transformation of traditional industries [9]. In the era of Industry 4.0, with the promotion of Cyber-Physical System (CPS), the popularization of smart devices, and the continuous upgrading of the IoT technology, the magnitude of industrial data is also increasing at an unprecedented rate. The huge amount of data in the industrial field covers a wide range, including manufacturing, promotion operations, and supply chain, which are the basis of "Industry 4.0" and industrial production in smart cities [10]. Given an industrial big data environment, various sensors monitor the operation of the entire process of mineral processing in real time, generating massive real-time machine data with extremely complex structures. Traditional human empirical knowledge and business process optimization theories have met bottlenecks in this environment, which restricts the further optimization of business processes. Therefore, by using the advantages of industrial big data and complex event processing technology, the manufacturing business processes can be understood in detail, which helps save costs, improve production efficiency, and improve product quality. In the meantime, this can find defects in business processes in time to ensure the long-term stable operation of optimized business process.

The rapid development of data also brings a series of new problems, such as energy consumption and operation problems. In particular, the development of cloud computing, big data, virtualization, software-defined network (SDN), network function virtualization, and other technologies has also brought new trends and challenges to data. As one of the new research directions of the next generation network, SDN has become the hottest research direction in the global network field. Massive industrial data has accumulated to a certain magnitude, reflecting in the amount and the complexity of information. These data are mainly characterized by multi-modality, high throughput, real-time updates, and dynamic features [11]. At present, the industrial data structure in a smart city environment is extremely complicated, and the phenomenon of data islands is common, resulting in low data utilization [12]. Therefore, how to take effective measures to analyze and utilize advanced industrial complex data is an urgent problem to be solved in industrial processes. Complex Event Processing (CEP) has received widespread attention in the development of process processing technology. It can perform real-time management and services through an event-driven model, thereby monitoring and making decisions for complex events [13].

Thus, this study explores advanced complex data analytics in smart cities by constructing industrial complex data models, extracting more valuable information from massive complex data, improving industrial production efficiency and product quality, and ensuring the stability of the industrial production process.

13.2 Related Works

With the passage of time and various transformation into the cities, they are converted into smart city. A key factor in the transformation process is the effective use of big data provided by these cities. Tang et al. [14] considered that data-intensive analysis was the major challenge faced by smart cities in the environment where various sensors were commonly deployed; therefore, to support the integration of many infrastructure components and services in smart cities, a layering distributed optic fiber gyroscope computing architecture confirmed the feasibility of the system in smart cities by identifying 12 different events. Mohammadi and Al-Fuqaha [15] proposed a three-level learning framework for smart cities from the perspective of machine learning, which provided different levels of knowledge abstraction to meet the service needs of smart cities; finally, good convergence control strategies confirmed the application value of machine learning in high-level smart city services. Enayet et al. [16] proposed a mobile-optimized resource allocation architecture (Mobi-Het) for real-time access, sharing, storage, processing, and analysis of big data in smart cities. In terms of reliability, Mobi-Het had higher efficiency, which was used to remotely monitor the execution of big data tasks. The effectiveness of the Mobi-Het system was shown in mobile big data applications in smart cities.

The premise of providing a series of intelligent services for smart cities is to collect and process information of various fields, thereby providing a complete service system through effective data analysis. In the industrial field, the application of intelligent equipment and terminals provides convenient data collection for industrial production and operation. However, at the same time, it has higher requirements for the analysis of industrial complex data. At present, many scholars have analyzed industrial multi-level complex data. Based on the characteristics of a large amount of data, high data correlation, deep integration, and dynamic integration in intelligent industrial manufacturing, Xu and Hua [17] proposed the ontology modeling and inference method according to the intelligent product line in the intelligent environment. In addition, the research on neural network production line prediction methods confirmed that the cloud-based deep learning between devices and three-dimensional self-organizing reconstruction mechanism could accelerate the production of smart factories. According to the properties of industrial big data processing, Yan et al. [18] proposed a new multi-source heterogeneous information structure framework, which was combined with the spatiotemporal characteristics to describe structured data and modeled invisible factors. Therefore, the production process became transparent; finally, the prediction and maintenance of the remaining life of key components of processing equipment in the era of Industry 4.0 were achieved. For the data analysis of complex industrial events, Yu and Zhao [19] utilized density peaks to quickly search and find algorithms and proposed clustering indexes; then, they established a multi-modal exponent discrimination model based on the index covariance matrix between and within subclasses. In addition, they proposed a multi-modal exponent discriminant analysis (MEDA) algorithm to diagnose faults with fuzzy subclass boundaries; finally, the Tennessee-Eastman process verified that the MEDA algorithm could effectively diagnose different types of faults in industrial production.

In summary, at present, for the industrial development in smart cities, massive amounts of complex information and data have participated in all aspects of industrial production and development. However, the above studies mostly focus on the analysis of industrial big data, while the research on the analysis of advanced complex industrial data is rarely reported. Therefore, this study explores the advanced complex data analytics in the industrial environment of smart cities and combines the industrial big data and complex event processing technology to process and analyze industrial big data, thereby realizing the joint improvement of industrial production efficiency and product quality.

13.3 Method

13.3.1 Architecture and Key Technologies of IIoT

A smart city is a combination of a variety of new information technologies, such as artificial intelligence, the IoT technology, and information grids, to achieve the common development of the information industry and smart city governance, thereby promoting urban economic and social development and improving the quality of life of residents [20]. Informationization is a key driving force in the development of human society and the promotion of smart cities. As a source of economic production and social change, informationization has promoted not only the development of new technology industries but also the innovative changes in industrialization. The foundation of smart city construction is the information construction of the city, which is not only a guarantee for sustainable urban development but also a core indicator for assessing the degree of smart city construction [21]. In a smart city, the information construction is mainly for the network environment and the data environment. The specific content is shown in Fig. 13.1. At the level of the network environment, it applies innovative information technologies, such as the Internet of Things, big data, AI recognition, and radio frequency. At the level of the data environment, it combines 5G and other emerging hot spot communication technologies.

With the development of Bluetooth, wireless networks, and Global System for Mobile Communications (GSM), data sharing and information exchange have become a reality. The information service system of a smart city is to integrate infrastructure into a super-micro network system management framework [22]. This system expands daily management control through high-level integration, which not only stores and analyzes data but also improves the efficiency of equipment performance monitoring, which is beneficial to equipment management and operation cost savings. The composition is shown in Fig. 13.2. Therefore, with the GIS automation platform, the information service system of the smart city is divided into two modules, i.e., the system control terminal and the ultra-structure of the infrastructure. The system control module is mainly composed of platforms, data management centers, and service clusters; the ultra-micro structure of the infrastructure refers to resources such as water, gas, and optical cables. The smart city information service system uses service-oriented architecture to realize

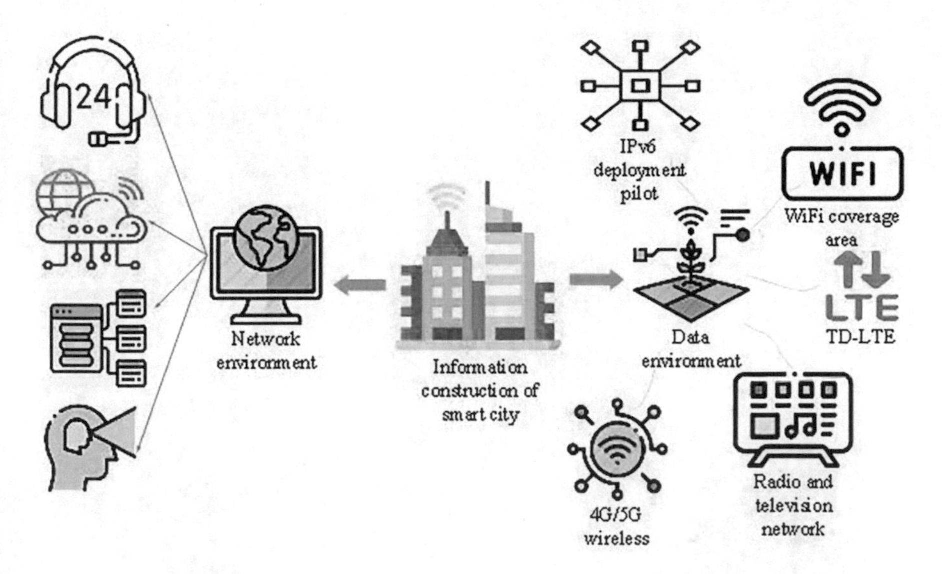

Fig. 13.1 Contents of smart city information construction

the interoperability and sharing of information for the massive data information involved in the city management process, which advances the interoperability between data and services [23].

13.3.2 Spectrum-Based SDN Deployment Algorithm

The network layering is simplified by dividing the network into data layer and control layer. In the data layer, it mainly includes dedicated SDN switches to meet the needs of rapid growth of traffic. In the control layer, the characteristics of SDN controller include: logic centralization and programmable ability, which can grasp the global information in the network. The open and unified interface is used for the interaction between the data layer and the control layer. The controller issues unified standard rules to the switch through the standard interface, and the switch only needs to perform corresponding actions according to these rules.

Based on the in-depth study of various SDN domain partition models and algorithms, a formal description of a general model of SDN domain partition problem is given by comprehensively considering various factors affecting SDN domain partition and controller deployment. If the network topology in the distributed cloud data is regarded as a graph, it can be represented by G(S,E). S represents the set of city nodes in the network topology, that is, the set of OpenFlow switches deployed in these city nodes; E represents the network connection between these city nodes, that is, the links across the Internet between OpenFlow switches. Then, the problem of controller domain partition is transformed into the problem of graph partition.

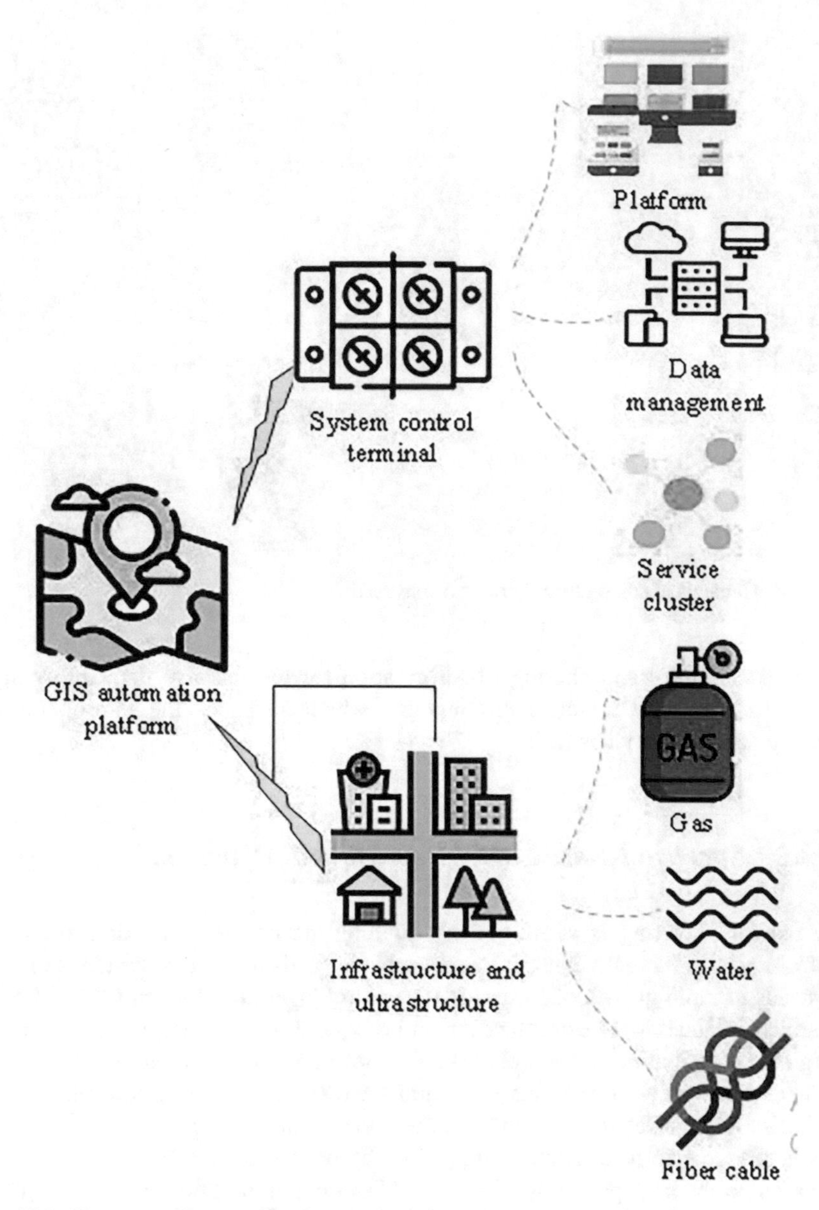

Fig. 13.2 The general framework of smart city information service system

The larger graph is divided into subgraphs and named SDN domain N_i. There is no similarity between the set of switches (S_i) in the ith SDN domain and the set of switches (S_j) in the jth SDN domain.

Let C represent the set of SDN controllers, that is, the SDN controller deployment scheme represents the similarity between controllers and switches. Then, the similarity matrix of graph $G(S, E)$ is $W = (w_{ij})_{i,j=1,2,\ldots,n}$. The partition method of topology graph is closely related to SDN domain load balancing technology. How to divide a topology graph into approximately equal k subgraphs and minimize the number of edges connecting different partitions. In other words, to get a relatively balanced SDN domain partition, the objective equation must be satisfied.

$$SDN_{out} = \sum_{i=1}^{k} \frac{\sum_{x \epsilon N, y \epsilon G-N_i} W_{xy}}{\sum_{x \epsilon N, y \epsilon G} W_{xy}} \tag{13.1}$$

After the partition problem of SDN domain is solved and the balanced cut set is obtained, the placement and deployment of controllers in each domain is urgent to be solved. After SDN domain division, the large distributed data center network across Wan is divided into several small subnets. Similar to the facility location problem, in each subnet, the deployment location of the master–slave controller should meet the minimum average delay of each node in the SDN domain to the controller, as shown in the below equation:

$$min \sum_{c_i \epsilon C} \sum_{s \epsilon N_i} dist(s, c_i) \tag{13.2}$$

C represents the set OF SDN controller placement scheme, c_i indicates the i-th controller in the set, and s suggests OpenFlow switch. $dist(s, c_i)$ refers to the shortest path from s to c_i. Therefore, the complex problem of controller deployment is transformed into dividing the large-scale network into SDN domains to get a relatively balanced operation domain, and then deploying the controller to the center of each SDN domain according to the principle of the shortest average delay. It not only ensures the load balance of the management domain but also considers the average delay and performance of the whole network.

13.3.3 Industrial Complex Events and Data Processing

In the era of smart production, the integration and analysis of multi-source data have become an essential element of the process industry. The advantage of CEP is to transform the data stream into an event stream, thereby detecting the abnormal behavior in the data. CEP uses the relationship between atomic events to build a complex event model and extracts valuable complex events from the event stream [24]. Taking the most representative petroleum, metallurgy, and chemical industries in the process industry as an example, a large amount of production data such as temperature, flow rate, and component pressure are generated every moment, while some data errors will inevitably exist in this process, which leads to an imbalanced measurement data.

The ERP/MES/PCS modern integrated manufacturing system model proposed by the United States for industrial automation systems is shown in Fig. 13.3.

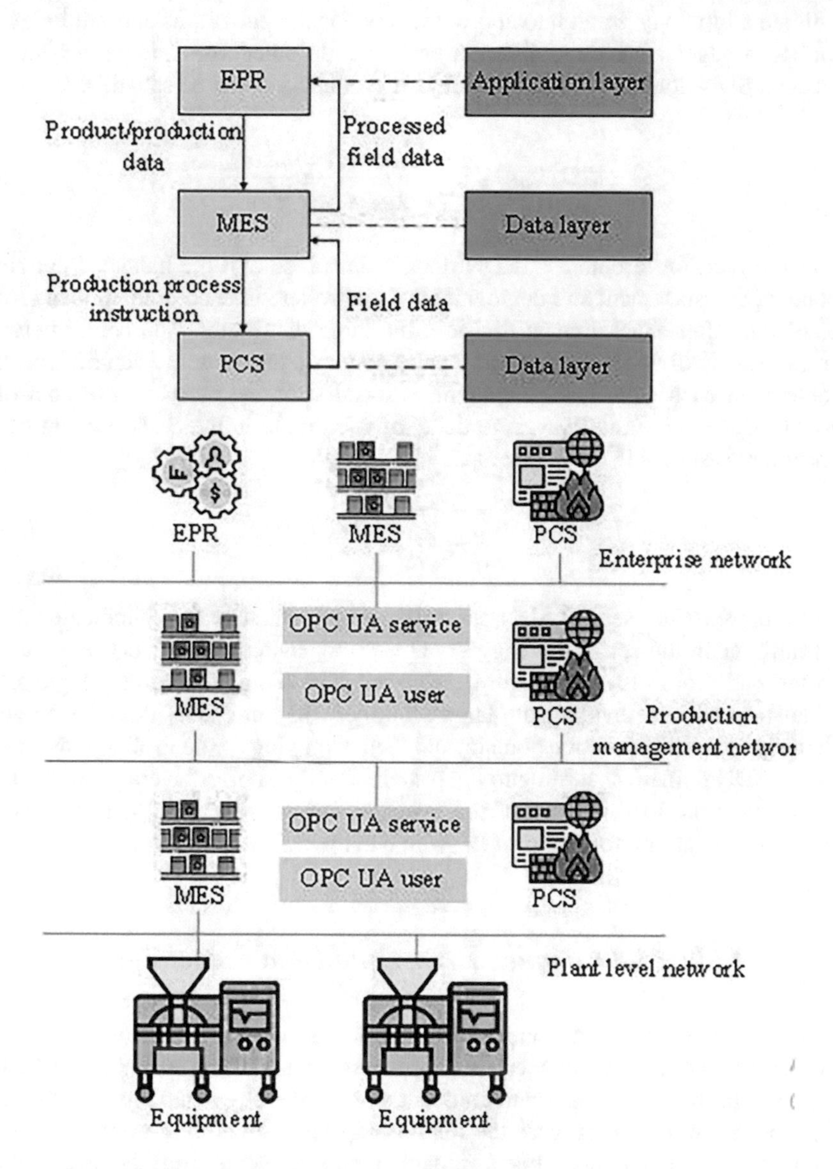

Fig. 13.3 ERP/MES/PCS three-layer architecture integrated manufacturing system

Among them, the Enterprise Resource Planning (ERP) is an overall resource optimization technology with financial analysis as the core, which emphasizes the planning of the enterprise; the Manufacturing Execution System (MES) is a production process optimization operation technique that aims at comprehensive production indicators, which emphasizes the plan execution; the Process Control System (PCS) is a technology with integrated equipment management and control as the core technology, which emphasizes on equipment control [25–27]. To achieve comprehensive enterprise automation, the middle-level production execution system (MES) is essential for the three-layer structure. The ERP/MES/PCS three-layer structure model in practical applications mainly includes dynamic data aggregation of industrial production equipment and virtual tanks. The ERP/MES/PCS three-layer architecture integrated manufacturing system is shown in Fig. 13.3.

This study establishes an industrial process optimization model based on complex events. Taking the mineral processing industry as an example, as a derivative link of autonomous beneficiation, the optimization of the flotation process flow has a huge impact on industrial production processes. A CEP model is constructed based on the business process of the flotation process. The complex and changeable characteristics of the flotation process are described; finally, the flotation business process is optimized. According to the CEP theory, a complex event is a combination of multiple atomic events based on a certain relationship. This study uses the Event to represent the atomic events decomposed during the flotation production process. Event1–Event10 are, respectively, the pulp flow, drug dose, aeration volume, pulp concentration, pulp study size, flotation level, foam movement speed, foam size ratio, average foam life cycle, and events where the concentrate grades change. M is used to indicate the changes in the state of the warehouse [28]. This study is based on two Petri nets that describe the relationship between events and conditions. Then, it builds a complex event processing model for the flotation business process, as shown in Fig. 13.4. After the initial event occurs, the token is continuously passed to the next event through the transition activity. Finally, when it passes to the Event10 event, a complex event occurs.

13.3.4 Independent Transportation Mode in Industrial Environment

Environment, safety, efficiency, and other issues have always been an important problem restricting the sustainable development of mining industry. To improve productivity, reduce costs, and cope with labor shortage, intelligent and unmanned technology is urgently needed to improve the industry capacity. The mine environment is bad, the location is remote and closed, and the operation of mining machinery is single, point-to-point transportation and repetitive operation, which is one of the most suitable scenes to realize autonomous driving. An autonomous transportation solution is provided to meet the challenges faced by customers in terms

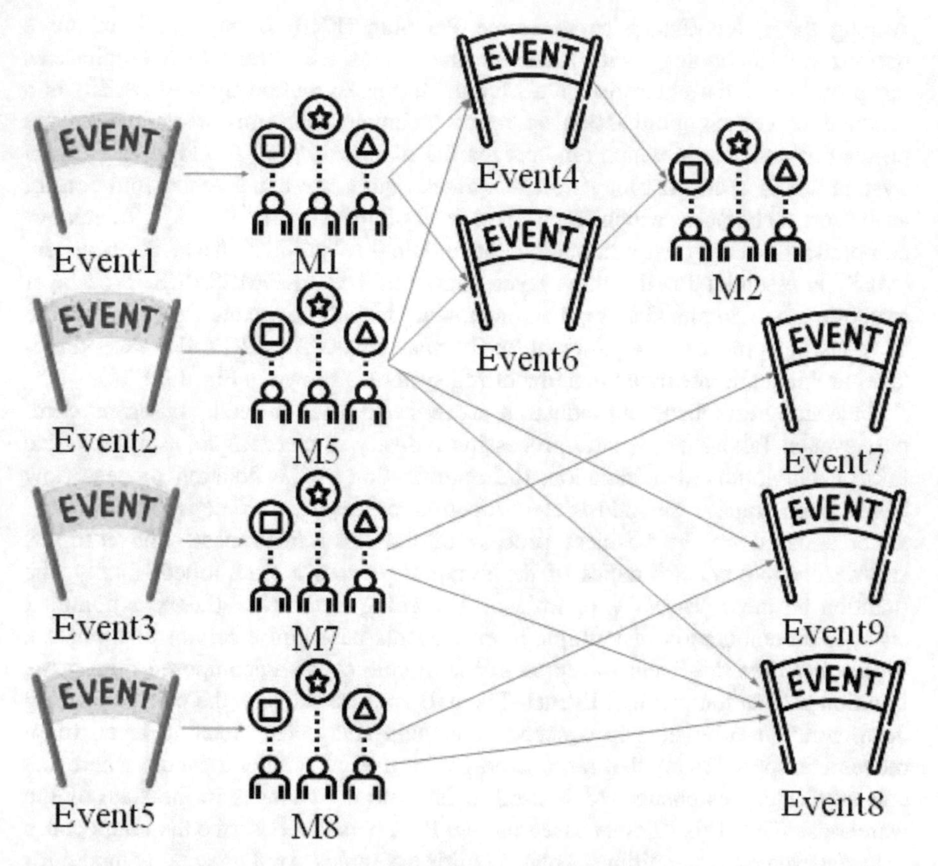

Fig. 13.4 Complex event processing model of flotation business process

of safety, reliability, and profitability. Users do not buy autonomous driving trucks or autonomous driving technology but purchase transportation services to improve work flexibility, efficiency, and productivity. Remote driving is a typical application scenario of unmanned transportation in opencast mines. In normal operation, the mine card, excavator, bulldozer, and other mechanical engineering vehicles, in the planned path, realize the cooperative operation under the autonomous driving mode. In case of failure or danger warning, the driver of the autonomous driving mine card will start the remote takeover mode in the control center, move the vehicle to a safe area, and send a warning to the surrounding vehicles. The original personnel mode of "one mine card and one driver" has also changed here, replaced by the commander mode of "one person in charge of multiple vehicles." Under the overall planning of the cluster scheduling system, manned vehicles and unmanned vehicles can cooperate to improve the production efficiency of the mine.

The progress of science and technology aims to make people's life more convenient. Intelligent airport is the best interpretation of this meaning. To achieve large-scale popularization of autonomous driving technology, a core problem that must be solved is how to "remove the safety officer," that is, without the supervision driver, all driving operations and surrounding environment monitoring work are completed by the system. The normal operation of airport autonomous driving is one of the milestones of unmanned driving toward large-scale commercialization. The driver driving vehicle is adopted, and the unmanned logistics vehicle does not need to consider the manpower. It can still operate in bad weather, to reduce the accident rate caused by human factors, ensure the safe transportation, improve the operation efficiency, and enhance the logistics reliability. In case of obstacles, the driverless system will make the best decision and path planning safely and efficiently according to the perception results of obstacles. Even in bad weather, the logistics link can still be smooth, bringing passengers a comfortable luggage service experience. Over the years, the unmanned logistics vehicle for transporting luggage and goods has been operating in the airport flight restricted area with the highest safety requirements. The application of unmanned driving technology can avoid the accident risk caused by great human errors.

For ports, the rise in shipping costs is caused by multiple factors, including the soaring demand for goods stimulated by global fiscal and monetary policies, port saturation, and too few ships, dockers, and truck drivers. Port operators are eager to improve the port operation efficiency and solve the problem of manpower shortage. Port autonomous driving is a typical closed scenario + low-speed operation scenario, in which autonomous driving takes the lead in commercialization. Autonomous driving port transport vehicles mostly use electric drive, and the control response time is shorter and can provide power for the autonomous driving system directly. Compared to the fuel vehicle, its efficiency is higher, and the electric driving vehicle is the application trend of the autonomous driving port. The L4 level of SAE (Society of Automotive Engineers) autonomous driving classification standard is generally used in the experiment and application of port unmanned container truck. In this level, the unmanned container truck can complete all driving operations without the driver's taking over and intervention; unmanned truck can autonomously perceive the driving environment and actively brake or switch lines in dangerous situations. If intelligent roadside equipment is appropriately introduced combined with the technical solutions of the autonomous driving company, it can significantly accelerate the large-scale mature commercial application of the unmanned container truck in the port.

13.3.5 Industrial Advanced Complex Data Analytics

In the industrial advanced complex data environment, the real-time monitoring of many sensors generates massive amounts of machine data, and the data structure is extremely complex. Traditional data analysis techniques have been unable to meet the current data processing requirements [29]. Therefore, constructing a CEP model in an industrial big data environment has certain feasibility for subsequent

Table 13.1 Comparison of the process industry and discrete industry

Process industry	Discrete industry
Maintain a stable and balanced production state, slow response speed	Ability to respond quickly to changes in requirements and plans
Serial production mode, low redundancy of equipment function	Parallel asynchronous production, high redundancy of equipment function
The production process is dynamic and continuous, and it is generally uninterrupted within the fixed cycle to ensure continuous feeding	Discrete production operation, the plan of course and fine production capacity is important
A complex process, strong coupling between control quantities and high switching cost	Independent control quantity, low switching cost, pay attention to production flexibility
The output and product quality are affected by many uncontrollable factors	Control the production schedule and quantity of spare parts according to the requirements to ensure the complete set and strong controllability of production

data analysis. Due to the different types of specific production objects, the types of industrial data also differ. The process industry and discrete industry have obvious differences in production logistics and supply chain operations. The comparison results of the two are shown in Table 13.1.

Based on the analysis of multi-level complex characteristics of process industry data, the emerging Flink stream processing framework, as a streaming execution engine for the data stream, not only performs fault-tolerant real-time analysis but also greatly simplifies the data processing process, which has obvious computational advantages in analyzing the large-scale industrial complex data [30]. Flink has a complex event processing library. The streaming computing model of Flink enables many functional features, such as state management, processing out-of-order data, and flexible windows. These functions are important to obtain accurate results from infinite data sets [31].

In the actual industrial production process, the production scheduling cycles at different levels are different. Assuming that the system is always in a quasi-steady state, the industrial measurement system in the actual industrial production process can be expressed as $G = (V, E)$, where the node-set V represents the meeting point of the reaction device and pipeline and the arc set X represents the pipeline connecting the units. The logistics data balance model at a single level can be expressed as

$$Ax + Bu = 0 \quad x\epsilon R^n, u\epsilon R^n \tag{13.3}$$

where X refers to the measured variable and u refers to the vector formed by the unmeasured variable. Two vertices (v_i, v_j) determine each arc. All variables determine the integration of flow in the time scale. If only spatial aggregation and disaggregation are considered, the set of nodes and arcs in the lower layer will be

$$U = \{V, X | VX = 0\} \tag{13.4}$$

$$V = [A, B] \; X = \begin{bmatrix} x \\ u \end{bmatrix} \tag{13.5}$$

Several nodes in the lower layer will be aggregated into one node in the upper layer. If the nodes i and j are merged, then:

$$f(U) = \{V, X | p_{i,j} VX = 0\} \tag{13.6}$$

The relationship matrix $p_{i,j}$ between the two layers can be expressed as

$$p_{i,j} = [0 \ldots 01^i 0 \ldots 01^j 0 \ldots]_{1 \times m} \tag{13.7}$$

The material balance equation is obtained:

$$PVX = 0 \tag{13.8}$$

If two arcs with the same vertex are to be merged, the measured values of arcs that can be combined without involving nodes will be combined:

$$f(U) = \{X_a + X_b + \cdots + X_c | X_a, X_b, \ldots, X_c \epsilon S_d\} \tag{13.9}$$

The material balance equation at this time is

$$V' X' = 0 \tag{13.10}$$

In the time scale, the structure of the model does not change. The lower-level model after the time scale expansion can be expressed as

$$V(X_1 + X_1 + +.. + X_c) = 0 \tag{13.11}$$

If it supposes that

$$X_{avg} = \frac{X_1 + X_1 + +.. + X_c}{c} \tag{13.12}$$

the following will be obtained:

$$cVX_{avg} = 0 \tag{13.13}$$

The upper-layer model after node combination can be expressed as

$$f(U) = \{V, X_{avg}|cp_{i,j}VX_{avg} = 0\} \tag{13.14}$$

$$p_{i,j} = [0\ldots01^i0\ldots01^j0\ldots]_{1\times m} \tag{13.15}$$

Then, the material balance equations at this time can be expressed as

$$cPVX_{avg} = 0 \tag{13.16}$$

The upper model of the arc combination becomes

$$f(U) = \{X_{avg,i} + X_{avg,j} + \cdots + X_{avg,k}|X_{avg,i} + X_{avg,j} + \cdots + X_{avg,k} \epsilon S_d\} \tag{13.17}$$

$$cV^{'}X^{'} = 0 \tag{13.18}$$

The application of dynamic data coordination can reduce uncertainties in process measurement. Data coordination methods based on Least Square (LS) estimation are more traditional but cannot meet the requirements of the LS algorithm in practical applications. To reduce the apparent error of the LS algorithm based on robust estimation under weak redundancy variables, the introduction of local redundancy is necessary [32]. Huber estimation is the maximum likelihood estimation of Huber function and is essentially a kind of maximum and minimum estimation, which lays the foundation for robust estimation of mapping [33]. The data processing method derived from improved Huber estimation in this study can be regarded as a Generalized Robust Least Square (GRLS) method. The probability density function estimated by Huber is expressed as

$$f(x) = \begin{cases} (1-\zeta)\frac{1}{\sqrt{2\pi exp[-\frac{1}{2}x^2]}} & |x| \leq c \\ (1-\zeta)\frac{1}{\sqrt{2\pi exp[\frac{c^2}{2}-c]}} & |x| > c \end{cases} \tag{13.19}$$

The loss function of Huber estimation is expressed as

$$\rho(v_i) = \begin{cases} \frac{v_i^2}{2\sigma_i^2} & |v_i| \leq c\sigma_i \\ c\frac{|v_i|}{\sigma_i} - \frac{c^2}{2} & |v_i| \leq c\sigma_i \end{cases} \tag{13.20}$$

The loss function of Huber estimation is expressed as

$$\phi(v_i) = \begin{cases} -c & v_i/\sigma_i < -c \\ v_i & -c \leq v_i/\sigma_i \leq c \\ c & v_i/\sigma_i > c \end{cases} \tag{13.21}$$

The weight function of Huber estimation is expressed as

$$w(v_i) = \begin{cases} 1 & z_i = |v_i/\sigma_i| \leq c \\ c\sigma_i sign(v_i)/v_i & z_i = |v_i/\sigma_i| > c \end{cases} \tag{13.22}$$

where c is a constant of Huber estimation, and the value of c determines the robustness and efficiency of Huber estimation; v_i represents the residual of the i-th variable, and σ_i represents the standard deviation of the i-th variable. The efficiency equation of Huber estimation can be expressed as

$$eff = \frac{1}{var} = \frac{[2\phi(c) - 1]^2}{2\phi(c) - 1 - 2c\phi(c) + 2c^2(1 - \phi(c))} \tag{13.23}$$

In addition to RLS, in the field of industrial data processing, there are some commonly used algorithms for dynamic data correction. As a method of measuring local similarity, correntropy has attracted wide attention due to its practicability and good robustness in the field of non-Gaussian signal processing, which has been continuously used in various fields of industrial data signal processing. However, the correntropy algorithm has the disadvantage of convergence performance. When the initial convergence position is far away from the optimal solution, the convergence speed becomes abnormally slow. Kalman filtering has two purposes of data prediction and correction. The prediction is to estimate the current state based on the state of the previous time, and the correction is to combine the estimated state and the observed state at the current time to estimate the optimal state. In addition, Adaptive Hampel filtering and Median Filter filtering are commonly used data correction methods, which will be compared with the GRLS algorithm proposed in this study.

13.3.6 Simulation Experiment of Industrial Complex Data Analytics

To explore the timeliness of complex event models in an industrial big data environment, this study tests the parallel application of complex event models. The experimental data in this study are from the confirmed data collected by the production equipment of the iron mine flotation plant between October 2018 and 2019. The collected data set is divided into three groups for testing, and the three sets of data experiments are processed by a parallel computing model for complex event processing. The three sets of data sets for large-scale chemical plants and metallurgical plants are selected for testing at the same time, and the correction effects of dynamic data of different algorithms in industrial data analysis are compared.

13.4 Results and Discussions

13.4.1 Comparison of Different Data Coordination Methods

The numerical case selection process in this study is a 5-node process, where W is used to represent the capacity. The numerical simulation results of different algorithms are shown in Fig. 13.5. As shown in Fig. 13.5, when there is no significant error, the comparison results of the W2 correction values of the LS algorithm and the GRLS algorithm coincide. When there is a short-term error, the error between the RLS algorithm and the actual value is larger. At this time, the W2 correction result of the GRLS algorithm is closer to the actual value.

When the error persists significantly, the correction results of W1 and W2 are shown in Fig. 13.6. The corrections of W1 and W2 are affected by the error. For W1, the correction results of the RLS algorithm before and after the improvement are basically the same after 42 h. For W2, the correction results of the two algorithms are the same after 40 h.

To further prove the efficiency of the algorithm, the RLS algorithm and GRLS algorithm are compared horizontally with the other four algorithms. The simulation duration is set to 200. When there is no significant error, except for Kalman filtering, the W2 correction results of the other five algorithms are the same, as shown in Fig. 13.7.

Assuming that there are four significant discontinuous errors, the comparison results of the W2 correction deviations of the six algorithms are shown in Fig. 13.8.

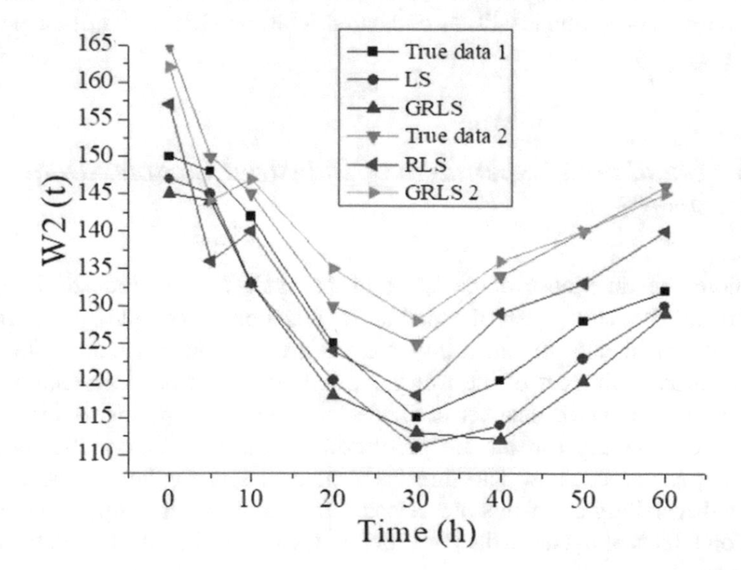

Fig. 13.5 Comparison of W2 correction results without significant errors

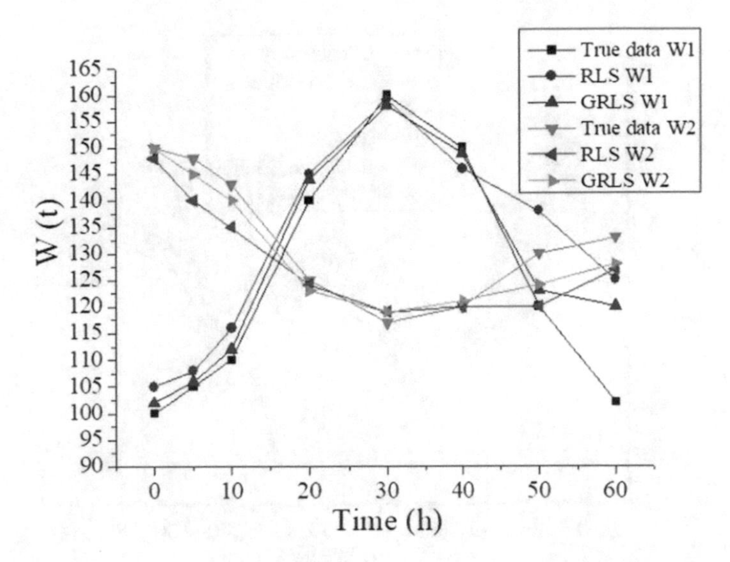

Fig. 13.6 Comparison of W1 and W2 correction results with persistent significant errors

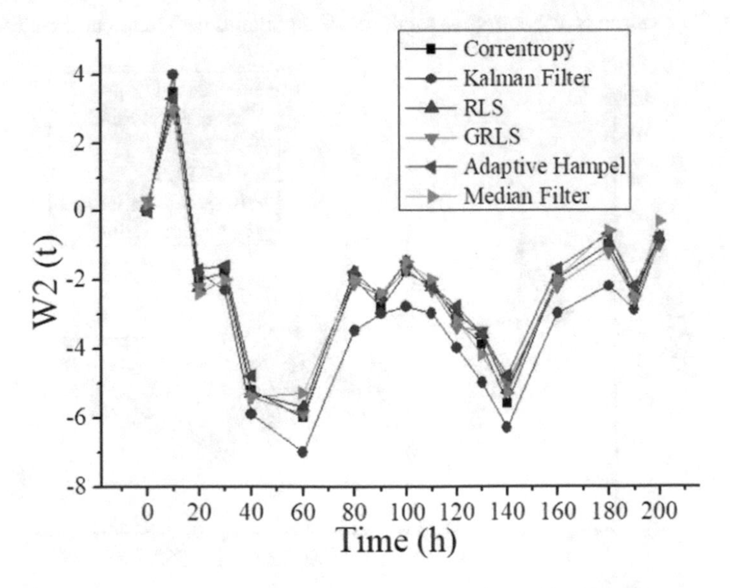

Fig. 13.7 Comparison of W2 correction results of six algorithms without significant errors

At this time, the results of the Kalman filtering independent significant error are the worst, and the W2 correction results of the other five algorithms are the same.

Assuming that there are significant continuous errors, the comparison results of the W2 correction bias of the six algorithms are shown in Fig. 13.9. When the interference lasts for more than three cycles, the W2 correction effect of the Kalman filtering becomes worse, and the effects of other algorithms are the same.

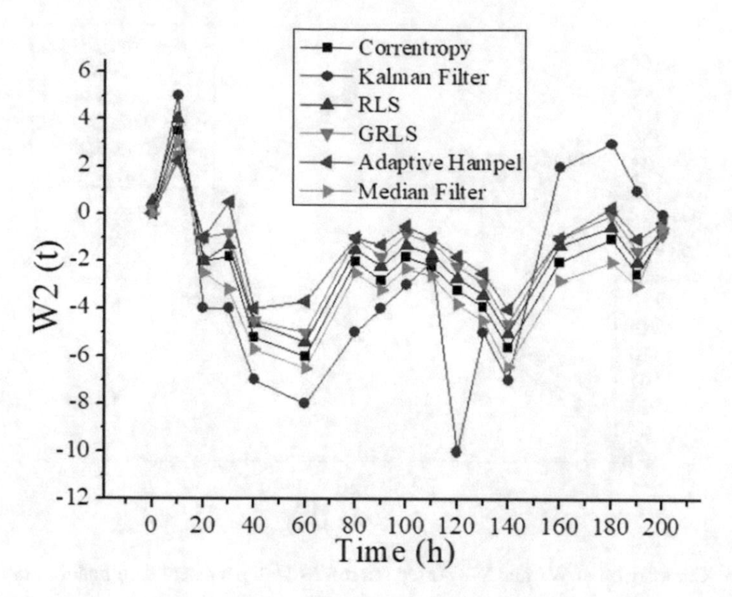

Fig. 13.8 Comparison of W2 correction results of six algorithms with transient significant errors

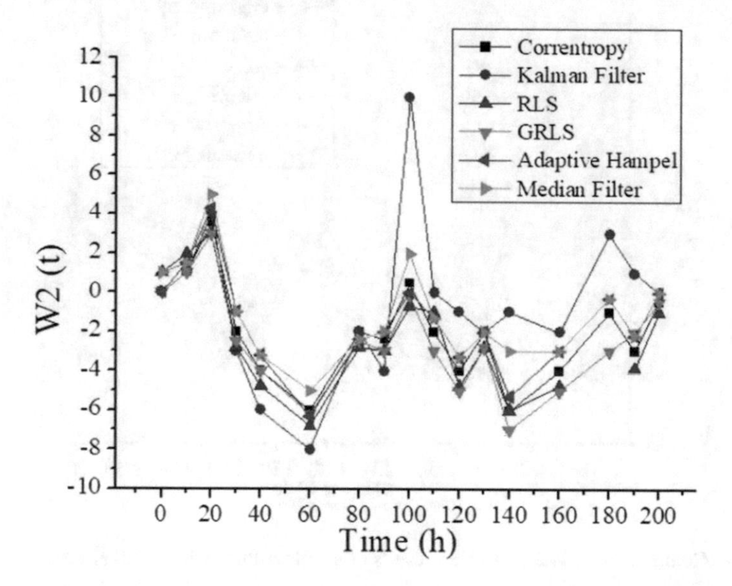

Fig. 13.9 Comparison of W2 correction results of six algorithms with persistent significant errors

It is confirmed that the GRLS algorithm, like most algorithms, has better robustness to significant error correction.

When the time of the interference persists for a long time, the Adaptive Hampel filtering and the Median Filter filtering reach the collapse point at different times in a cycle, as shown in Fig. 13.10. When interference persists, the robustness of Adaptive Hampel filtering will fluctuate, and sometimes not all significant errors can be detected, as shown in Fig. 13.11. The robustness of the GRLS algorithm is still stable when significant errors persist for a long time, which indicates that the performance of the GRLS algorithm is better.

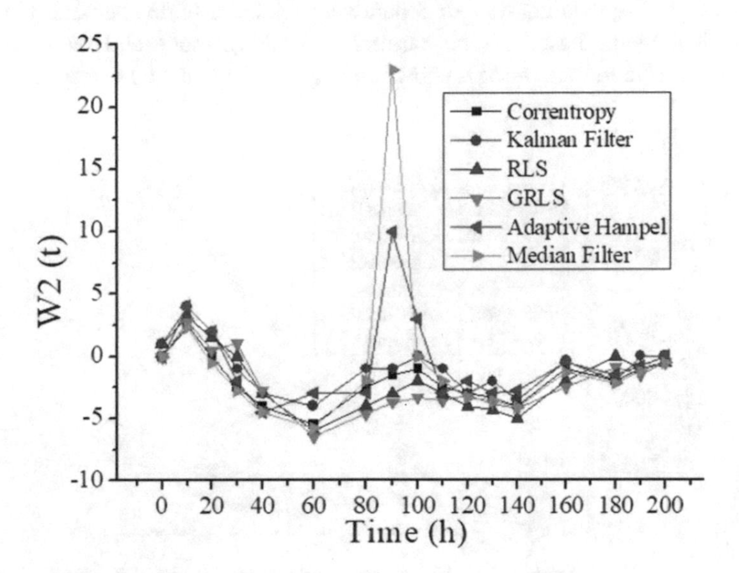

Fig. 13.10 Comparison of W2 correction bias of six algorithms with significant errors persisting for a long time

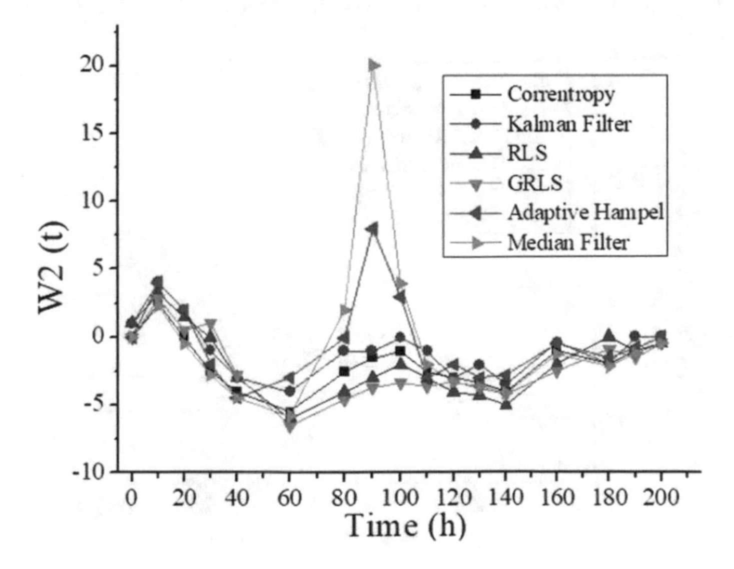

Fig. 13.11 Comparison of W2 correction bias stability of six algorithms

13.4.2 Analysis of the Operation Effect of Complex Event Mode Based on Industrial Data

The comparison between the data analysis results based on complex event model and the single machine operation results of the iron mine flotation plant is shown in Fig. 13.12. The data analysis of 3 data sets for large-scale chemical plants and metallurgical plants based on the parallel computing mode and the comparison results of the stand-alone mode is shown in Figs. 13.13 and 13.14, respectively.

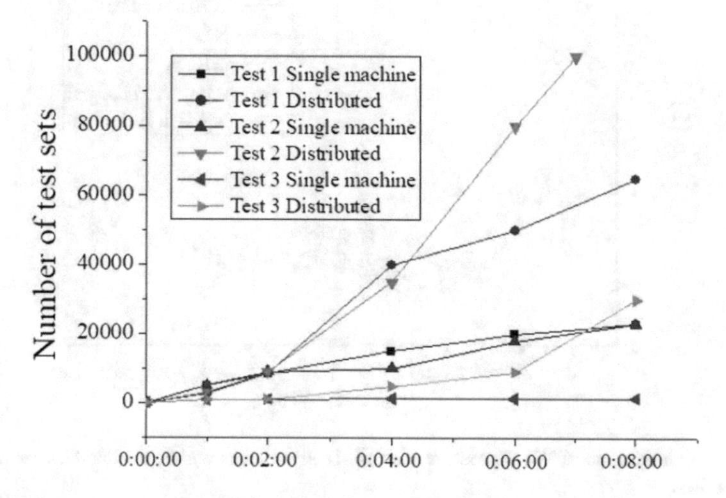

Fig. 13.12 Operation effect of complex event mode in iron mine flotation plant

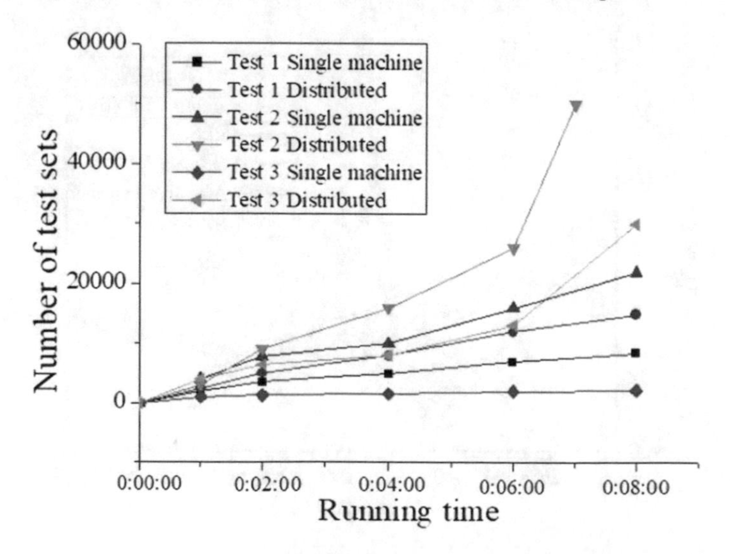

Fig. 13.13 Operation effect of complex event mode in a chemical plant

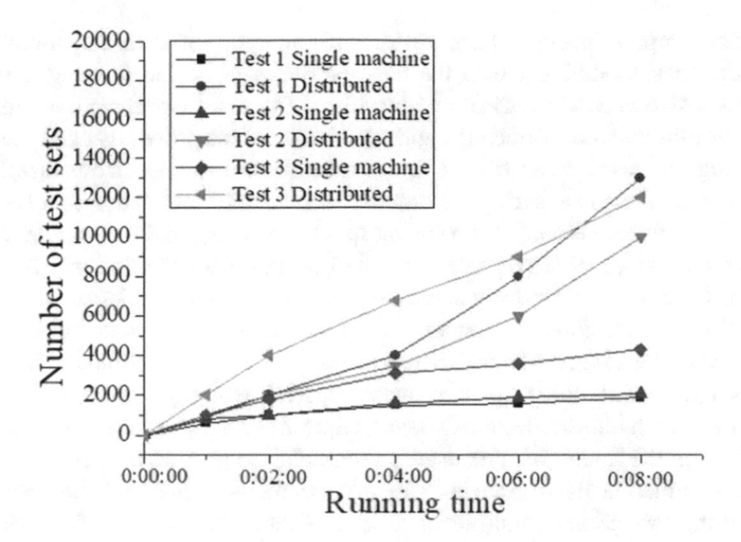

Fig. 13.14 Operation effect of complex event mode in metallurgical plant

In the initial stage, the data processing speed of the single machine environment is faster than that of the parallel environment; after some time, the data processing speed of the parallel environment is accelerated. In a parallel environment, as the number of data increases, the data processing speed will also increase, while the data processing capacity in a single machine environment will affect the data processing speed.

Complex event processing technology is an event-driven data processing technology, which can transform the business process with complex data structure and many influencing factors into event-driven processing. By building a complex event detection model, the controlled parameters that affect the industrial production process are integrated. From a data perspective, the complex event detection model is used to integrate multimodal data, mine the potential of the data, transfer the data into an event stream, and, thus, optimize the flotation process by the event-driven model. In the actual industrial production process, it becomes more meaningful to perceive production fluctuations through event changes.

13.5 Conclusions

With the advancement of information technology, smart cities based on IoT and artificial intelligence technologies have gradually developed. In this era, the upgrading of traditional industrial manufacturing processes is imminent. In the process of industrial transformation, the processing of industrial big data and complex events is a key link in optimizing industrial processes. Therefore, this study builds an industrial complex data model and explores the analytics of advanced complex data in smart cities. Combining with the characteristics of online filtering, this study proposes a GRLS data coordination algorithm. The results confirm that the GRLS has better robustness in correcting significant errors. This proves that complex event processing technology can extract more valuable information from massive and complex data, thereby ensuring the stability of industrial production processes. The topological structure is analyzed by using spectrum theory, and the similarity matrix is mined by using matrix perturbation and eigengap theory to find out its inherent characteristics and to adaptively determine the number of SDN domains.

For the decomposition of complex data of industrial systems and multi-level data correction in the context of smart cities, although a relatively complete framework has been proposed, there are still many special working conditions in actual production. Each industrial site will have unique and detailed problems to be solved. Therefore, in the future, the proposed process will be applied to specific industrial sites to summarize its advantages and disadvantages. This study systematically explores the correction of multi-level complex data in the industrial field. However, the amount of industrial data and data complexity will continue to increase in the future. This study has not discussed the non-linear problems in actual industrial production, which will be the topic of the subsequent study.

References

1. Khatoun, R., & Zeadally, S. (2016). Smart cities: Concepts, architectures, research opportunities. *Communications of the ACM, 59*(8), 46–57.
2. Aujla, G. S., & Jindal, A. (2020). A decoupled blockchain approach for edge-envisioned IoT-based healthcare monitoring. *IEEE Journal on Selected Areas in Communications, 39*(2), 491–499.
3. Habibzadeh, H., Kaptan, C., Soyata, T., et al. (2019). Smart city system design: A comprehensive study of the application and data planes. *ACM Computing Surveys (CSUR), 52*(2), 1–38.
4. Luna-Reyes, L. F. (2018). The search for the data scientist: Creating value from data. *ACM SIGCAS Computers and Society, 47*(4), 12–16.
5. Calyam, P., & Ricart, G. (2018). Research and infrastructure challenges for applications and services in the year 2021. *ACM SIGCOMM Computer Communication Review, 46*(3), 1–5.
6. Singh, M., Aujla, G. S., Singh, A., Kumar, N., & Garg, S. (2020). Deep-learning-based blockchain framework for secure software-defined industrial networks. *IEEE Transactions on Industrial Informatics, 17*(1), 606–616.
7. Aujla, G. S., Prodan, R., & Rawat, D. B. (2021). Big data analytics in Industry 4.0 ecosystems. https://doi.org/10.1002/spe.3008
8. Schleicher, J. M., Vögler, M., Dustdar, S., et al. (2016). Enabling a smart city application ecosystem: Requirements and architectural aspects. *IEEE Internet Computing, 20*(2), 58–65.
9. Zhang, K., Ni, J., Yang, K., et al. (2017). Security and privacy in smart city applications: Challenges and solutions. *IEEE Communications Magazine, 55*(1), 122–129.
10. Zhu, C., Shu, L., Leung, V. C. M., et al. (2017). Secure multimedia big data in trust-assisted sensor-cloud for smart city. *IEEE Communications Magazine, 55*(12), 24–30.
11. Xiao, Z., Fu, X., & Goh, R. S. M. (2017). Data privacy-preserving automation architecture for industrial data exchange in smart cities. *IEEE Transactions on Industrial Informatics, 14*(6), 2780–2791.

12. Menouar, H., Guvenc, I., Akkaya, K., et al. (2017). UAV-enabled intelligent transportation systems for the smart city: Applications and challenges. *IEEE Communications Magazine, 55*(3), 22–28.
13. Sharma, P. K., Rathore, S., & Park, J. H. (2018) DistArch-SCNet: Blockchain-based distributed architecture with Li-Fi communication for a scalable smart city network. *IEEE Consumer Electronics Magazine, 7*(4), 55–64.
14. Tang, B., Chen, Z., Hefferman, G., et al. (2017). Incorporating intelligence in fog computing for big data analysis in smart cities. *IEEE Transactions on Industrial Informatics, 13*(5), 2140–2150.
15. Mohammadi, M., & Al-Fuqaha, A. (2018). Enabling cognitive smart cities using big data and machine learning: Approaches and challenges. *IEEE Communications Magazine, 56*(2), 94–101.
16. Enayet, A., Razzaque, M. A., Hassan, M. M., et al. (2018). A mobility-aware optimal resource allocation architecture for big data task execution on mobile cloud in smart cities. *IEEE Communications Magazine, 56*(2), 110–117.
17. Xu, X., & Hua, Q. (2017). Industrial big data analysis in smart factory: Current status and research strategies. *IEEE Access, 5*, 17543–17551.
18. Yan, J., Meng, Y., Lu, L., et al. (2017). Industrial big data in an industry 4.0 environment: Challenges, schemes, and applications for predictive maintenance. *IEEE Access, 5*, 23484–23491.
19. Yu, W., & Zhao, C. (2018) Online fault diagnosis in industrial processes using multimodel exponential discriminant analysis algorithm. *IEEE Transactions on Control Systems Technology, 27*(3), 1317–1325.
20. Boukerche, A., Coutinho, R. W. L., & Loureiro, A. A. F. (2019). Information-centric cognitive radio-based networking for content distribution in smart cities. *IEEE Network, 33*(3), 99.
21. Kai, C., Li, H., Xu, L., et al. (2018). Energy-efficient device-to-device communications for green smart cities. *IEEE Transactions on Industrial Informatics, 14*(4), 1542–1551.
22. Ta-Shma, P., Akbar, A., Gerson-Golan, G., et al. (2017). An ingestion and analytics architecture for IoT applied to smart city use cases. *IEEE Internet of Things Journal, 5*(2), 765–774.
23. Ma, M., Preum, S. M., Ahmed, M. Y., et al. (2019). Data sets, modeling, and decision making in smart cities: A survey. *ACM Transactions on Cyber-Physical Systems, 4*(2), 1–28.
24. Shah, S. A., Seker, D. Z., Rathore, M. M., et al. (2019). Towards disaster resilient smart cities: Can internet of things and big data analytics be the game changers? *IEEE Access, 7*(1), 91885–91903.
25. Usman, M., Jan, M. A., He, X., et al. (2019). A survey on big multimedia data processing and management in smart cities. *ACM Computing Surveys (CSUR), 52*(3), 1–29.
26. Hwang, J. Y., An, J. G., Aziz, A., et al. (2019). Interworking models of smart city with heterogeneous internet of things standards. *IEEE Communications Magazine, 57*(6), 74–79.
27. Prandi, C., Mirri, S., Ferretti, S., et al. (2017). On the need of trustworthy sensing and crowdsourcing for urban accessibility in smart city. *ACM Transactions on Internet Technology (TOIT), 18*(1), 1–21.
28. Zhao, L., Wang, J., Liu, J., et al. (2019). Routing for crowd management in smart cities: A deep reinforcement learning perspective. *IEEE Communications Magazine, 57*(4), 88–93.
29. He, Y., Yu, F. R., Zhao, N., et al. (2017). Software-defined networks with mobile edge computing and caching for smart cities: A big data deep reinforcement learning approach. *IEEE Communications Magazine, 55*(12), 31–37.
30. Lu, X., Chen, B., Chen, C., et al. (2018). Coupled cyber and physical systems: Embracing smart cities with multistream data flow. *IEEE Electrification Magazine, 6*(2), 73–83.
31. Gharaibeh, A., Salahuddin, M. A., Hussini, S. J., et al. (2017). Smart cities: A survey on data management, security, and enabling technologies. *IEEE Communications Surveys & Tutorials, 19*(4), 2456–2501.

32. Behrisch, M., Streeb, D., Stoffel, F., et al. (2019). Commercial visual analytics systems-advances in the big data analytics field. *IEEE Transactions on Visualization & Computer Graphics, 25*(10), 3011–3031.
33. Santana, E. F. Z., Chaves, A. P., Gerosa, M. A., et al. (2017). Software platforms for smart cities: Concepts, requirements, challenges, and a unified reference architecture. *ACM Computing Surveys (Csur), 50*(6), 1–37.

Chapter 14
A Meta Sensor-Based Autonomous Vehicle Safety System for Collision Avoidance Using Li-Fi Technology

Amil Roohani Dar, Munam Ali Shah, and Mansoor Ahmed

14.1 Introduction

Approximately 1.35 million people every year are cut short due to road traffic crashes. Around 20 to 50 million people suffer with severe injuries with many experiencing debilities because of their injury. According to World Health Organization (WHO) most of the deaths for children & adults amongst ages 5–29 years were the road traffic injuries. To reduce fatalities and injuries from road traffic crashes WHO acts as a team with partners to be responsible for technical support to countries. Main cause of fatalities and injuries is the rear end collision, and this is 70% out of all vehicle's collisions [1]. Another report according to authors in [2] is that 1.078 million injuries in the USA only are due to rear end collisions. So, there is a need of efficient collision avoidance system in vehicles to reduce the death rate published by WHO. The collision avoidance systems use different tools and technologies to handle the situation of running vehicles. The well-known available communications technologies are (ZigBee, Bluetooth, Li-Fi, and Wi-Fi). A short comparison is discussed in given Table 14.1.

Vehicles in Vehicular Ad hoc Network (VANET) can connect and communicate with each other and share the information regarding necessary actions using the technologies depicted in Table 14.1. The VANET architecture and interaction is shown in Fig. 14.1. VANET avoids the possibilities of an accident and offers a wireless communication between moving vehicles [4–6], as visible in Fig. 14.1.

The authors in [8] described that multimedia applications demand is high. So, people are concerned in applications that could be incorporated with VANET. Wireless Fidelity (Wi-Fi) and Light Fidelity (Li-Fi) are one of the most promising

A. R. Dar · M. A. Shah (✉) · M. Ahmed
Department of Computer Science, COMSATS University Islamabad, Islamabad, Pakistan
e-mail: mshah@comsats.edu.pk; mansoor@comsats.edu.pk

S. Garg et al. (eds.), *Intelligent Cyber-Physical Systems for Autonomous Transportation*, Internet of Things, https://doi.org/10.1007/978-3-030-92054-8_14

Table 14.1 Comparison of wireless communication technologies [3]

Attributes	Wi-Fi	Li-Fi	Bluetooth	ZigBee
Mode operation	Radio waved	Light waves	Short wavelength radio waves	Radio waves
Distance covered	32 m	10 m	–	2.4 GHz
Transmission speed	150 Mbps	1 Gbps	25 Mbps	250 kbit/s

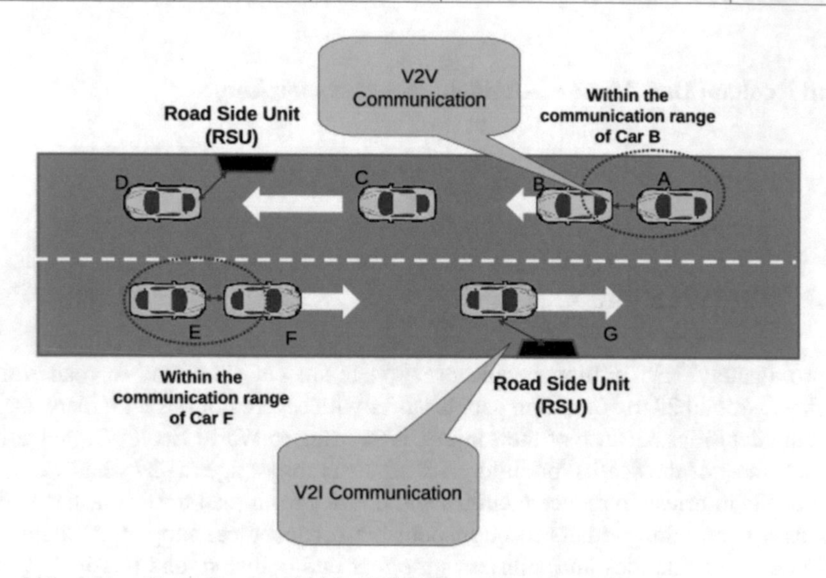

Fig. 14.1 Ad hoc Network using Vehicles (VANETs) [7]

technology to create vehicle communications. Prof. Harald Haas on July 2011 at TED Global Talk introduced the Li-Fi which is a relatively recent technology that refers to a Visible Light Communication (VLC) system which uses light as a pathway for transfer of data at high speeds" [3]. Wi-Fi is a wireless technology that uses radio frequency and enables devices to interact with full speed via a wireless message. Dedicated Short Range Communication (DSRC) was designed from Wi-Fi standard to meet the requirements of V2V communication [9]. In this system, we use Li-Fi technology for message communication between vehicles as seen in Fig. 14.2.

Many researchers have been proposed solutions for collisions avoidance between V2V based on Li-Fi and Wi-Fi technologies. In available literature we did not find any solution to avoid collision using internal sensors monitoring. In [10], the authors proposed rear end collision avoidance controller based on proportional–integral–derivative. Another research was proposed by authors in [11] for vehicles rear end collision avoidance using linear quadratic optimal control technique. In [12], the authors used the physical, environmental, and mental factors to reduce the chances of accident. The authors in [13] analyzed and discuss the road and weather conditions factor in collision occurrence. The authors in [14] also used the environmental factors: road and weather conditions in rear end crash avoidance.

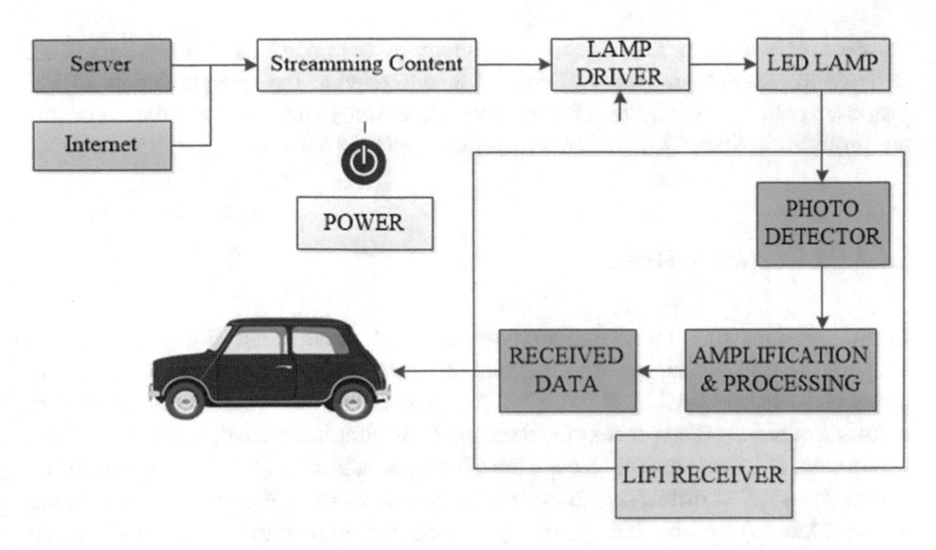

Fig. 14.2 Li-Fi technology

Driver's characteristics can be added in reducing accident chances and increase the flexibility of algorithm. The authors in [15] proposed an algorithm in which they pass the characteristics of driver in the proposed algorithm & showed the significant improvement. Driver's characteristics are also important in decision making because the Warning thresholds can be improved by adding driver experience [16]. There are many implemented and proposed models exist in current research about how to use Li-Fi technology in accident avoidance or collision avoidance. As spectrum of light frequency is vast, it is very advantageous to adopt it in a short-range wireless communication [17, 18].

Collisions happen when an AV contacted other road traffics entities. The main idea to avoiding collision is either change the direction or apply brake to stop AV through deceleration. Our focus is on the second option. So, it becomes logical to focus on AV braking system. The control decisions, such as increasing or decreasing of the speed and the parking, can be made by an autonomous system, according to automation and perception results [19, 20].

There are two types of sensors in AVs:

- Internal sensors (speed sensor, brake sensor, etc.)
- External sensors (radar, camera, etc.).

However, there are inherent safety and security challenges for AVs. If anyone component or sensor of AV is failed or attacked, then internal system of AV will be disrupted. After this, "brain" of AV may be or surely issue the wrong command, and it will directly compromise the safety of road entities [21]. There is need of serious attention towards AVs internal system to control the chances of collision.

The main aim of this chapter is to highlight the issues of internal sensor failure of AVs and to provide a system that can detect any failure in advance and can avoid collision of the AVs. The rest of the chapter is organized as follow: Sect. 14.2 reviews the related studies. In Sect. 14.3, we provide the design details of the proposed collision free system for the AVs. Case Study and other simulation details are provided in Sect. 14.4. We conclude the chapter in Sect. 14.5.

14.2 Literature Review

In the existing literature huge amount of work have been done in collision avoidance-based algorithms and techniques. These proposed solutions are mostly based on Li-Fi and Wi-Fi technologies. External sensors of vehicles collect the data from surroundings then process the data and controller initiate a command after this processing. Internal sensors accept the command and finally vehicle perform the action. We studied different papers and found that most of the work is on external sensors data collection, but no one proposed any algorithm for internal sensor condition checking. Because it is possible that collision can occur due to the failure of internal sensor working. Li-Fi technology is part of our research work for V2V communication for collision avoidance. So, we are presenting some research works for the importance of this technology. The author in [22] described a cost-effective solution for V2V communication using Li-Fi technology. Their purpose was only used the transmitting and receiving circuit for simulation and implementation and described how LEDs can eliminate the need of complex network and protocols. The authors in [23] said that communication using Li-Fi technology can greatly enhance the driver safety in V2V environment. A real time vehicle unidirectional communication test using visible light communication was performed. The authors in [22] described the importance of Li-Fi technology in Intelligent Transportation System (ITS). The proposed work elaborates that in near future data sending and receiving through Li-Fi technology can help in accident avoidance.

The authors in [24] also proposed that VLC is cost-effective solution for V2V communication. The proposed idea was to use brake lamp for information sending so the following vehicle can avoid from collision. The authors in [25] discussed Li-Fi technology as an alternative solution due to the challenges of latency and bandwidth in near future. Comparison table from their research provide us the speed of Li-Fi, Wi-Fi, and WIMAX. But they did not perform experiments for their comparison between these technologies. The authors in [26] proposed and implemented the concept of inter vehicles communication. The concept was to convey message to next vehicle about driving style or fuel leakage. The authors in [27] described Li-Fi as ecofriendly technology which is healthy and proposed a V2V communication framework using Li-Fi. Authors in proposed a different approach for V2V communication using LED as transmitter at the sender side and CMOS [28] image sensor for receiving the signal at receiver side. Currently they did not use this to control V2V collision avoidance. The authors in [29] proposed a system to avoid collisions by detecting driver condition. For the said purpose they used alcohol sensor and sleeping sensor to detect the driver condition. If results are positive, the message is communicated using Li-Fi technology. The authors in [30] presented novel scheme for V2V front and rear collision avoidance using Li-Fi technology. They presented two scenarios, i.e., first send message to rear vehicle and second send message to front vehicle at T-junction.

The authors in [31] proposed a new V2V collision avoidance method using ZigBee. This technique has several drawbacks listed by authors, e.g., expensive, fail in heavy rain, snow, etc. Arduino with ultrasonic sensor and Li-Fi module-based framework was proposed for smart car for collision detection [32]. The ultrasonic sensor was used to measure the distance between vehicles. The authors used the Li-Fi for data transmission between V2V. The authors in [3] discussed the advantages and disadvantages of Li-Fi. This paper demonstrated the use of Li-Fi in controlling the speed of motors. The authors in [33] provide an overview of proposed technologies for collision prediction. In this work authors proposed the use of Li-Fi and Wi-Fi in V2V communication for collision avoidance. Two scenarios were discussed considering a front and rear vehicle and vehicles at the T-junctions. No implementation, results, and comparison provided in their work. According to authors in [34] Li-Fi is a must have feature by most of the manufacturer of next generation cars. They proposed the use of already available feature of automatic high beam controller and an acquisition camera to read the beam fluctuations at night.

The authors in [34] proposed a prototype model for V2V communication for accident prevention. The proposed system on facing any obstacle applies the brake and this information is transmitted to the following vehicle. Implementation was performed successfully but no results or comparison is provided. According to author their proposed system is simple, low cost and friendly in accident prevention. Emerging technologies are the part of intelligent transportation system now. Despite these technologies, collision avoidance is still an open challenge. The authors in [35] tried to overcome the fatigue driving for collision avoidance. The proposed system assesses the driver drowsiness using machine learning model and trigger the alarm. Now the trending topics in the field of transportation are V2V &V2I communication for effective information transfer in real time systems. The author in this paper [9] discussed different wireless communication technologies. They proposed a framework where every vehicle must have communication device with Li-Fi and Wi-Fi. At the end authors suggested that these two complementary communication methods can be used in various ways in future in intelligent transportation system. Though, they have implemented the proposed model, they did not provide any results and comparisons. In [36], the authors proposed Li-Fi based architecture for collision avoidance at both front and rear ends. Authors provided only theoretical knowledge. No implementation and results are discussed. Table 14.2 summarizes the V2V Collision Avoidance.

Table 14.2 V2V collision avoidance

Sr. no.	Year	Purpose	Tools and technology used	Internal sensor failure	References
1	2015	Proposed for collision avoidance	Li-Fi and Proteus	No	[20]
2	2016	Data transmission in V2V	Li-Fi	No	[19]
3	2016	V2V communication for collision avoidance	Li-Fi and Arduino pro mini	No	[21]
4	2016	Communicating warning between V2V	Li-Fi, micro controller, MMA7361L	No	[22]
5	2017	Used as data transmission medium in VANET	Li-Fi	No	[23]
6	2017	V2V communication only	Li-Fi and micro controller	No	[37]
7	2017	V2V communication	Li-Fi	No	[24]
8	2017	Accident avoidance in automobiles	Li-Fi and Raspberry PI, CMOS image sensor	No	[25]
9	2017	Prevent road accidents	Li-Fi, ultrasonic sensor, alcohol sensor and eye blink sensor	No	[27]
10	2018	Collision avoidance	Li-Fi, Proteus	No	[27]
11	2018	V2V collision avoidance	ZigBee, 8051 micro controller	No	[28]
12	2018	Collision detection system	Li-Fi and ultra sonic sensor and Arduino	No	[38]
13	2019	V2V communication only	Li-Fi and Arduino	No	[3]
14	2019	Crash avoidance system for V2I	–	No	[30]
15	2020	Car to car communication	Li-Fi, automatic high beam controller and acquisition camera	No	[31]
16	2020	V2V accident avoidance	Li-Fi, Arduino UNO, Arduino MEGA, LED, Ultrasonic, GSM, and GPS	No	[32]
17	2020	V2V collision avoidance	Li-Fi, Ultrasonic Sensor, Infrared (IR) Sensor	No	[33]
18	2020	V2V communication	Wi-Fi and Li-Fi	No	[39]
19	2021	V2V collision avoidance	Li-Fi, LCD, Ultrasonic, Arduino, and Webcam	No	[34]

14.3 Proposed Framework

In this section, we provide a detailed design overview of the proposed system which helps in the collision avoidance amongst the AVs on the road. The proposed system is comprised of different network layers such as cloud, fog, and IoT layers. Cloud computing provides fastest computing services over the Internet. We are proposing this technology in our framework because with the use of cloud computing technologies, AVs on roads can communicate with each other to prevent accidents. AVs can update maps, traffic information, and route the passengers to their destination quickly. This technology is now part of automotive IoT industry. Data collected in cloud can be analyzed for further improvements in AVs behavior on roads. The fog computing layer is also used in the proposed framework which helps to reduce the traffic of cloud server. Figure 14.3 presents a typical workflow

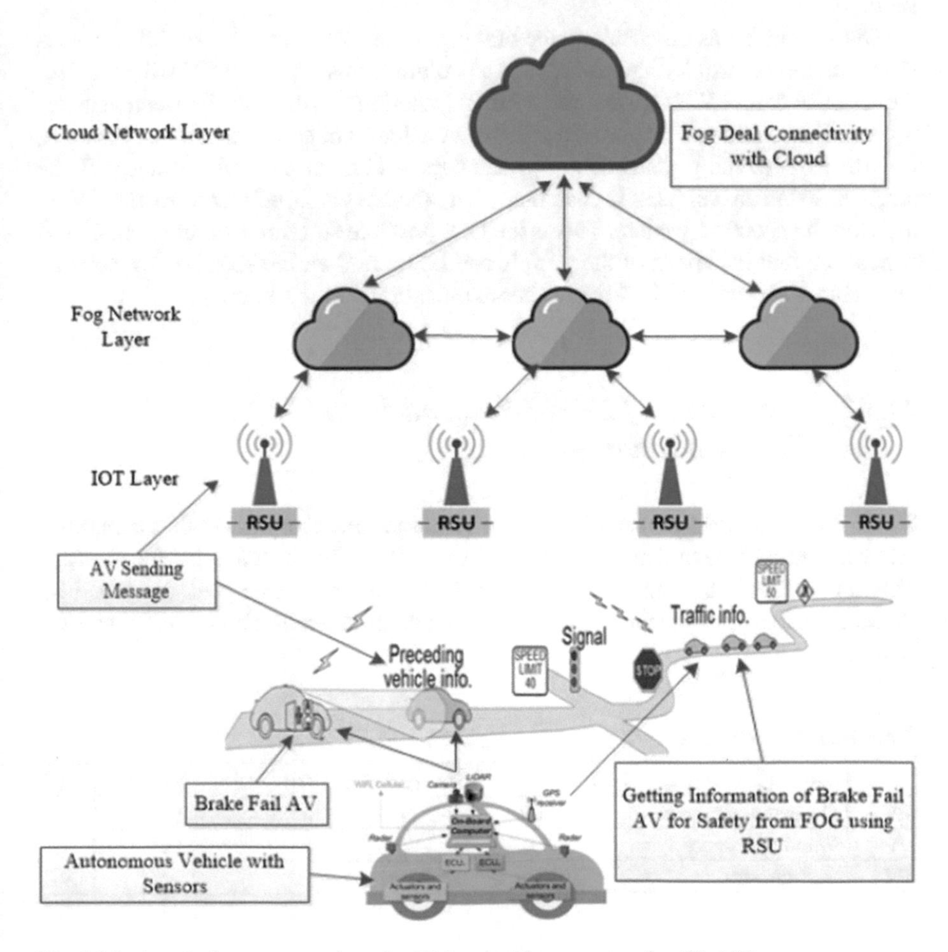

Fig. 14.3 A typical meta sensor-based collision avoidance system for AVs [40]

of an AV environment [40]. It can be observed that when an AV Meta Sensor detects the brake status as failed, it sends message to the fog layer using the roadside unit (RSU). The brake failure message in an AV is passed to the LV with Li-Fi to avoid collision. An FV cannot apply brakes so a LV can avoid the accident with increase in its speed and by maintaining the safe distance.

14.3.1 Proposed System Flowchart

The proposed system working is based on Meta Sensor status. If the status is "on," this indicates that the braking system of the AV is being monitored. If the brake fails, a failure indicator is generated, and the system rechecks the braking system status after some timer. This allows the braking system to recover itself from the failure state.

The system keeps on checking the braking of the AV. If the "brake fail" status is true, then the control will go to distance calculation function. The FV will calculate the distance from LV. There are numerous calculations on distance between vehicles for safety operations. For our research, we consider 5 m as the dangerous distance, 6–10 m as a warning distance and greater than 11 m is the safe distance. If the distance between vehicles is less than 5 m, the message will pass to the LV to increase the speed of vehicle. This is the best possible solution to avoid collision. It is possible that the speed of the vehicle can be more than the speed of LV vehicle. See Table 14.3 and Fig. 14.4 for further explanation and working.

14.3.2 Description of LV & FV Scenario and Li-Fi Communication

Figure 14.5 describes the working of two Arduino microcontrollers that communicate when Meta Sensor detects the brake failure. It can be observed that the message from FV will only be sent to the LV when there is only an emergency. Figure 14.6 shows the use of Li-Fi technology for message communication between FV and LV vehicles.

Table 14.3 V2V Distance

	1 m	2 m	3 m	4 m	5 m	6 m	7 m	8 m	9 m	10 m	>10
FV	Dangerous distance				LV						
FV	Warning distance									LV	
FV	Safe distance										LV

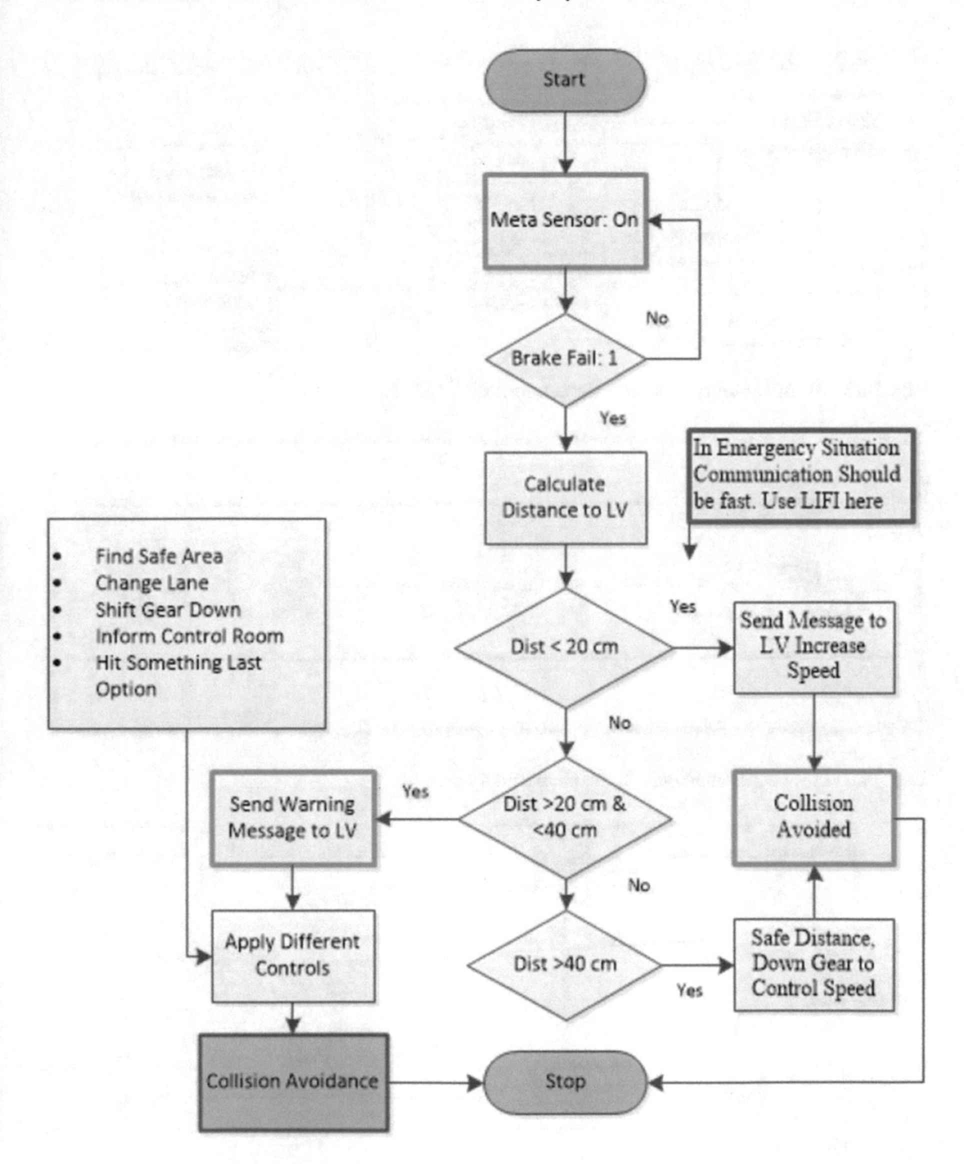

Fig. 14.4 Flowchart of the proposed system for collision avoidance in AVs

14.3.3 Case Study of the Need of the Meta Sensor

In this section, we discuss the critical role of actuator which we call controller phase in AV. The actuator ensures the desirable behavior and steady state response. The actuator control part comes at the end in AV as seen in figure below. It is very important to know the state of AV internal components which perform the action on the command issued by controller. Now think of that brake component is waiting for

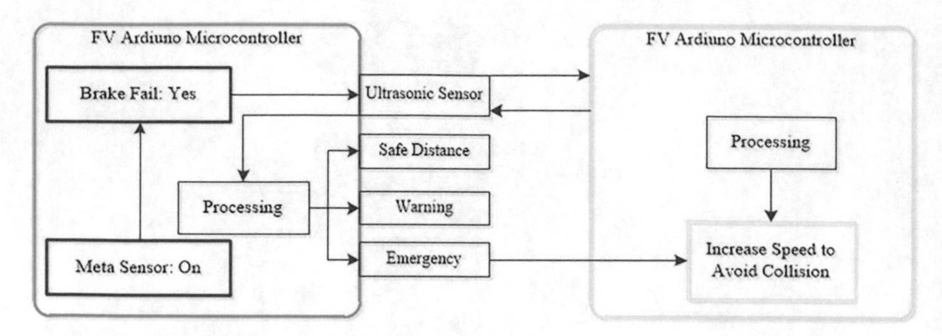

Fig. 14.5 FV & LV with Arduino Microcontroller

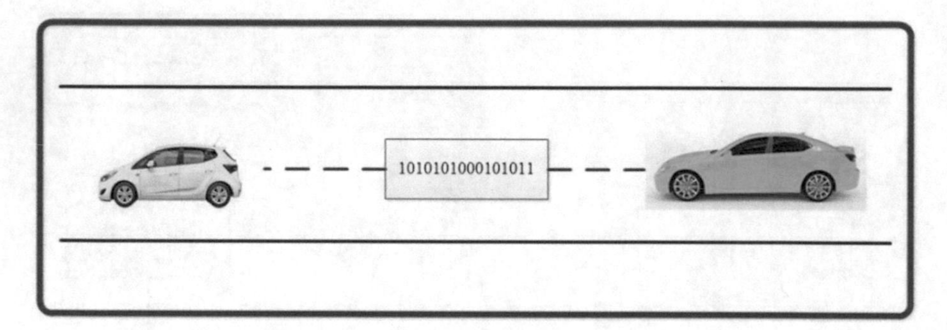

Fig. 14.6 Li-Fi communication in the proposed system

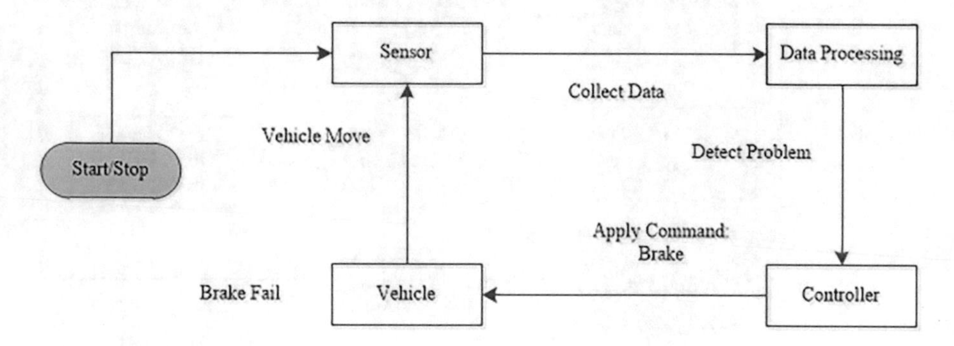

Fig. 14.7 Autonomous vehicle control process

apply brake from controller. When the controller issue applies command and brake component is not applying the brake. Vehicle will move forward due to brake failure and collisions chances are high as the driver can see in Fig. 14.7. We try to explain this in scenario here. In Fig. 14.8 we proposed the position of Meta Sensor.

Scenario If an AV or human driver is travelling at the speed of 120 km/h and suddenly it detected a human or anything else in front of him. Surely, it will try to apply brake to avoid accident. This stage is very critical when the driver applies

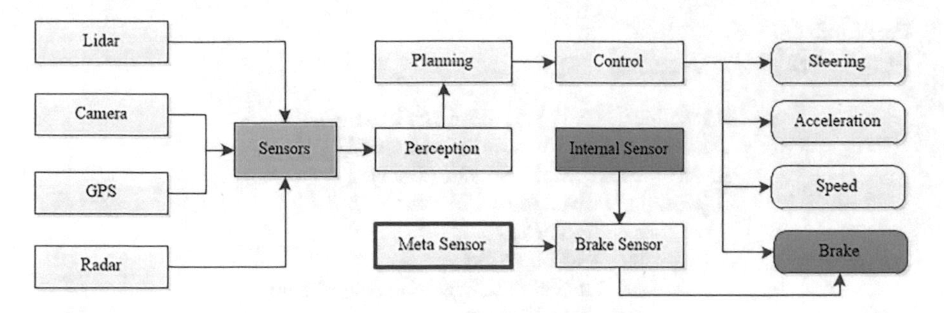

Fig. 14.8 AV system with meta sensor position

the brake, and the driver determines that there is a brake failure. Here, the chances of an accident or a collision are at peak values. To avoid from this critical situation, the proposed Meta Sensor-based solution which will intimate the AV or human driver about the condition of internal sensor working.

14.3.4 Importance of Brake Sensor

The modern AVs are equipped with built-in technology that allows the driver to become aware of the status and condition of the vehicle at any moment. The status and condition of brake system of the vehicle has too much importance that keeps the driver safe in sudden danger situation. Here, the brake sensor works and informs the driver that whether or not the brake is working because this particular sensor is attached to break pad itself.

14.3.5 Meta Sensor

The Meta sensor works by checking the condition of sensors working under his control as seen in Fig. 14.8. To overcome the scenario discussed in case study, we apply a Meta sensor over brake sensor. This sensor helps us in knowing the current condition of brake before it is too late. Characteristics observed by Meta Sensor:

- Brake quality
- Maintenance history
- Oil status
- Brake lifetime
- Speed of brake

```
Procedure ():
        While (True)
        {
                BS_Value-≥(Get Value from Brake sensor)
                Data = Calculate Current Situation of BS_Value
                Result = Brake_Meta_Sensor (Data)
                If (Result == Chances of Collision)
                        a.   Hooter On
                        b.   Red Light On
                        c.   Collect data and send to Fog/Cloud
                        d.   Auto Pilot On
                             If (Speed is very high)
                                     •   Reduce Speed
                                     •   Update the RSU
                             If (Leading Vehicle in Danger Zone)
                                     •   Inform the Leading Vehicle to Change Lane
                                     •   Leading Vehicle Increase Speed
                             If (Gear Status is high)
                                     •   Shift Gear to Lower
                             If (Brake Oil is Low)
                                     •   Find the Maintenance
                             If (Speed is Low)
                                     •   Park AV in Safe Lane
                Else
                        Control back to the AV.
                End if
        }
```

Fig. 14.9 Brake meta sensor algorithm

14.3.6 Proposed Algorithm

Figure 14.9 shows algorithm checks the input values from brake sensor and passes these values to the Brake Meta Sensor system. If chances of accident are high then it will apply given options otherwise, it will execute normal behavior.

14.4 Simulation and Results

For simulation and results, we used the NetLogo 6.2.0 simulation model [41]. Most of the computers can run NetLogo. It is a multi-agent programmable modeling environment. NetLogo web version is also available for researchers to run simulations.

Figure 14.10 shows the proposed system interface in NetLogo. We can set number cars, acceleration, and deceleration speed according to requirements. Figures 14.11 and 14.12 shows the result when brake failure is detected by Meta

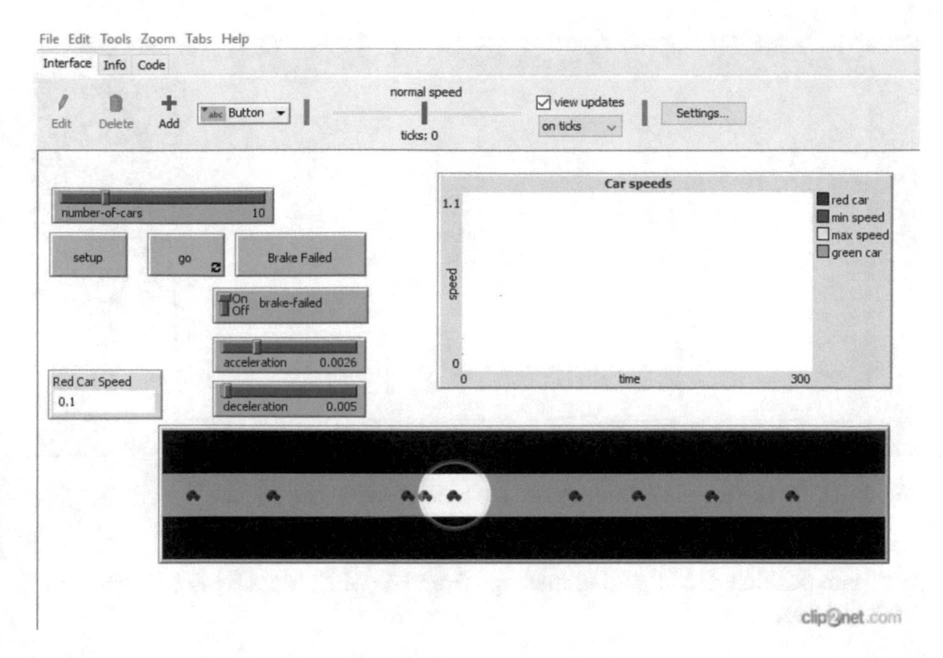

Fig. 14.10 Proposed system emulation interface using NetLogo

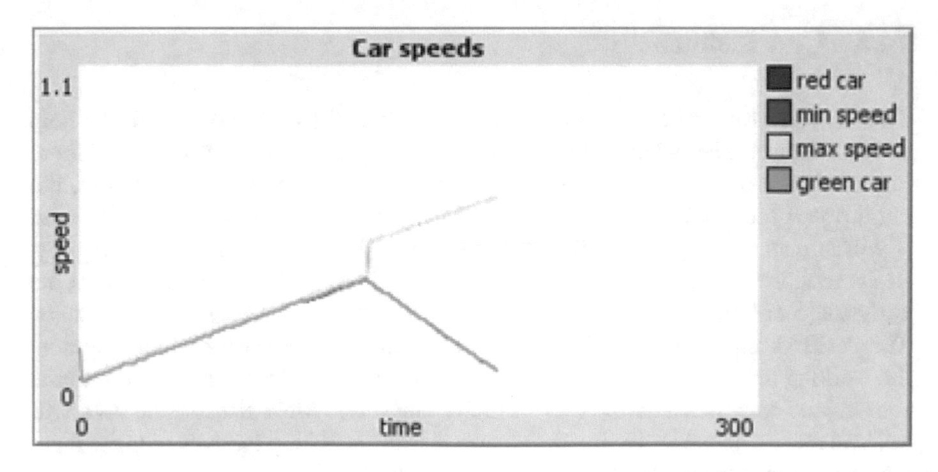

Fig. 14.11 The LV increases and FV decreases the speed when a brake failure is detected

Sensor. Yellow line in graph of Figs. 14.11 and 14.12 is for LV whereas green line is for FV. Green line drop shows the declaration in speed of FV because brake is failed, and it cannot stop immediately. The decrease in green line shows FV will stop when approaching speed to zero. Increase in Yellow line in results shows the increase of speed of LV to avoid collision. Because, if LV does not increase the speed, then chances of accident are very high. Increase in Yellow line shows LV

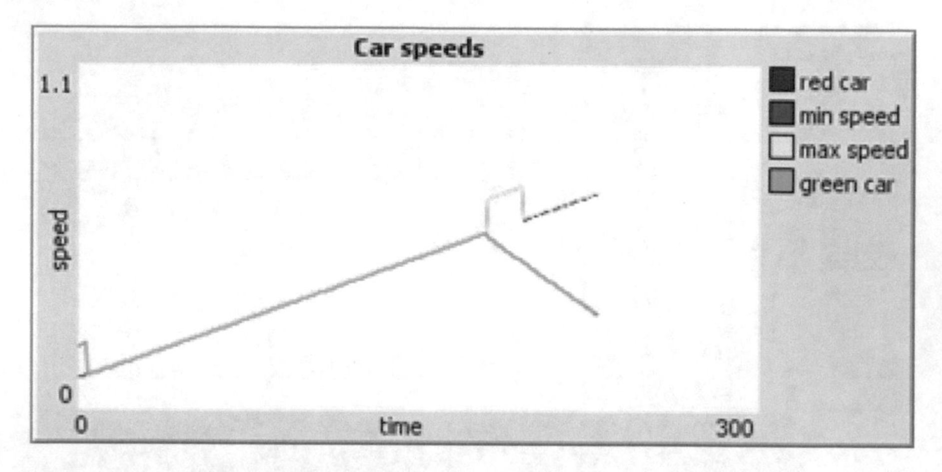

Fig. 14.12 LV increase and FV decrease in speed when brake failure detected

is in safe position due to getting message from FV when Meta Sensor detected the brake failure.

14.5 Conclusions

This chapter provided a new concept of collision avoidance in AVs with the help of Meta Sensor. Hardware and software failure is possible at any stage. But there is a need of a mechanism which can detect this issue in advance. Otherwise, the entities will remain in danger on roads. Usually, when an AV brake is failed then it will inform the fog layer using RSU to update the other AVs in the proximity. In the proposed system, an AV with Meta Sensor installed in it can get alert in the early stage of brake failure & can overcome the situation as illustrated in results. This research might be helpful for AV vendors to embed a sensor which monitors the working and condition of other sensors. This sensor can be applied to internal sensors and to external sensors as well. By using this Meta Sensor, the AVs will become more capable in dealing the emergency and will help the AVs to make them trustworthy in safe travelling and collision free driving.

References

1. Meng, Q., & Qu, X. (2012). Estimation of rear-end vehicle crash frequencies in urban road tunnels. *Accident Analysis & Prevention, 48*, 254–263.
2. Chen, C., Zhang, G., Tarefder, R., Ma, J., Wei, H., & Guan, H. (2015). A multinomial logit model-Bayesian network hybrid approach for driver injury severity analyses in rear-end crashes. *Accident Analysis & Prevention, 80*, 76–88.

3. George, R., Vaidyanathan, S., Rajput, A. S., & Deepa, K. (2019). Li-Fi for vehicle to vehicle communication–A review. *Procedia Computer Science, 165*, 25–31.
4. Sheikh, M. S., & Liang, J. (2019). A comprehensive survey on VANET security services in traffic management system. *Wireless Communications and Mobile Computing, 2019*. https://doi.org/10.1155/2019/2423915
5. Dua, A., Sharma, P., Ganju, S., Jindal, A., Aujla, G. S., Kumar, N., & Rodrigues, J. J. P. C. (2018). RoVAN: A rough set-based scheme for cluster head selection in vehicular ad-hoc networks. In: *2018 IEEE Global Communications Conference (GLOBECOM)* (pp. 206–212). IEEE.
6. Garg, S., Singh, A., Kaur, K., Aujla, G. S., Batra, S., Kumar, N., & Obaidat, M. S. (2019). Edge computing-based security framework for big data analytics in VANETs. *IEEE Network, 33*(2), 72–81.
7. Abdulshaheed, H. R., Yaseen, Z. T., Salman, A. M., & Al_Barazanchi, I. (2020). A survey on the use of WiMAX and Wi-Fi on Vehicular Ad-Hoc Networks (VANETs). *IOP Conference Series: Materials Science and Engineering, 870*(1), 012122.
8. Akbar, M. S., Khan, M. S., Khaliq, K. A., Qayyum, A., & Yousaf, M. (2014). Evaluation of IEEE 802.11 n for multimedia application in VANET. *Procedia Computer Science, 32*, 953–958.
9. Yogarayan, S., Razak, S. F. A., Azman, A., Abdullah, M. F. A., Raman, K. J., Muthu, K. S., & Ibrahim, S. Z. (2020). A comprehensive study of vehicle communication framework in Malaysia. *Journal of Physics: Conference Series, 1502*(1), 012012.
10. Moon, S., Moon, I., & Yi, K. (2009). Design, tuning, and evaluation of a full-range adaptive cruise control system with collision avoidance. *Control Engineering Practice, 17*(4), 442–455.
11. van den Berg, J., Wilkie, D., Guy, S. J., Niethammer, M., & Manocha, D. (2012). LQG-obstacles: Feedback control with collision avoidance for mobile robots with motion and sensing uncertainty. In: *IEEE International Conference on Robotics and Automation* (pp. 346–353).
12. Razzaq, S., Riaz, F., Mehmood, T., & Ratyal, N. I. (2016). Multi-factors based road accident prevention system. In: *2016 International Conference on Computing, Electronic and Electrical Engineering (ICE Cube)* (pp. 190–195).
13. Malin, F., Norros, I., & Innamaa, S. (2019). Accident risk of road and weather conditions on different road types. *Accident Analysis & Prevention, 122*, 181–188.
14. Shangguan, Q., Fu, T., & Liu, S. (2020). Investigating rear-end collision avoidance behavior under varied foggy weather conditions: A study using advanced driving simulator and survival analysis. *Accident Analysis & Prevention, 139*, 105499.
15. Zhang, Y. J., Du, F., Wang, J., Ke, L. S., Wang, M., Hu, Y., Yu, M., Li, G. H., & Zhan, A. Y. (2020). A safety collision avoidance algorithm based on comprehensive characteristics. *Complexity, 2020*. https://doi.org/10.1155/2020/1616420
16. Ashraf, I., Hur, S., Shafiq, M., & Park, Y. (2019). Catastrophic factors involved in road accidents: Underlying causes and descriptive analysis. *PLoS One, 14*(10), e0223473.
17. Takai, I., Harada, T., Andoh, M., Yasutomi, K., Kagawa, K., & Kawahito, S. (2014). Optical vehicle-to-vehicle communication system using LED transmitter and camera receiver. *IEEE Photonics Journal, 6*(5), 1–14.
18. Wang, J., Zou, N., Dong, W., Kentaro, I., Zensei, I. H. A., & Namihira, Y. (2012). Experimental study on visible light communication based on LED. *The Journal of China Universities of Posts and Telecommunications, 19*, 197–200.
19. Davis, L. C. (2020). Optimal merging into a high-speed lane dedicated to connected autonomous vehicles. *Physica A: Statistical Mechanics and Its Applications, 555*, 124743.
20. Millard-Ball, A. (2018). Pedestrians, autonomous vehicles, and cities. *Journal of Planning Education and Research, 38*(1), 6–12.
21. Cui, J., & Sabaliauskaite, G. (2018). US: An unified safety and security analysis method for autonomous vehicles. In: *Future of Information and Communication Conference* (pp. 600–611).

22. Al Abdulsalam, N., Al Hajri, R., Al Abri, Z., Al Lawati, Z., & Bait-Suwailam, M. M. (2015). Design and implementation of a vehicle to vehicle communication system using Li-Fi technology. In: *2015 International Conference on Information and Communication Technology Research, ICTRC 2015* (pp. 136–139). https://doi.org/10.1109/ICTRC.2015.7156440
23. Mohammed, D., Bourzig, D. K. D., Abdelkim, M., & Mokhtar, K. (2017, May 15). Digital data transmission via Visible Light Communication (VLC): Application to vehicle to vehicle communication. In: *4th International Conference on Control Engineering and Information Technology, CEIT 2016*. https://doi.org/10.1109/CEIT.2016.7929059
24. Mamatha, K. R., & Pavithra, S. (2018). Visible light communication in intelligent transportation system for I2V and V2V mode. *International Research Journal of Engineering and Technology*. www.irjet.net
25. Agyemang, J. O., Kponyo, J. J., & Mouzna, J. (2017). Light fidelity (Li-Fi) as an alternative data transmission medium in VANET. In: *2017 European Modelling Symposium (EMS)* (pp. 213–217).
26. Gurav, M. S. S., Ghatage, A. F., Nandgave, S. S., & Kole, R. S. (2019). Inter-vehicle communication system using Li-Fi technology. *Journal of Telecommunication Study, 4*(2), 1–5.
27. Kulkarni, S., Darekar, A., & Shirol, S. (2017). Proposed framework for V2V communication using Li-Fi technology. In: *2017 International Conference on Circuits, Controls, and Communications (CCUBE)* (pp. 187–190).
28. Anitha, R., Bharathi, S., Jayalakshmi, J., Nancy, J., & Thiruppathi, M. (2017). Accident avoidance by using Li-Fi technology in automobiles. *Asian Journal of Applied Science and Technology, 1*(2). www.ajast.net
29. Singh, P. (n.d.). VANET based intelligent transportation system.
30. Aleem Jamali, A., Kumar Rathi, M., Hakeem Memon, A., & Das, B. (2018). Collision Avoidance between vehicles through Li-Fi based communication system. *IJCSNS International Journal of Computer Science and Network Security, 18*(12), 81.
31. Jahnavi, M. (2018). Vehicle to vehicle communication for collision avoidance. *International Journal for Research in Applied Science and Engineering Technology, 6*(5), 1380–1386. https://doi.org/10.22214/ijraset.2018.5227
32. Krishnan, P. (2018). Design of collision detection system for smart car using Li-Fi and ultrasonic sensor. *IEEE Transactions on Vehicular Technology, 67*(12), 11420–11426. https://doi.org/10.1109/TVT.2018.2870995
33. Kavyapriya, S. (2019). Review paper on vehicle to vehicle communication for crash avoidance system. *Journal of Electrical & Electronic Systems, 8*(295), 2332–2796.
34. Sivakumar, S., Alagumurugan, A., Baala Vignesh, G., & Dhanush, S. (2020). Accident analysis and avoidance by V2V communication using LI-FI technology. *International Research Journal of Modernization in Engineering Technology and Science, 730*. www.irjmets.com
35. Li, Y. (2020). Anti-fatigue and collision avoidance systems for intelligent vehicles with ultrasonic and Li-Fi sensors. In: *2020 3rd IEEE International Conference on Information Communication and Signal Processing, ICICSP 2020* (pp. 203–209). https://doi.org/10.1109/ICICSP50920.2020.9232054
36. Yuvaraju, M., Benson Mansingh, P. M., & Sekar, G. (2021). A LI-FI based collision avoidance system for vehicles using visible light communication. In: *Proceedings of the 7th International Conference on Electrical Energy Systems, ICEES 2021* (pp. 114–116). https://doi.org/10.1109/ICEES51510.2021.9383736
37. Nachimuthu, S., Pooranachandran, S., & Aarthi, B. S. (2016). Design and implementation of a vehicle to vehicle communication system using Li-Fi technology. *International Research Journal of Engineering and Technology*. www.irjet.net
38. Santhosam, I. B. (2017). VANET based Intelligent TransportationSystem using Li-Fi technology. *IOSR Journal of Electronics and Communication Engineering, 12*(02), 30–32. https://doi.org/10.9790/2834-1202013032

39. Spahiu, C. S., Stanescu, L., Brezovan, M., & Petcusin, F. (2020, October 27). Li-Fi technology feasibility study for car-2-car communication. In: *Proceedings of the 2020 21st International Carpathian Control Conference, ICCC 2020*. https://doi.org/10.1109/ICCC49264.2020.9257263
40. Cui, J., Liew, L. S., Sabaliauskaite, G., & Zhou, F. (2019). A review on safety failures, security attacks, and available countermeasures for autonomous vehicles. *Ad Hoc Networks, 90*, 101823.
41. Tisue, S., & Wilensky, U. (2004). NetLogo: Design and implementation of a multi-agent modeling environment. In: *Proceedings of the Agent 2004 Conference on Social Dynamics: Interaction, Reflexivity and Emergence, Chicago, IL*. Center for Connected Learning and Computer-Based Modeling. Northwestern University, Evanston, IL.

Part V
Security Perspective in Intelligent Transportation Systems

Chapter 15
Secure Information Transmission in Intelligent Transportation Systems Using Blockchain Technique

Anju Devi, Geetanjali Rathee, and Hemraj Saini

15.1 Introduction

The pervasive computing technique is growing exponentially over the years. In addition, the engagement of IoT in various fields such as home automation, industrial application, and agriculture is on heavy side [1]. The self-automation and self-organization network integrated with IoT systems give rise to the engagement of various data communication nodes [2, 3]. The involvement of automated systems while sharing, exchanging, and managing of information within IoT devices pose various issues in the network that may lead to verify each and every communication node before transmission process.

Further, the involvement of various computing techniques such as big data, Artificial Intelligence (AI), cloud based services create and gather huge amount of information via phones, PCs, or other traditional approaches [4, 5]. However, the IoT based systems entirely rely on communicating-return procedure for an efficient processing and communicating of information among devices. Since, the data honesty and reliability are the significant concerns, especially a single fraud and dishonest with the shared information or leakage of data may lead to security and privacy explorer [6].

A. Devi · H. Saini
Jaypee University of Information Technology, Waknaghat, India

G. Rathee (✉)
Department of Computer Science and Engineering, Netaji Subhas University of Technology, New Delhi, India

S. Garg et al. (eds.), *Intelligent Cyber-Physical Systems for Autonomous Transportation*, Internet of Things, https://doi.org/10.1007/978-3-030-92054-8_15

15.1.1 Motivation and Research Objective

Security of network is considered as a consequence aspect while working over LAN, internet, and other online systems. Whenever it comes to setup the network for communication process, data security is assured at the higher precedence. The security of data transmitted through network is very critical because of number of hackers and invaders [3, 7]. Further, the data security in Internet-of-Vehicles (IoV) while sharing/exchanging the information has attracted wide attention in various applications. In vehicular transportation system organizations are afraid to fully adopt intelligent IoV systems because of various security breach and threats [8]. Numbers of researchers have proposed cryptographic security schemes in IoV using various encryption and decryption issues. The existing systems have their own limitations while managing the key overhead, key storage, complexity, cost delay, and involvement of third party may further invoke number of network security threats [9–12]. In order to prevent from these limitations, blockchain technique nowadays have been emerged as a most popular and advantageous mechanism to ensure a secure, transparent, and efficient information transmission. Fortunately, blockchain system ensures a secure, decentralized, self-organizing, and temper-resistant technique using token-based application such as shared economic, digital currency, consensus, etc. [13]. Blockchain provides and enhances the data management in a temper-resistant and distributed manner [14, 15]. The aim of this paper is to propose a secure information transmission process using blockchain technique in intelligent vehicular systems during the data sharing and information exchanging among various vehicular devices. The blockchain technique ensures a transparent and secure information transmission in real-time scenarios. Further, the proposed phenomenon is simulated against various security measures such as DoS, data alteration parameters in IoV systems through MATLAB environment.

The remaining structure of the paper is organized as follows. The number of blockchain technique used in various real-time applications for ensuring a secure information transmission proposed by various scientists is discussed in Sect. 15.2. In addition, a secure information transmission mechanism using blockchain technique in IoV is described in Sect. 15.3. Further, the validation of proposed scheme over various simulation results is discussed in Sect. 15.4. Finally, the conclusion of the paper along with future directions is detailed in Sect. 15.5.

15.2 Related Work

Various authors have proposed blockchain based solution [16, 17]. Some of them are depicted in Table 15.1. Liu et al. [12] have proposed a blockchain based data security mechanism in MANETs. The proposed mechanism used a route request discovery process to control the forwarding requests. In addition, the proposed mechanism avoided the collusion threats such as spoofing and cooperative reporting

Table 15.1 Performance comparison with the existing works based on accuracy using same dataset

Authors	Technique	Mechanism
Zhang et al. [18]	A secure architecture of VANET	The proposed architecture used three different layers, namely edge, perception, and service layers, to ensure a secure message transmission
Zhaofeng et al. [19]	Blockchain based trusted management	The authors have used a blockchain based matrix multichannel data security and privacy for encrypting the information using smart contracts
Nehe et al. [20]	The limitations of ensuring the data security	All the transmitted transactions and information are secured through digital signatures along with a tamper-proof
Asare et al. [21]	Authentication mechanism	The authors have used GOST algorithm to verify and authenticate the transmitted information
Zhang et al. [22]	Blockchain based centralized storage mechanism	The proposed model consists of three servers including cloud, HIoT, and blockchain
Harshini et al. [23]	Patient centric blockchain	The proposed mechanism ensured a secure record sharing, billing, and identification of thefts through blockchain mechanism

by using digital signatures. The consensuses mechanisms avoided the forking and ensured decentralized and increased efficiency among participating entities. The proposed mechanism is analyzed over various measured simulation results. Zhang et al. [18] have proposed a secure architecture of VANET that is based upon edge computing and blockchain technique. The proposed architecture used three different layers, namely edge, perception, and service layers, to ensure a secure message transmission. The edge and service layers ensured the computing resources and cloud based techniques to ensure the data security.

Zhaofeng et al. [19] have proposed a blockchain based trusted management system to ensure data security in the network. The authors have used a blockchain based matrix multichannel data security and privacy for encrypting the information using smart contracts. The proposed mechanism is simulated by ensuring the availability, security, and flexibility in the network while transferring the data with security in the network. Nehe et al. [20] have discussed the limitations of ensuring the data security through blockchain techniques. All the transmitted transactions and information are secured through digital signatures along with a tamper-proof. Each block is hashed and linked to chain of blocks using blockchain concept. Asare et al. [21] have proposed an authentication mechanism using blockchain technique in IoT systems while ensuring the data integrity. The authors have used GOST algorithm to verify and authenticate the transmitted information. The proposed results claimed the security of communicating data with integrity among nodes in the network. Zhang et al. [22] have proposed a blockchain based centralized storage mechanism by reducing the bandwidth occupied by data transmission in the network. The proposed model consists of three servers including cloud, HIoT, and blockchain.

The authors have analyzed the proposed mechanism with improved efficiency and data characteristics by improving the efficiency in the network. Harshini et al. [23] have proposed a patient centric blockchain based security approach to ensure the records integrity in the network. The proposed mechanism ensured a secure record sharing, billing, and identification of thefts through blockchain mechanism. The proposed mechanism implemented the model using smart contracts while exchanging the data among each other. In addition, the authors have highlighted its future scope and practicality to implement blockchain in healthcare systems.

15.3 Proposed Approach

In the proposed mechanism, the data is stored and retrieved between data creator and data user in the encrypted form [24]. All the information or transaction is recorded into the Ganache blockchain. As shown in Fig. 15.1, six various entities are used as Ganache blockchain, data creator, data user, Myetherwallet, solidity, and IPFS algorithm.

Data Creator It is responsible to deploy and create smart contracts by encrypting them and upload the files into IPFS by accessing various policies to data users.

Data User It is used to access the file from IPFS with the help of the data creator and decrypt the ciphertext of the encrypted file.

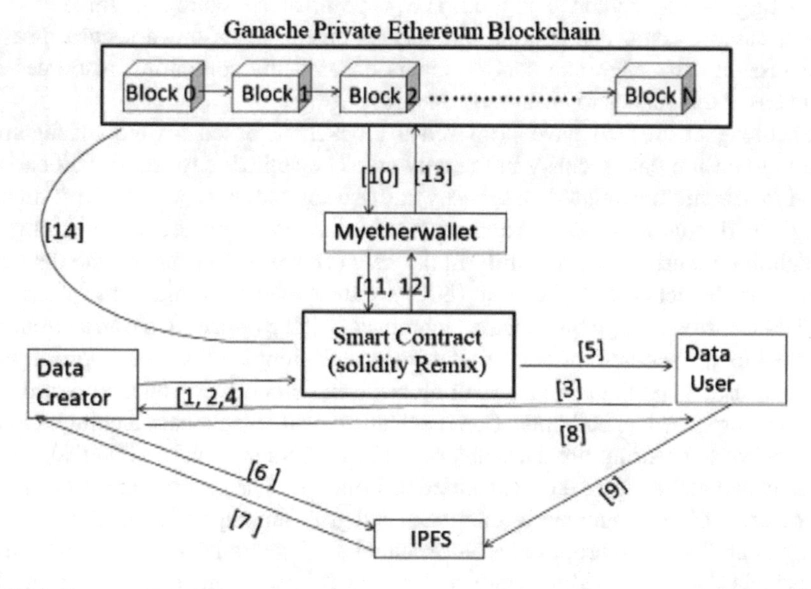

Fig. 15.1 Proposed blockchain mechanism

Solidity It is used to create the smart contract in solidity language and compile it with an online Ethereum Virtual Machine.

Myetherwallet It is software that is used to online connect the ether wallet.

Ganache Deploy and run the contracts by storing and retrieving the data.

15.3.1 IPFS Algorithm

The following steps are needed in order to store the file into encrypted form via data creator.

1. Data creator deployed a smart contract in solidity remix IDE named CPABE.
2. Data creator stores the ciphertext, hash, and file ID.
3. The data user sends an access request to the data creator.
4. The data creator encrypts the keys within the timing to access through the data user and stores them into a smart contract.
5. Data user obtains the ciphertext of the secret key and timing from the smart contract.
6. The data creator uploads the file onto the IPFS.
7. In addition, it gives the IPFS code to the data creator.
8. The data creator sends the IPFS address to the data user securely.
9. Data user downloads the encrypted files from IPFS.
10. Further, it deploys and interacts with the smart contract through Myetherwallet within the private (Ganache) Ethereum blockchain.
11. Deploy the smart contract through the Byte code within the blockchain.
12. Interact with the contract through the ABI code within the blockchain.
13. The number of transactions is shown within the hash, gas used, contract address, etc.
14. Same accounts with address show on Ganache and solidity with ethers.

As shown in Fig. 15.2, smart contracts create an online solidity remix and compile the contract. Data creator stores the smart-contracts on IPFS. Initially check the availability of all the ethers. If they are present in the network then blocks are successful created otherwise there is an error while generating or creating the block in the network. After creating the blocks, smart contracts are deployed in the network that further used to connect the online ether wallet via ABI and Byte code. Further, deployed smart contract will interact with Ganache through Myetherwallet by creating and using some hash, transactions and gas while performing transmission in the network.

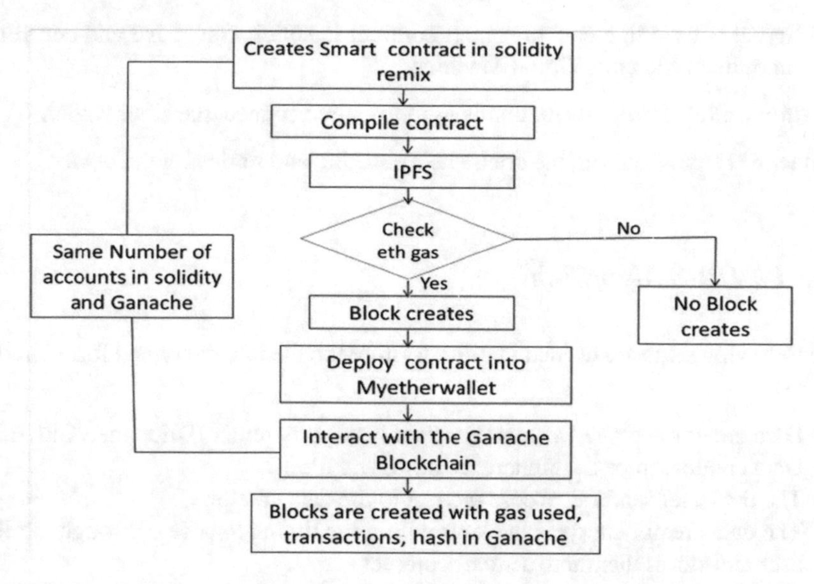

Fig. 15.2 Flowchart of proposed mechanism

15.4 Results and Analysis

The result analysis and discussion is done over various simulating graphs. All the simulated graphs are used to deploy the contracts using Ethereum through Myetherwallet using the used gas. Figure 15.3 shows the contracts that are deployed through online Myetherwallet with a gas limit that creates automatic and see on to the blockchain. Figure 15.4 shows the number of blocks that is created with the gas used along with the smart contract.

Figure 15.5 shows the transaction of every block with the transaction holder's address. All the values are encrypted within the hash value by the Keccak algorithm. Figure 15.6 shows the contract creation timing, gas, cost in ethers, and USD of existing and proposed approaches.

Figure 15.7 represents the comparisons against proposed and existing mechanisms over Ethereum using Ganache blockchain mechanism that defines the amount of time required to process the contract. The proposed mechanism outperforms existing approach that takes less amount of contract generation due to Keccak algorithm used to encrypt the message.

15.5 Conclusion

This paper has proposed a blockchain based data security mechanism in smart based systems. The proposed approach developed a smart consensus mechanism

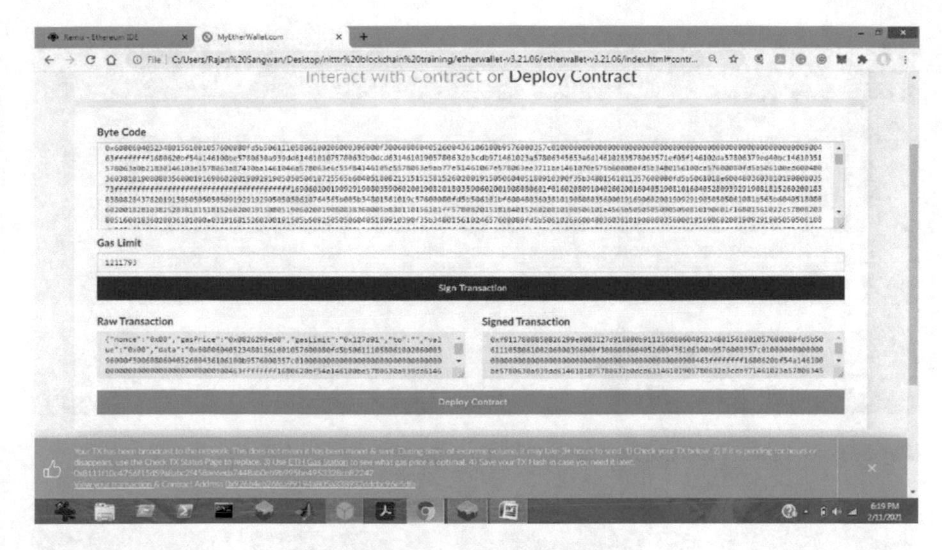

Fig. 15.3 Deploy the contract through Myetherwallet on Ganache

Fig. 15.4 Deploy the contract through Myetherwallet on Ganache

Fig. 15.5 Number of blocks is created with the gas used

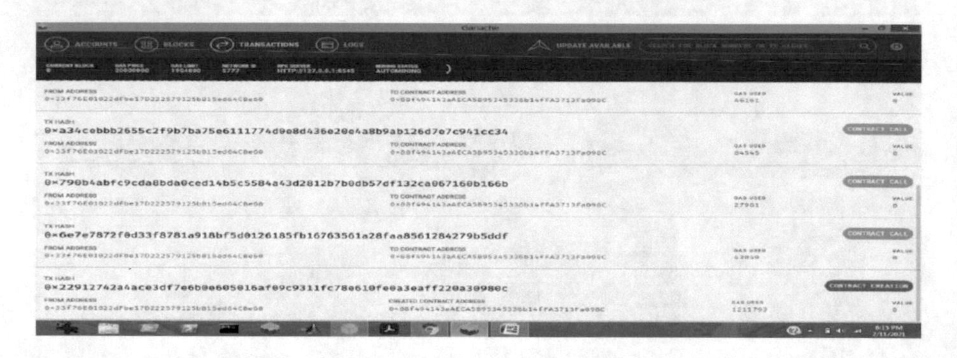

Fig. 15.6 Contracts deployed and interact in Ganache Private Ethereum

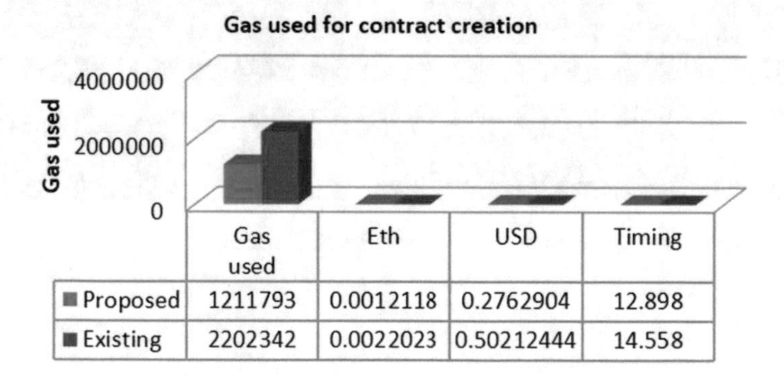

	Gas used	Eth	USD	Timing
Proposed	1211793	0.0012118	0.2762904	12.898
Existing	2202342	0.0022023	0.50212444	14.558

Fig. 15.7 Gas used during contract creation

while ensuring the integrity of data in the network. The proposed approach efficiently ensured the data management, resource management, and integrity of the information while sharing through cryptographic mechanism. The proposed approach is verified and validated against various security measurements such as contract creations over encrypted algorithms against existing scheme. In addition, the number of attacks that can be generated by reading the ideal nodes' patterns can be illustrated and develops a secure mechanism using blockchain that can be considered for the future directions.

References

1. Hansmann, U., Merk, L., Nicklous, M. S., & Stober, T. (2013). *Pervasive computing handbook*. Springer Science & Business Media.
2. Rathee, G., Sharma, A., Saini, H., Kumar, R., & Iqbal, R. (2019). A hybrid framework for multimedia data processing in IoT-healthcare using blockchain technology. *Multimedia Tools and Applications, 79*(15), 9711–9733.
3. Rathee, G., Garg, S., Kaddoum, G., & Choi, B. J. (2020). A decision-making model for securing IoT devices in smart industries. *IEEE Transactions on Industrial Informatics, 17*(6), 4270–4278.
4. Campolo, A., Sanfilippo, M. R., Whittaker, M., & Crawford, K. (2017). Ai now 2017 report.
5. Holzinger, A., Biemann, C., Pattichis, C. S., & Kell, D. B. (2017). What do we need to build explainable ai systems for the medical domain? Preprint, arXiv:1712.09923.
6. Fischer, R., Edward Halibozek, M. B. A., & Walters, D. (2012). *Introduction to security*. Butterworth-Heinemann.
7. Shrivastava, S., & Jain, S. (2013). A brief introduction of different type of security attacks found in mobile ad-hoc network. *International Journal of Computer Science & Engineering Technology (IJCSET), 4*(3).
8. Rathee, G., Sharma, A., Iqbal, R., Aloqaily, M., Jaglan, N., & Kumar, R. (2019). A blockchain framework for securing connected and autonomous vehicles. *Sensors, 19*(14), 3165.
9. Chen, W., Chen, Y., Chen, X., & Zheng, Z. (2019). Toward secure data sharing for the IoV: A quality-driven incentive mechanism with on-chain and off-chain guarantees. *IEEE Internet of Things Journal, 7*(3), 1625–1640.
10. Yaga, D., Mell, P., Roby, N., & Scarfone, K. (2019). Blockchain technology overview. Preprint, arXiv:1906.11078.
11. Dinh, T. T. A., Liu, R., Zhang, M., Chen, G., Ooi, B. C., & Wang, J. (2018). Untangling blockchain: A data processing view of blockchain systems. *IEEE Transactions on Knowledge and Data Engineering 30*(7), 1366–1385.
12. Liu, G., Dong, H., Yan, Z., Zhou, X., & Shimizu, S. (2020). B4SDC: A blockchain system for security data collection in MANETs. *IEEE Transactions on Big Data*. https://doi.org/10.1109/TBDATA.2020.2981438
13. Aujla, G. S., Singh, M., Bose, A., Kumar, N., Han, G., & Buyya, R. (2020). BlockSDN: Blockchain-as-a-service for software defined networking in smart city applications. *IEEE Network, 34*(2), 83–91.
14. Aujla, G. S., Singh, A., Singh, M., Sharma, S., Kumar, N., & Choo, K.-K. R. (2020). BloCkEd: Blockchain-based secure data processing framework in edge envisioned V2X environment. *IEEE Transactions on Vehicular Technology, 69*(6), 5850–5863.
15. Singh, M., Aujla, G. S., & Bali, R. S. (2020). A deep learning-based blockchain mechanism for secure Internet of drones environment. *IEEE Transactions on Intelligent Transportation Systems*. https://doi.org/10.1109/tits.2020.2997469

16. Singh, M., Aujla, G. S., & Bali, R. S. (2020). ODOB: One drone one block-based lightweight blockchain architecture for internet of drones. In: *IEEE INFOCOM 2020-IEEE Conference on Computer Communications Workshops (INFOCOM WKSHPS)* (pp. 249–254). IEEE.
17. Aujla, G. S., Barati, M., Rana, O., Dustdar, S., Noor, A., Llanos, J. T., Carr, M., Marikyan, D., Papagiannidis, S., & Ranjan, R. (2020). COM-PACE: Compliance-aware cloud application engineering using blockchain. *IEEE Internet Computing, 24*(5), 45–53.
18. Zhang, X. D., Li, R., & Cui, B. (2018). A security architecture of VANET based on blockchain and mobile edge computing. In: *2018 1st IEEE International Conference on Hot Information-Centric Networking (HotICN)* (pp. 258–259). IEEE.
19. Zhaofeng, M., Xiaochang, W., Jain, D. K., Khan, H., Hongmin, G., & Zhen, W. (2019). A blockchain-based trusted data management scheme in edge computing. *IEEE Transactions on Industrial Informatics, 16*(3), 2013–2021.
20. Nehe, M., & Jain, S. A. (2019). A survey on data security using blockchain: Merits, demerits and applications. In: *2019 International Conference on Recent Advances in Energy-efficient Computing and Communication (ICRAECC)* (pp. 1–5). IEEE.
21. Asare, B. T., Quist-Aphetsi, K., & Nana, L. (2019). Nodal authentication of IoT data using blockchain. In: *2019 International Conference on Computing, Computational Modelling and Applications (ICCMA)* (pp. 125–1254). IEEE.
22. Zhang, G., & Xie, J. (2019). Blockchain-enabled security-aware applications in home internet of thing. In: *2019 International Conference on Communications, Information System and Computer Engineering (CISCE)* (pp. 559–565). IEEE.
23. Harshini, V. M., Danai, S., Usha, H. R., & Kounte, M. R. (2019). Health record management through blockchain technology. In: *2019 3rd International Conference on Trends in Electronics and Informatics (ICOEI)* (pp. 141–1415). IEEE.
24. Wang, S., Wang, X., & Zhang, Y. (2019). A secure cloud storage framework with access control based on blockchain. *IEEE Access, 7*, 112713–112725.

Chapter 16
Privacy-Preserved Mobile Crowdsensing for Intelligent Transportation Systems

Qinyang Miao, Hui Lin, Jia Hu, and Xiaoding Wang

16.1 Introduction

With the advent of the era of "intelligent connection," all things are interconnected. Intelligent transportation provides a new paradigm for alleviating traffic problems and has become a hot research field for relevant people [1, 2]. In the context of intelligent transportation, mobile crowdsensing still faces the following key challenges. The data collectors do not trust each other, and the unreliable collectors are easy to forge or modify the data. Secondly, honest collectors worry about the privacy leakage caused by the collected data, but they have no enthusiasm to collect data. Therefore, in the process of data collection to achieve data privacy protection and ensure that the data collector to provide reliable and effective data are two important aspects of our research.

Cloud computing plays an important role in storage and data aggregation. Mobile edge computing can provide the required services nearby by using computing, storage, and application platform in the device source or data source, and users can initiate requests on the edge side, so that they can quickly respond and complete crowdsensing. However, the privacy security problem of data aggregation in the distributed scene has not been well solved.

Q. Miao · H. Lin · X. Wang
College of Computer and Cyber Security, Fujian Normal University, Fuzhou, China

Engineering Research Center of Cyber Security and Education Informatization, Fujian Province University, Fuzhou, Fujian, China
e-mail: 18291972715@163.com; linhui@fjnu.edu.cn; wangdin1982@fjnu.edu.cn

J. Hu (✉)
University of Exeter, Exeter, UK
e-mail: j.hu@exeter.ac.uk

S. Garg et al. (eds.), *Intelligent Cyber-Physical Systems for Autonomous Transportation*, Internet of Things, https://doi.org/10.1007/978-3-030-92054-8_16

Recently, the rise of federated learning provides a strong technical support to solve the privacy security of data aggregation in distributed environment. Federated learning not only solves the problem of privacy and security, but also relieves the computing burden of traditional centralized devices by allowing data collectors to provide training models instead of the original data. On the other hand, blockchain is a new application mode of computer technology such as distributed data storage, point-to-point transmission, consensus mechanism, and encryption algorithm. The blockchain is essentially a decentralized database. It is a series of data blocks associated with cryptographic methods. Each data block contains a batch of Bitcoin network transaction information, which is used to verify its information. Because the blockchain has the characteristics of decentralization, openness, independence, security, anonymity, etc., it can be used as a reliable and safe data storage mechanism, especially for the crowdsensing. In intelligent transportation, due to the cross regional and mutual distrust of vehicles, how to ensure that each owner provides reliable data needs to be solved. In this chapter, we use federated learning technology and blockchain technology to propose a mobile crowdsensing mechanism based on teamwork. The contributions of this article are summarized as follows:

1. We transform the aggregation of raw data into the aggregation of data models. On this basis, combined with federated learning, we propose a team-based crowdsensing mechanism that can protect the privacy of team members.
2. Since the crowdsensing tasks are completed in the form of a team, the team leader can manage and supervise the team members. In addition, we introduce the blockchain to record the execution process of the crowdsensing task, so as to ensure that the team members provide accurate data model, thus guaranteeing the efficiency of the completion of crowdsensing tasks.
3. We have introduced a differential privacy mechanism. Each team member adds differential privacy noise to disturb the model during the process of submitting the data model, thereby further protecting the privacy of the team members. In addition, in order to improve the accuracy of data model aggregation, we propose a reasonable reward and punishment mechanism, which can provide corresponding rewards and punishments based on the data model provided by team members.
4. The experimental results show that the proposed scheme can achieve efficient crowdsensing while protecting the privacy of team members.

The rest of this article is organized as follows. Section 16.2 covers related work. We introduced the system model in Sect. 16.3. The details of the proposed strategy are detailed in Sect. 16.4. The experiment is described in Sect. 16.5. We summarized this paper in Sect. 16.6.

16.2 Related Work

The mobile crowdsensing in intelligent transportation systems is of great significance, thereby a lot of excellent works have been proposed. Zhao et al. [3] implemented a reputation management based on privacy protection and a prevention mechanism for user malicious behavior. In [4], a quorum sensing mechanism called blockchain is designed to achieve location privacy protection. In [5], Shen et al. combined machine learning technology and blockchain technology to design a new privacy protection mechanism. In [6], deep learning technology is applied to information hiding of embedded sensors by Liang et al., so as to realize user privacy protection. In [7], Hu and others combined data credibility and privacy protection to design an enhanced crowdsourcing mechanism that can resist internal attacks. In [8], Liu et al. considered the privacy of participants, the randomness of perception tasks, and the cost of the platform, and designed a privacy protection framework that minimizes the trade-off of system stability and data aggregation. In [9], Xiong et al. use machine learning and game theory to realize privacy-protected mobile crowdsensing. In [10], Huang et al. combined a dynamic incentive mechanism with the blockchain to realize the perception and verification of mobile edge users while incentivizing users to provide reliable data. Heterogeneous privacy protection was first studied by Wang et al. [11]. Zhang et al. [12] consider how to motivate users while protecting their privacy while considering a limited budget. He et al. [13] proposed the setting of privacy preference and added it to the design of incentive mechanism to realize privacy protection with privacy preference. Nie et al. [14] use Stackelberg game mechanism to design the incentive mechanism in privacy protection. In [15], multi-stage programming is used to design incentive mechanisms to encourage user participation. Zhan[16] combines deep reinforcement learning and Stackelberg game to design the incentive mechanism in mobile crowdsensing. Qlearning[17] is used in the design of dynamic pricing in privacy protection. Liu et al. [18] combined game theory and Q learning to design an incentive mechanism based on user privacy protection.

Existing work can realize user privacy protection in mobile crowdsensing, but there are still the following challenges: (1) How to change the data aggregation method in mobile crowdsensing to achieve user privacy protection; (2) How to combine blockchain with crowdsensing to protect users' privacy while ensuring that users provide real data. In response to the above problems, this chapter proposes a crowdsensing mechanism based on federated learning, which can effectively improve the quality of mobile crowdsensing and protect user privacy.

16.3 System Model

In this chapter, we transforms the data aggregation of the traditional mobile crowdsensing into the aggregation of data models to protect the privacy of data owners. Among them, users who are willing to aggregate relevant data, will respond

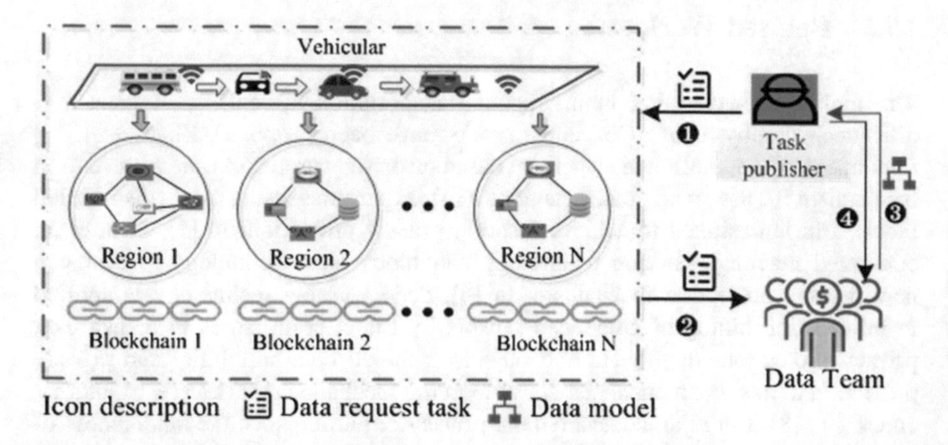

Fig. 16.1 The architecture of the proposed strategy

to tasks in the form of a team, and the team leader guides and supervises the aggregation of the data model within the team. In addition, after the final aggregated data model is generated, the model will be stored on the blockchain and provided to the next data publisher of the same task. The architecture of the solution proposed in this chapter is shown in Fig. 16.1, including the following four processes:

1. The data requester in each region issues the mobile crowdsensing tasks, that is, the task of identifying buildings and obstacles in the intelligent transportation environment;
2. The data owners respond to the crowdsensing tasks in the form of a team;
3. The team trains the data model and delivers the model to the data requester;
4. The data requester provides corresponding task rewards.

In this chapter, we consider the following security threats.

1. Dishonest team members may provide false or malicious data models, resulting in low availability of the final data aggregation model. In addition, some untrustworthy team members may temporarily withdraw from the training process of the data model, which will affect the quality of the final data model.
2. The task publisher may be interested in the privacy of team members. If the team only provides the data model that has not been protected by privacy protection to the task issuer, the task issuer can infer the privacy data of the data owner with a certain probability. For these two security threat models, this chapter supervises the training process of the data model through the team leader, and protects the privacy of team members by adding noise to the data model.

16.4 Crowdsensing Based on Federated Learning

16.4.1 Data Aggregation with Team Participation

In this chapter, in order to ensure the high-quality completion of data request tasks, we propose a new blockchain registration method, in which each data collector is registered as a team member and responds to the crowdsensing tasks. The trust relationship between them and the efficiency of completing tasks have also been improved to a certain extent. When the task publisher releases the perception task, the team that meets the requirements responds to the task. It is an efficient and mutually beneficial way for multiple participants to complete the assigned tasks through cooperation. First, the team sponsor initiates the team to aggregate information, and sets the aggregate response deadline, work tasks, and corresponding requirements for members. Of course, membership requirements are based on specific tasks to be solved, such as number and ability requirements. The process of signing the agreement between the data team leader and the task publisher is as follow in Algorithm 4.

Algorithm 4 Signing the contract

Require: Data sharing task T_s, completion time T, task requirements R_e, task rewards T_r
Ensure: Signed contract
A team of users respond to the task
The team leader and the data requester sign the protocol with their private keys P_k.
Crowdsensing platform acts the supervisor of protocol implementation
Contract signed.

16.4.2 Data Model Aggregation Process

Encrypting data is the way most existing methods perform data aggregation. However, in actual data aggregation scenarios, encrypting the original data will result in a significant reduction in the availability of the final aggregated data. With the increasing demand for data aggregation in a distributed environment, aggregating data models is a safer and more efficient way than data aggregation. Aggregation in this way not only achieves the purpose of data aggregation, but also protects the privacy of data collectors.

In this chapter, we consider the following mobile crowdsensing scenarios. When the perception task is released by the task publisher through the blockchain, the data-collecting node receives the task in the form of a team, that is, the team leader decides whether to accept the task. Once the task is accepted, the team members perform federated learning to train a global data model. The specific data model aggregation scheme is summarized as follows:

1. The task initiator initiates a mobile crowdsensing task, which includes task details, rewards, and timestamps. The task is signed by the task initiator through its private key. When the task issuer publishes the task to the blockchain, the identity of the task issuer will be verified. Then the system will retrieve whether such request has been processed before, and if there is a record, it will directly return the query result, otherwise it will post the task.
2. The user responds to the task in the form of a team, where the team leader decides whether to accept the task, and then the team members perform federated learning to train a global model, that is, each team member first trains the local data model, and then sends the model to the team leader. The team leader aggregates each local model and sends it back to each team member. This process is repeated until the maximum training time is reached. The training process is shown in Fig. 16.2.
3. All aggregations will be regarded as aggregation transactions, and these transactions will be packaged into blocks by transaction record nodes to generate aggregate transaction records.
4. The team members perform the consensus process, that is, team members compete for the opportunity to write transactions into the block, and the node with the accounting rights broadcasts its block to other data nodes for verification. After the verification is passed, the block is added to the blockchain.

Using the form of a team to achieve data model aggregation not only solves data privacy and security issues, but also improves the effectiveness of the aggregation model. Since each team member can be managed by the team leader, this further enhances the reliability of data model aggregation. In addition, through a team to respond to crowdsensing tasks, this is because we consider that there is a certain

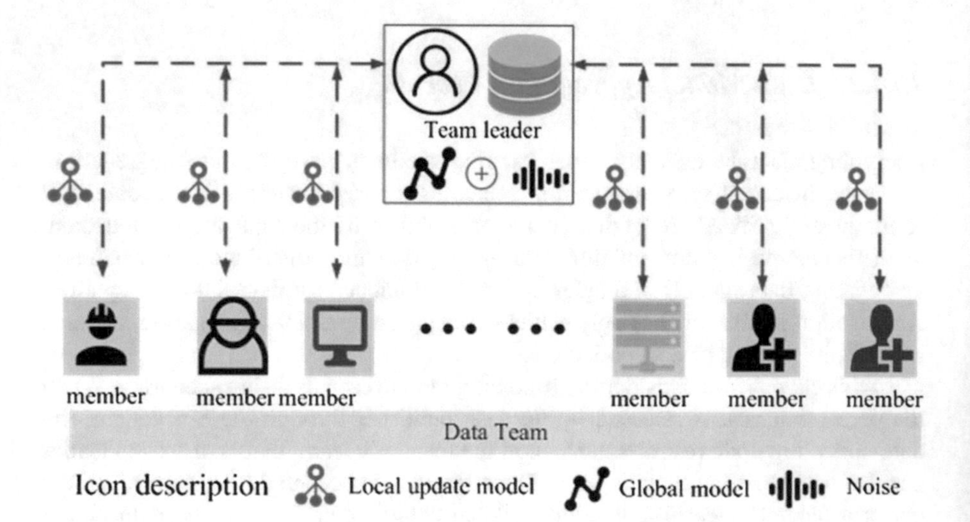

Fig. 16.2 Model training process

degree of mutual trust between team members, but there is no such trust relationship between task publishers and team members. Therefore, in order to protect the privacy of team members, in the process of federated learning, each team member adds differential privacy noise to the data model provided by itself, which ensures that the data model aggregated by team leaders is also privacy-protective. The federated learning algorithm with differential privacy is summarized in following Algorithm 5:

Algorithm 5 Federated learning with privacy preservation

Require: crowdsensing task D_r, team D_T
Ensure: Model G_m
  for each member $D_t \in D_T$ **do**
    Team leader generates the initial model m_i
    while the training time reaches the maximum rounds **do**
      Each team member trains the local data model according to local data and add the corresponding noise for model disturbance
      The team leader aggregates the data models and return the aggregated model to all team members
      Team members continue to train local data models
    end while
  end for
  Obtain the final aggregated model G_m
  The team leader uploads the G_m to the task publisher

16.4.3 Team Credit Management

After the aggregation process of federated learning, we can obtain the local data model and the final aggregated data model provided by the team members. If the accuracy of the final aggregated model on the test set meets the requirements, the team will receive the corresponding credit reward, otherwise the entire team will be punished. In addition, only by quantifying the contribution of each team member and rewarding and punishing their contribution can the enthusiasm of the team members be improved. In the process of federated learning, the greater the contribution of a team member, the greater the weight of the member's local model in the final global model.

Assuming that the accuracy required by the task publisher is A_r, the test accuracy of the final aggregation model must be greater than the required accuracy A_r. We divide credit rewards in the following ways.

1. If the accuracy rate of m members in the team exceeds A_r, and the remaining $n-m$ members do not achieve this accuracy rate, only the m members mentioned above will be rewarded. The rewards for each member are as follows:

$$R_i = \frac{acc_i - A_r}{\sum_{i=1}^{m}(acc_i - A_r)} \cdot R_{total}, \tag{16.1}$$

where acc_i is the test accuracy of member i, R_{total} is the total reward.

2. If the test accuracy of the final aggregation model is lower than the required accuracy, but the test accuracy of m members of the team is greater than the required accuracy, and the other $n - m$ members are less than the required accuracy, only the above $n - m$ members will be punished. The specific fines are calculated as follows:

$$P_i = \frac{acc_i - A_r}{\sum_{i=1}^{n-m}(acc_i - A_r)} \cdot P_{total}, \tag{16.2}$$

where P_{total} is the total punishment.

Since each team has two roles, one is the team leader and the other is the team member. Among them, team members are responsible for training the local data model and handing it over to the team leader for aggregation; and the team leader needs to decide whether to accept the task, supervise the team members to train the data models and provides the model aggregation, and eventually submits the final aggregated model to the task publisher. This indicates that the team leader have a lot of responsibilities compared to team members. Thereby, for successfully completed tasks, the team leader must get more rewards, while for failed tasks, the team leader must also get more punishments. Only in this way we can achieve fair rewards and punishments based on individual contributions, thereby incentivizing the team to provide better data models.

16.5 Performance Evaluation

16.5.1 Simulation Setup

The experiment in this chapter is completed on a computer equipped with Windows 7 system, which is equipped with an Intel Core i7 processor with a CPU frequency of 6.4 GHz. We use the python programming language to verify the effectiveness of the proposed scheme. This experiment uses the MNIST data set, which comes from the National Institute of Standards and Technology (NIST). The training set consists of handwritten numbers from 250 different people, of which 50% are high school students and 50% are from the staffs of the Census Bureau. The test set is also the same proportion of handwritten digital data, but it is guaranteed that the test set and the training set do not intersect. The MNIST data set has a total of 70,000 images, of which 60,000 are the training set and 10,000 are the test set. Each picture is composed of 28*28 handwritten digital pictures of 0–9. Each picture is in the form of white text on a black background. The black background is represented by 0, and the white text is represented by a floating point number between 0–1. The closer to 1, the whiter the color.

16.5.2 Experimental Results

16.5.2.1 Model Training

Figures 16.3 and 16.4 each show the loss of the strategy proposed in this chapter during the training of the local data model and the accuracy curve of the federated learning data model aggregation. By observing Fig. 16.4, we find that when the number of local training rounds reaches 60, the loss curve area is stable. This shows the efficiency of our proposed team-based data model training. By observing Fig. 16.3, we find that as the number of federated learning rounds increases, the training accuracy gradually increases, and finally converges to an accuracy of about 97%.

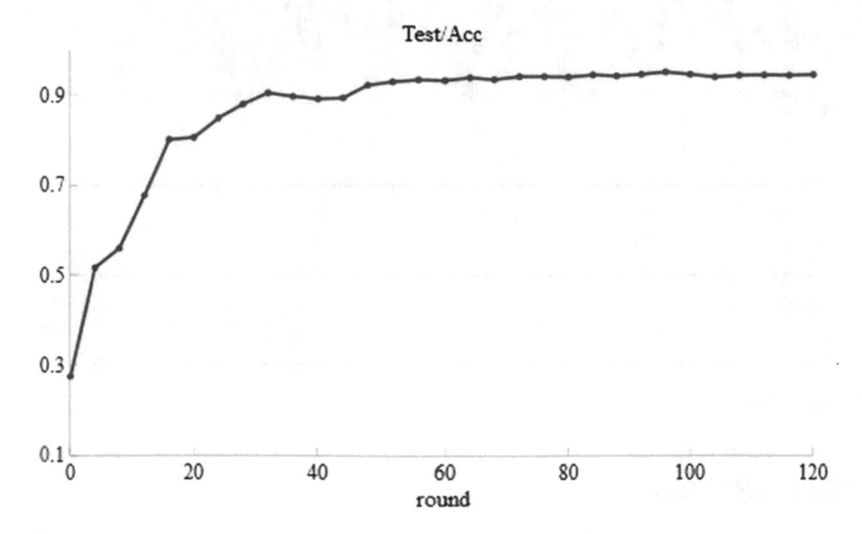

Fig. 16.3 Test accuracy

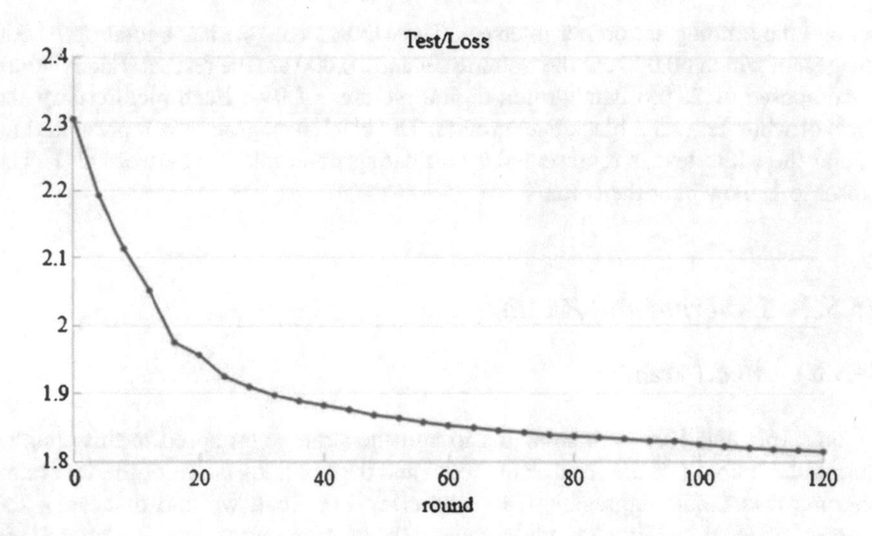

Fig. 16.4 Test loss

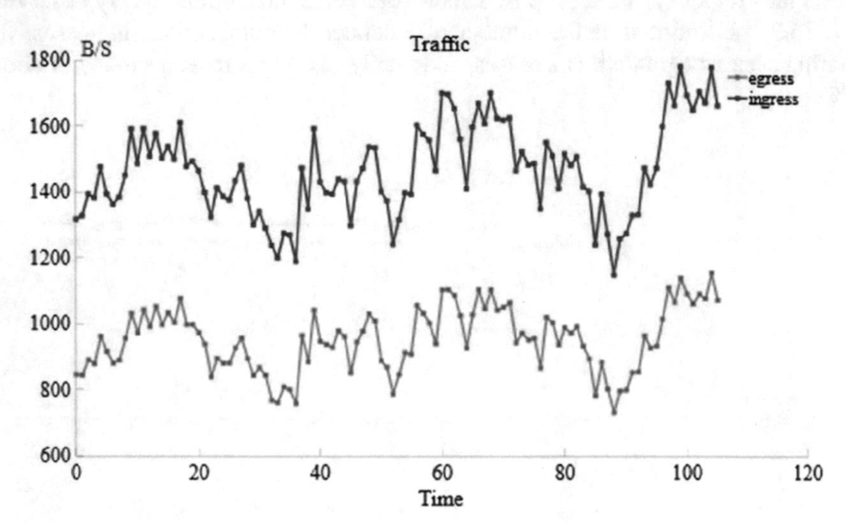

Fig. 16.5 Traffic condition

16.5.2.2 System Performance

In this chapter, we build a Hyperledger Fabric to verify the performance of the proposed strategy. According to the specific plan proposed above, we have written related smart contracts to compete for the right to bookkeeping based on the contributions of team members. In addition, we set a block to be generated every four seconds. Figures 16.5, 16.6, 16.7, and 16.8, respectively, show the performance

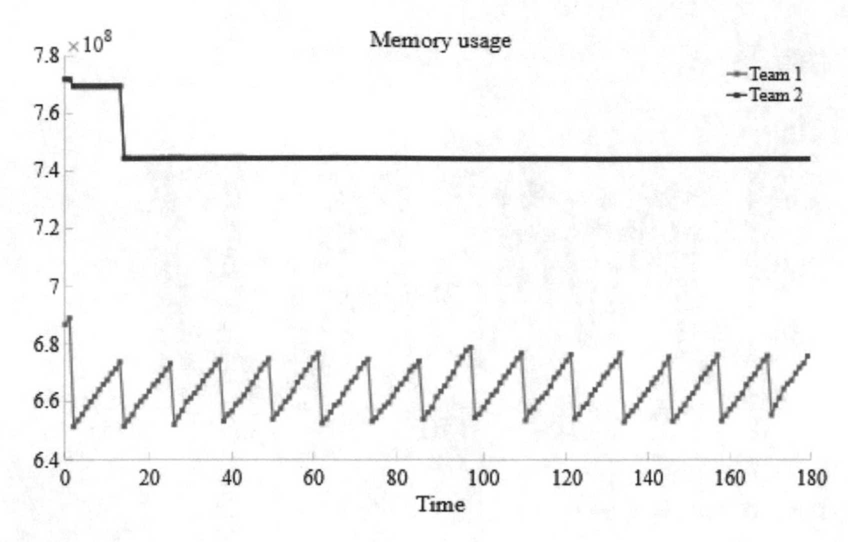

Fig. 16.6 Memory usage

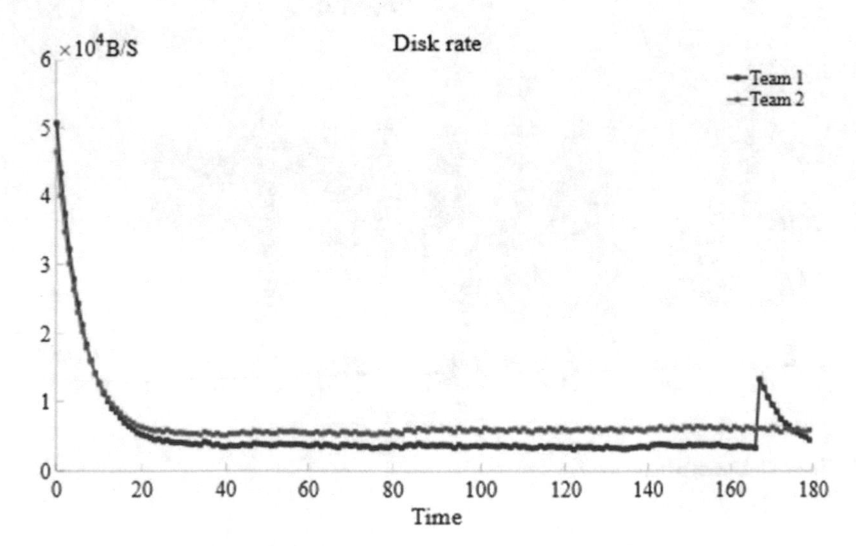

Fig. 16.7 Disk usage

of our proposed strategy in terms of traffic conditions, memory usage, disk usage, and CPU usage of the proposed strategy.

By observing Fig. 16.5, we find that the traffic conditions fluctuate with time. In addition, the highest egress and ingress reached 1800B/S and 1200B/S, respectively, which means that the strategy proposed in this chapter has good load balancing.

As shown in Figs. 16.6 and 16.7, different teams have different memory and disk usage. In addition, with the increase of time, the utilization rate of disks by each

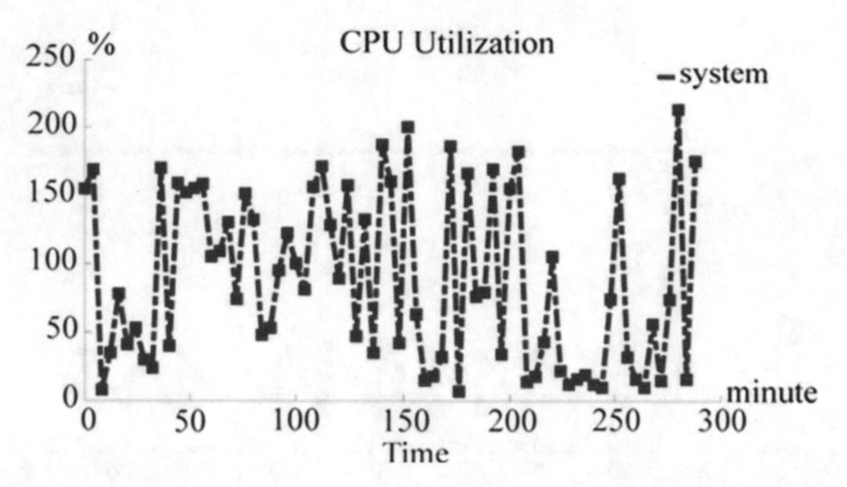

Fig. 16.8 CPU usage in region 1

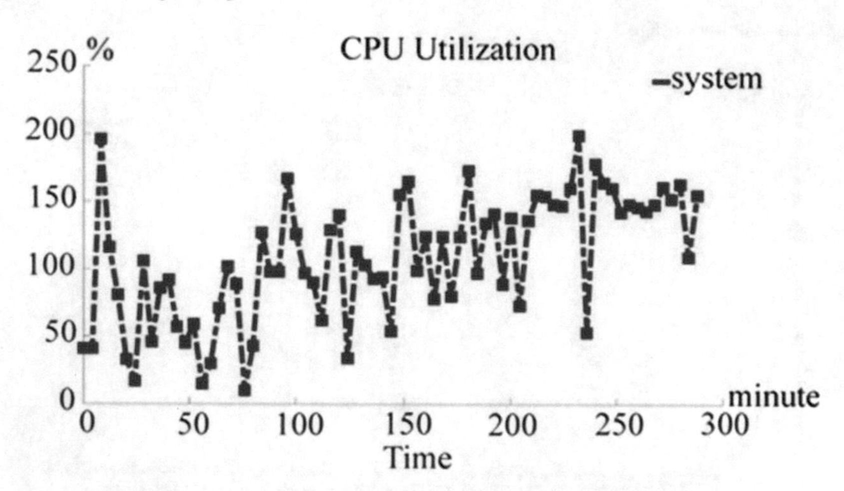

Fig. 16.9 CPU usage in region 2

team has dropped significantly. Figures 16.6 and 16.7 show the advantages of the strategy proposed in this chapter in resource utilization.

Observing Figs. 16.8 and 16.9, the CPU utilization of different teams performing crowdsensing tasks in different regions is significantly different. Although the CPU utilization rate fluctuates greatly over time, the highest CPU utilization rate is not more than 200%, which means that the strategy we proposed will not cause a huge impact on the team members' own equipment for training local data models.

16.6 Conclusion

In the chapter, we propose a security mobile crowdsensing mechanism based on federated learning in the context of intelligent transportation. Specifically, this mechanism allows users to complete mobile crowdsensing tasks in the form of a team. In addition, the team leader aggregates the data models provided by the team members through federated learning and supervises the team members in the process. The team members further protect their privacy by adding disturbing noise to the data model. Finally, the reward for task completion and the punishment for task failure are allocated based on the contribution of each team member (that is, the accuracy of the data model provided by each member), so as to motivate team members to provide reliable and true data and finally realize the data aggregation process Security and reliability in the system. In the case of preventing inference attacks, the usability of the model after disturbance can be guaranteed. Compared with the traditional scheme, the proposed data aggregation scheme has better results in terms of privacy protection and data quality improvement. Experimental results show that the proposed strategy performs well in road condition recognition accuracy for intelligent transportation.

References

1. Aujla, G. S., Singh, A., Singh, M., Sharma, S., Kumar, N., & Choo, K.-K. R. (2020). BloCkEd: Blockchain-based secure data processing framework in edge envisioned V2X environment. *IEEE Transactions on Vehicular Technology, 69*(6), 5850–5863.
2. Singh, A., Aujla, G. S., & Bali, R. S. (2020). Intent-based network for data dissemination in software-defined vehicular edge computing. *IEEE Transactions on Intelligent Transportation Systems*. https://doi.org/10.1109/TITS.2020.3002349
3. Zhao, K., Tang, S., Zhao, B., & Wu, Y. (2019). Dynamic and privacy-preserving reputation management for blockchain-based mobile crowdsensing. *IEEE Access, 7*, 74694–74710.
4. Yang, M., Zhu, T., Liang, K., Zhou, W., & Deng, R. H. (2019). A blockchain-based location privacy-preserving crowdsensing system. *Future Generation Computer Systems, 94*, 408–418.
5. Shen, M., Tang, X., Zhu, L., Du, X., & Guizani, M. (2019). Privacy-preserving support vector machine training over blockchain-based encrypted IoT data in smart cities. *IEEE Internet of Things Journal, 6*(5), 7702–7712.
6. Liang, Y., Cai, Z., Yu, J., Han, Q., & Li, Y. (2018). Deep learning based inference of private information using embedded sensors in smart devices. *IEEE Network, 32*(4), 8–14.
7. Hu, J., Lin, H., Guo, X. C., & Yang, J. (2018). DTCS: An integrated strategy for enhancing data trustworthiness in mobile crowdsourcing. *IEEE Internet of Things Journal, 5*(6), 4663–4671.
8. Liu, Y., Feng, T., Peng, M., Guan, J., & Wang, Y. (2020). DREAM: Online control mechanisms for data aggregation error minimization in privacy-preserving crowdsensing. *IEEE Transactions on Dependable and Secure Computing*. https://doi.org/10.1109/TDSC.2020.3011679
9. Xiong, J., Ma, R., Chen, L., Tian, Y., Li, Q., Liu, X., & Yao, Z. (2020). A personalized privacy protection framework for mobile crowdsensing in IIoT. *IEEE Transactions on Industrial Informatics, 16*(6), 4231–4241.
10. Huang, J., Kong, L., Dai, H., Ding, W., Cheng, L., Chen, G., Xi, J., & Zeng, P. (2020). Blockchain based mobile crowd sensing in industrial systems. *IEEE Transactions on Industrial Informatics, 16*(10), 6553–6563.

11. Wang, X., He, J., Cheng, P., & Chen, J. (2019). Privacy preserving collaborative computing: Heterogeneous privacy guarantee and efficient incentive mechanism. *IEEE Transactions on Signal Processing, 67*(1), 221–233.
12. Zhang, Z., He, S., Chen, J., & Zhang, J. (2018). REAP: An efficient incentive mechanism for reconciling aggregation accuracy and individual privacy in crowdsensing. *IEEE Transactions on Information Forensics and Security, 13*(12), 2995–3007.
13. He, J., Cai, L., Cheng, P., Pan, J., & Shi, L. (2018). Distributed privacy-preserving data aggregation against dishonest nodes in network systems. *IEEE Internet of Things Journal, 6*(2), 1462–1470.
14. Nie, J., Luo, J., Xiong, Z., Niyato, D., & Wang, P. (2019). A Stackelberg game approach toward socially-aware incentive mechanisms for mobile crowdsensing. *IEEE Transactions on Wireless Communications, 18*(1), 724–738.
15. Cao, B., Xia, S., Han, J., & Li, Y. (2020). A distributed game methodology for crowdsensing in uncertain wireless scenario. *IEEE Transactions on Mobile Computing, 19*(1), 15–28.
16. Zhan, Y., Liu, C. H., Zhao, Y., Zhang, J., & Tang, J. (2020). Free market of multi-leader multi-follower mobile crowdsensing: An incentive mechanism design by deep reinforcement learning. *IEEE Transactions on Mobile Computing, 19*(10), 2316–2329.
17. Zhang, M., Chen, J., Yang, L., et al. (2019). Dynamic pricing for privacy-preserving mobile crowdsensing: A reinforcement learning approach. *IEEE Network, 33*(2), 160–165.
18. Liu, Y., Wang, H., Peng, M., Guan, J., Xu, J., & Wang, Y. (2020). DeePGA: A privacy-preserving data aggregation game in crowdsensing via deep reinforcement learning. *IEEE Internet of Things Journal, 7*(5), 4113–4127.

Index

S. Garg et al. (eds.), *Intelligent Cyber-Physical Systems for Autonomous Transportation*, Internet of Things, https://doi.org/10.1007/978-3-030-92054-8

Printed in the United States
by Baker & Taylor Publisher Services